Functional Connections of Cortical Areas

Functional Connections of Cortical Areas

A New View from the Thalamus

S. Murray Sherman and R. W. Guillery

The MIT Press
Cambridge, Massachusetts
London, England

MIT Press books may be purchased at special quantity discounts for business or sales promotional use. For information, please email special_sales@mitpress.mit.edu or write to Special Sales Department, The MIT Press, 55 Hayward Street, Cambridge, MA 02142.

This book was set in Syntax and Times Roman by Toppan Best-set Premedia Limited, Hong Kong. Printed and bound in the United States of America.

Library of Congress Cataloging-in-Publication Data

Sherman, S. Murray.
Functional connections of cortical areas : a new view from the thalamus / S. Murray Sherman and R. W. Guillery.
  p. ; cm.
Includes bibliographical references and index.
ISBN 978-0-262-01930-9 (hardcover : alk. paper)
I. Guillery, R. W. II. Title.
[DNLM: 1. Thalamus—physiology. 2. Cerebral Cortex physiology. 3. Neural Pathways—cytology. 4. Neural Pathways—physiology.   WL 312]
616.8—dc23
2012046926

10 9 8 7 6 5 4 3 2 1

The front cover shows a photograph of a section of the mouse brain containing the primary and secondary somatosensory cortex at the top and various subcortical structures below, including the posterior medial nucleus of the thalamus (which contains the tiny blue patch). The red and yellow zones in the cortex show regions of elevated neuronal activity as does the blue zone in the thalamus. In this experiment, layer 5 of primary somatosensory cortex was directly activated by delivery of glutamate (red patch on left in cortex), and this led to activity in the posterior medial nucleus, shown as the faint small patch of blue in the thalamus. This was followed by activity in the secondary somatosensory cortex (yellow patch immediately to the right of the red of the primary somatosensory cortex). Further to the right of this activity in the second somatosensory area the figure shows later, fainter activity in the third somatosensory area. This photograph is taken from figure 3a of Theyel et al. (2010b), where the dependence of the corticocortical transmission upon the thalamic relay is demonstrated. See chapter 5.1.2 for further details.

Brief Contents

Preface   xi
Abbreviations   xvii

1   An Introduction to Thalamocortical Pathways   1

2   Cell and Synaptic Properties   17

3   The Basic Organization of Cortex and Thalamus   49

4   Classification of Afferents in Thalamus and Cortex   83

5   First and Higher Order Thalamic Relays   119

6   The Dual Nature of the Thalamic Input to Cortex   141

7   Linking the Body and the World to the Thalamus   179

8   The Inputs to the Cortex from the Thalamus and the Cortical Descending Outputs   195

9   Thalamocortical Links to the Rest of the Brain and the World   217

References   245
Index   275

Contents

Preface   xi
Abbreviations   xvii

1   An Introduction to Thalamocortical Pathways   1
1.1   An Overall View of the Thalamus and Cortex in Relation to the Rest of the
       Brain   1
1.2   Thalamocortical Connections   7
1.3   Tracing the Message   11
1.4   An Overall Summary of Some of the Major Points   14

2   Cell and Synaptic Properties   17
2.1   Intrinsic Cell Properties   17
2.1.1 Cable Properties   17
2.1.2 Passive Membrane Conductances   22
2.1.3 Active Membrane Conductances: Thalamic Neurons   23
2.1.4 Active Membrane Conductances: Cortical Neurons   36
2.2   Synaptic Properties   37
2.2.1 Ionotropic and Metabotropic Receptors   37
2.2.2 Short-Term Plasticity   41
2.2.3 Synaptic Integration   46
2.2.4 Summary and Conclusions   46

3   The Basic Organization of Cortex and Thalamus   49
3.1   The Cortex   49
3.1.1 The Cortical Areas   49
3.1.2 Localization of Function in the Cortex   54
3.1.3 Structurally Distinct Cortical Areas and Their Functions   57
3.2   The Thalamus   60
3.2.1 The Thalamic Nuclei and Their Inputs   61

3.2.2 The Cells of the Thalamus   69
3.2.3 Thalamic Connectivity Patterns   76
3.2.4 An Overall View of Thalamic Inputs   79
3.3   The Thalamic Reticular Nucleus   80
3.4   Outstanding Questions   82

4   Classification of Afferents in Thalamus and Cortex   83
4.1   Classifying Glutamatergic Afferents   85
4.1.1 Terminology: Class 1 and Class 2 Glutamatergic Inputs   85
4.1.2 Properties of Class 1 and Class 2 Glutamatergic Inputs   86
4.1.3 Possible Functional Implications for Class 1 and Class 2 Inputs   96
4.1.4 Variation within Classes   100
4.1.5 Known Distributions of Class 1 and Class 2 Inputs in Thalamus and
        Cortex   104
4.1.6 Possible Differences in Properties at Different Terminals of an Axon   113
4.2   Nonglutamatergic Afferents   113
4.3   Concluding Remarks   115
4.4   Outstanding Questions   117

5   First and Higher Order Thalamic Relays   119
5.1   Evidence for Distinguishing First and Higher Order Thalamic Relays   119
5.1.1 Anatomical Evidence   119
5.1.2 Physiological Evidence   121
5.2   Some Differences between First and Higher Order Relays   122
5.2.1 Percentage of Class 1 Synapses   122
5.2.2 Sources of Afferent Input   122
5.2.3 Thalamocortical Relationships   123
5.2.4 Cellular Responses to Modulatory Inputs   123
5.2.5 Burst versus Tonic Response Modes   124
5.3   Developmental and Evolutionary Differences   124
5.3.1 Phylogeny   124
5.3.2 Ontogeny   125
5.4   Role of Higher Order Relays in Corticocortical Communication   126
5.4.1 Direct versus Transthalamic Corticocortical Pathways   126
5.4.2 Significance of Routing the Indirect Pathway through Thalamus   132
5.5   Concluding Remarks   139
5.6   Outstanding Questions   139

6   The Dual Nature of the Thalamic Input to Cortex   141
6.1   A Brief View of the Phylogenetic Origins of Thalamocortical Inputs   141
6.2   Driver/Class 1 Afferents to the Thalamus Are Branching Axons   144

6.2.1 The Afferents to the First Order Relays   144
6.2.2 The Afferents to the Higher Order Relays   164
6.3   Efference Copies and Forward Models   169
6.4   Overall Conclusions about the Branching Driver Axons   173
6.5   Outstanding Questions   177

7   **Linking the Body and the World to the Thalamus   179**
7.1   Introduction   179
7.2   The Inputs to Thalamus   181
7.2.1 The Inputs to the First Order Thalamic Relays   181
7.2.2 The Inputs to the Higher Order Thalamic Relays   191
7.3   Outstanding Questions   193

8   **The Inputs to the Cortex from the Thalamus and the Cortical Descending Outputs   195**
8.1   Early Studies of Thalamocortical Relationships   195
8.2   More Recent Views of the Thalamocortical Pathways   199
8.2.1 Thalamic Afferents to the Temporal Lobe   200
8.2.2 Thalamic Nuclei That Innervate More than One Cortical Area   201
8.3   The Topography of Thalamocortical Projections   202
8.4   Different Types of Thalamocortical Projection   206
8.5   Cortical Outputs   211
8.5.1 The Corticotectal Pathways   211
8.5.2 The Corticopontine Pathways to the Cerebellum   212
8.5.3 The Corticostriatal Pathways   213
8.5.4 Direct Corticobulbar Pathways   214
8.6   Outstanding Questions   215

9   **Thalamocortical Links to the Rest of the Brain and the World   217**
9.1   A Brief Overview   217
9.1.1 First Order and Higher Order Relays   217
9.1.2 Drivers and Modulators   217
9.1.3 The Message and Its "Meaning"   218
9.1.4 The Thalamic Gate   218
9.1.5 Motor Branches of the Thalamic Afferents   219
9.1.6 Cortex Acts through Lower, Subcortical Motor Centers   219
9.1.7 The Functional Capacities of Cortical Areas   219
9.2   First Order and Higher Order Relays   220
9.3   Identifying the Drivers and Modulators   223
9.4   Reading the Message   226
9.5   The Thalamic Gate   228

9.6   The Motor Branches   230
9.7   Cortical Areas Act through Their Connections with Lower Motor Centers   234
9.8   The Functional Capacities of Cortical Areas   236
9.9   Comparing Two Models of Thalamocortical Functional Relationships   237
9.10 Examples of How the Functions of Particular Pathways Can Be Analyzed in
      Terms of the Organizational Principles Summarized So Far   239
9.11 Relating Thalamocortical Connectivity Patterns to Action, Perception, and
      Cognition   240

References   245
Index   275

Preface

This book follows two earlier books by the same two authors: the first one on the thalamus (Sherman and Guillery, 2001) and a second edition reaching somewhat beyond the thalamus to include more of the relationships to cortex (Sherman and Guillery, 2006). We decided that we now needed a distinct book, heading in a different direction but based on large parts of the same thalamocortical foundations. To some extent the title indicates this: we started with the ground rules that we had been able to define for the thalamus and followed their implications for the thalamic inputs to the connections of the cortical areas and their outputs. That is, we argued our way from the thalamus to the rest of the brain and arrived at new ways of thinking about the relationship of the brain to the world, to cognition, and to behavior. However, this is not a book about the brain and the mind. It is a collection of thoughts about neuroscience—the structure and functions of nerve cells and their interconnections. Our earlier thoughts about the thalamus, revisited and somewhat extended in the first three chapters, have led us now to look more closely at how the neural circuits of the brain relate to our actions and our perceptions, how not only people but also many other complex or even quite simple organisms relate to the world.

Two major points have become clear to us as we planned, wrote, and argued about the contents of this book. One, which runs through much of this book and is explored in detail in later chapters, was that a functional and structural analysis of the neural circuits that connect thalamus and cortex leads us beyond these neural centers to lower centers of the brain and through them to the body and the world. That is, not only the cortex as a whole but also each part of the cortex individually, each of the cortical areas that are often treated as distinct organs, is closely linked to the body and the world and relates closely to the way in which we act and think. Each receives inputs from the thalamus, and each has outputs to subcortical centers. This second view of cortical areas all having their own links to subcortical centers and to the body was a recognition that the view of cortex now dominating much contemporary research needs to be challenged. This currently widely accepted view is one that sees the cortex as a large collection of separate, functionally distinct areas that interact with each other in complex hierarchies and pass messages about the body and world to and fro among each other, eventually toward centers that can store memories or produce actions. In this view, the inputs from the body enter through the sensory areas, and the motor areas provide the outputs. The subcortical outputs from the sensory and intermediate areas do not play a part in this schema, nor do the thalamic inputs to higher cortical areas. There are several reasons why such a challenge is needed, and for us they arose directly out of our thoughts about the thalamus.

One element that is missing from this current view of corticocortical communication concerns the distinction between drivers and modulators. This distinction played a significant role in our earlier books, and it is revisited in chapter 4. It was crucial for understanding the thalamus and is proving equally relevant for understanding the functioning of the direct connections between cortical areas (Covic and Sherman, 2011; DePasquale and Sherman, 2011). Drivers carry messages for relay, whereas modulators modify the way in which the messages are relayed. The latter greatly outnumber the former in the thalamus and to a significant extent in the early stages of cortical processing. We have no relevant information about the proportions at higher levels. The nature of the message itself that is passed to the cortex and from one cortical area to another is important for understanding how the functions of any one cortical area are generated from their inputs, but this is unknown for most of cortex. For corticocortical communication, especially at higher levels, we know nothing about the distinctions between drivers and modulators, and we have very little information about the nature of the message. That is, we know about a great many of the connections in terms of their anatomical links, but we have no information about what they do.

A second element that plays essentially no role in most current views of cortical functions is the presence of transthalamic corticocortical connections, which were revealed once it was recognized that there are thalamic nuclei that relay messages from one cortical area to another. These were called higher order relays in our earlier books. These transthalamic links are pathways providing a route additional to the direct corticocortical pathways that dominate current thinking. They are reconsidered in chapter 5. They have two important features that the direct corticocortical pathways lack. One is that the messages have to pass through the thalamic gate, which can transmit a message with reasonable accuracy, block a message, or transmit it in a distorted form with a high signal-to-noise ratio for attracting attention. The condition of the thalamic gate that controls these different actions depends on a balance of excitatory and inhibitory inputs at the thalamic relay cell, and this can vary depending upon the particular relay and on many currently undefined conditions of the organism and the environment. We explore the functions of this gate to some extent but recognize that this is a key area where we need to learn more about how thalamocortical transmission for any one thalamic nucleus under any one set of particular conditions is controlled and how that control relates to action and perception.

Another feature present in the transthalamic pathway and missing from the direct corticocortical pathway, considered further in the next paragraph, is that the messages that come to the thalamus from the cortex are like the messages that travel along all thalamocortical pathways in that they transmit information to cortex about ongoing motor instructions in addition to the more generally recognized messages that they carry about activity at their origin in sensory receptors or other neural centers. We consider the actions of the thalamic gate and the motor copies that tell one cortical area about motor instructions being issued by another cortical area in chapters 5 and 6. They play a role in cortical processing that is almost entirely unexplored in the current literature.

A crucial feature that grew out of our earlier thoughts concerns the fact that the drivers that bring messages to the thalamus for relay to cortex all have branches that innervate motor centers. This raises an issue that puzzled us (and colleagues) for a long time and that is considered in chapter 6. Detailed consideration of the functional implications of these branching axons has led us to treat the driver afferents to the thalamus in a new light. We now suggest that these important inputs to the thalamus for relay to cortex should all be regarded as carrying copies of the motor instructions. Such a view allows recognition of the fact that these axons carry more than one type of message; not only do they bring to cortex information from peripheral centers or lower levels of the nervous system about the world and the body, but they also bring information about the motor instructions that these centers are currently generating and that will be executed after the afferent (sensory) messages have reached cortex. That is, they provide a view of future actions.

In chapters 7 and 8, we look at the way in which thalamus relates to cortex and cortex relates to the rest of the nervous system. We argue that, on the one hand, every cortical area receives inputs from the thalamus and that these inputs bring information about events in the world and in the rest of the nervous system to each area. On the other hand, every cortical area also has the capacity to influence the lower motor outputs through one or another of phylogenetically old centers relevant for the control of behavior. These include not only the spinal cord and the brain stem but also the basal ganglia, the tectum, and the cerebellum. That is, to understand the functions of any one cortical area, one needs to ask questions not only about the conditions under which the cells in that area become active or questions about how that area may be connected to other cortical areas. These are currently the two most common approaches: defining complex behavioral or perceptual situations, then showing which areas become more active in any one such situation, and subsequently exploring the corticocortical links of the active area. We argue that one also needs to ask about the nature of the thalamic inputs to that area and the motor outputs coming from that area. One needs to know which axons forming the connections are drivers and which are modulators, and when one has identified the drivers it will become of interest to define the nature of the message that is being transmitted. To understand the functions of the phylogenetically most recent, highest cortical centers concerned with cognitive functions, it will be necessary to learn how they are linked to the phylogenetically oldest parts of the nervous system that relate to the body and the world.[1]

These considerations raise a lot of unanswered questions about the thalamus, the cortex, and the rest of the brain. It is fashionable to ask, and we are often asked, what are the hypotheses that we are testing? We mention several in the course of the book, but the key consideration is about the questions we are asking about currently unexplored aspects of how thalamus and cortex relate to the rest of the brain. As in our previous books, we provide a short list of currently unanswered questions at the ends of chapters 3 to 8. These are not intended as questions that students should be able to answer in an exam but are questions that need to be answered if we wish to understand how the brain relates to the body and the world. It will be obvious to all that the questions we list represent but a fraction of the questions raised in the book and waiting to be attacked by the next generation of neuroscientists.

This book, as its two predecessors, is planned for individuals with a reasonable background in neuroscience at the undergraduate or graduate level. It should be of interest to anyone concerned with thinking about what the central nervous system may be doing. We have tried to let each chapter stand on its own so that it can be read independently of the other chapters. This has involved some repetition of subjects treated in one chapter in others. Chapter 9 provides an overview, but it has to be stressed that the detailed arguments supporting the views summarized in chapter 9 are in the earlier chapters. We

---

1. Evans (1982) has written, "Conscious experience results when 'the internal states which have a content by virtue of their phylogenetically more ancient connections with the motor system also serve as input to the concept-exercising and reasoning system." Here, a philosopher has anticipated one of the major conclusions of our book. We thank Andy Clark for citing this wonderfully percipient statement.

have avoided abbreviations as far as possible; they are a horrible feature of many current publications that often make comprehension difficult and usually do not save much ink. The few abbreviations that we use throughout the book are listed on page xvii at the beginning of the book.

We owe thanks to many colleagues who have discussed some of the issues we raise in the book or have read one or another of the chapters. These include Richard Boyd, Kevin Cheng, Christian Hansel, Jason MacLean, David Freedman, Peggy Mason, Kouichi Nakamura, Leslie Osborne, Luis Populin, Cliff Ragsdale, and Maria Wai. Robert Prior of MIT Press encouraged and advised us and helped us to learn what the publishers need. We thank the authors whose illustrations we have used for permission for their use, particularly Professor H. Ojima for sending us unlabeled high-definition versions of the images used in figure 3.7. Finally, we thank Marjorie Sherman for proofreading the entire manuscript.

Abbreviations

We have, as far as possible, avoided the use of abbreviations. Here, we list the ones that are commonly used by neuroscientists and widely recognized for complex names. They are the following:

A1   primary cortical auditory area A1

A2   secondary cortical auditory area A2

AMPA (R,S)-α-amino-3-hydroxy-5-methyl-4-isoxazolepropionic acid

EPSP(C)   excitatory postsynaptic potential (current)

GABA   γ-aminobutyric acid

GAD   glutamic acid decarboxylase

IPSP(C)   inhibitory postsynaptic potential (current)

NMDA N-methyl-D-aspartate

PSP   postsynaptic potential

S1   primary cortical somatosensory area S1

S2   secondary cortical somatosensory area S2

S3   cortical somatosensory area S3

V1   primary cortical visual area V1

V2   secondary cortical visual area V2

V4   cortical visual area V4

1 An Introduction to Thalamocortical Pathways

1.1 An Overall View of the Thalamus and Cortex in Relation to the Rest of the Brain

In this chapter, we introduce some of the major functional and structural observations and interpretations that provide the basis of this book. These are more fully developed and related to each other in the last six chapters of the book. In the first three chapters, we introduce some basic facts and concepts about the structural and functional relations of thalamus and cortex and about neural and synaptic properties in general so that a reader with limited neuroscience knowledge can have enough information to follow the material in the last six chapters, in which we introduce a number of distinct observations of the thalamocortical pathways that lead to a view of thalamic and cortical functions substantially different from a currently widely accepted view prevalent in much current literature and in most textbooks. We summarize some of the major differences later in this chapter and in figures 1.1 and 1.2. These two figures illustrate only the main points in which the two views differ. Many details have been left out for the sake of clarity and simplicity. The figures show some of the major connections that thalamus and cortex establish with each other and with the rest of the brain. They are discussed in more detail later in this chapter and more fully in the rest of the book. However, these figures do not illustrate other major differences, which are more difficult to illustrate and which concern the functional properties of the pathways shown in the figures and the extent to which they differ from one another and can change depending on local conditions. These are briefly described in section 1.2 and fully considered in chapters 4 and 5. Nor do the figures illustrate the important connections that outputs of the cerebral cortex establish with the rest of the brain except to indicate that they involve very much more than the outputs of the motor cortex (see figure 1.2).

There is one other crucial point that is easily recognized when sensory or motor pathways near the periphery are under consideration but is often lost

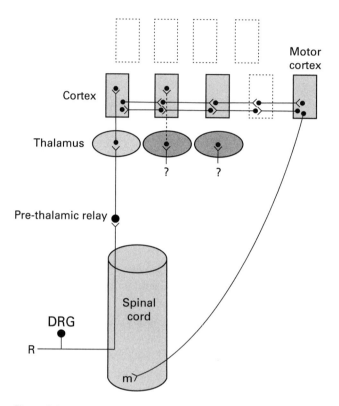

Figure 1.1
Schematic view of the major thalamic and cortical connections as seen in most current accounts. The sensory input from the spinal nerves shown here is representative of sensory inputs in general. Three thalamic nuclei are shown as ovals, and nine cortical areas are shown as rectangles. Only the thalamocortical and the corticofugal connections of the cortex are shown; cortical areas not shaded would have rich corticocortical connections, which are not illustrated here. Only the driver (Class 1) inputs, which transmit messages from the world and from other neural centers (see text), are shown in this figure and in figure 1.2. Modulatory connections, which do not transmit messages but modulate the way that messages are transmitted or processed, are not shown. In this schema, the information about events in the world once it reaches the cortex is thought to be processed through the several cortical areas, passing through a complex pattern of parallel and hierarchical feedforward and feedback connections, only some of which are shown in the figure, eventually reaching either areas for memory storage (not shown in the figure) or to the motor cortex for a motor output to lower centers and a contribution to the control of behavior. It is important to note that, on this view, once information reaches cortex through a primary thalamic relay, it is processed entirely within cortex through corticocortical connections without reference to any subcortical structures. There is a single entry point for each sensory modality for cortical processing (primary sensory cortex) and a single exit point (motor cortex). This view of thalamocortical connections can be found in most contemporary textbooks and is in contrast to the view on which this book is based, which is schematized in figure 1.2. DRG, dorsal root ganglion; m, lower motor center; R, receptor.

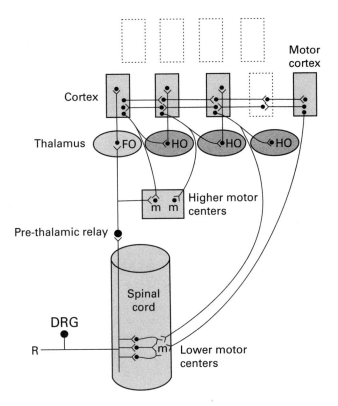

Figure 1.2
Schematic view of the major thalamocortical and corticofugal connections as viewed in the current account. Conventions and abbreviations as in figure 1.1, except that the unshaded cortical areas would also be in receipt of thalamic input and have corticocortical connections as well as corticofugal outputs to motor centers. The four major differences between figure 1.1 and figure 1.2 are the following: (1) Whereas some thalamic nuclei in both figures are shown as receiving ascending inputs for relay to the cortex from lower centers (FO in figure 1.2), a major part of the thalamus, commonly referred to as "association nuclei" that receive few or no such first order inputs, is shown with question marks for inputs in figure 1.1 and with inputs from cortex in figure 1.2. (2) Figure 1.2 defines two functionally distinct thalamic relays (first order, FO, and higher order, HO) that are not present in figure 1.1. (3) Figure 1.2 also provides a novel transthalamic route for corticocortical communication. (4) Figure 1.2 also shows all of the inputs to the thalamus, the ascending inputs from the sensory periphery to the first order relays as well as the corticothalamic inputs to the higher order relays as sending motor branches to lower motor centers. This indicates that essentially all cortical areas have access to motor outputs and shows that the messages that reach the thalamus are copies of messages that are on the way to motor centers, a relationship that is considered more fully in the text and in chapter 6. FO, first order thalamic relay; HO, higher order thalamic relay.

sight of for other parts of thalamus and cortex. This concerns the nature of the messages that are transmitted by particular pathways, which are about something specific happening in the world or in the nervous system, and the equally crucial question as to whether the pathways transmit a message at all or instead serve to modify the way in which a message is transmitted or processed. We have previously called the former the drivers and the latter the modulators. Messages transmitted over significant distances depend on action potentials that can be read as a code representing events in the world or in other parts of the brain. The nature of such a message is changed (elaborated, related to other messages) to a greater or lesser extent as it is passed from one neural center to another, but so far as we know, it always represents events in the world or in the brain, and one challenge that has not been met yet for large parts of the brain is to determine first whether a particular pathway either serves to transmit these messages (is a driver) or else serves to modify the way in which the messages are transmitted (is a modulator). Then if it is a driver, it is crucial to define the nature of the message (that is, to read the code), and if it is a modulator, the messages that are modulated need to be identified as well as the nature of the modulation. We stress that the distinction outlined in this paragraph provides a key to the different actions of drivers and modulators. We do not argue that modulators transmit no information, but rather that they do not transmit information about specific events in the world or the brain that can be interpreted by the recipient cell in terms of their source and their meaning. That is, the drivers carry an interpretable pattern of action potentials that comes from an identifiable pathway, often called a "labeled line."

We will treat the vertebrate nervous system as a series of interconnected structures arranged as a hierarchical series. The lowest is the peripheral nervous system, which links the brain to the body and the world, and the highest is the cerebral cortex. The peripheral nervous system contributes afferents, which bring messages to the central nervous system from sensory receptors, and efferents, which carry messages from the central nervous system to muscles and secretory organs. We are mainly concerned with the somatic nervous system and will be dealing primarily with messages that come from the skin, the muscles, and the joints and from the distance receptors in the eye and ear and are passed to the thalamus for relay[1] to the cerebral cortex. We include the taste receptors and the vestibular receptors but have

---

1. *Relay* used as a verb or a noun is often taken to mean a machine-like, often electronic, unerring passage of a message, although the main dictionary definition in the Oxford and Webster dictionaries is to a set of fresh horses substituted for tired ones. We regard the thalamic relay as more like a relay inn, where the horses would be changed, the travelers might eat a meal, spend a night on their own or with others, and leave refreshed and perhaps significantly changed.

little to say about the autonomic or vegetative nervous system, either its afferents from the viscera or its efferents to the gut, gut derivatives, and glands. This is not because we regard this part of the nervous system as unimportant for understanding thalamus and cortex, but because at present we have a full program in addressing the somatic afferent and efferent nerves that play a key role in current knowledge of thalamus and cortex and much less is known of the visceral role of these two structures. We also leave out the olfactory receptors, because olfactory information is relayed via the olfactory bulb to the phylogenetically old paleocortex and has no immediate thalamic relay. How specific patterns of olfactory inputs travel from paleocortex to the phylogenetically newer neocortex that is linked to the thalamus has not yet been clearly charted.

In the central nervous system, the lowest and phylogenetically oldest centers are in the spinal cord and brain stem. They receive afferents from the periphery, send efferents to the periphery, and play a role in simple reflex reactions like the withdrawal response to painful stimuli or the stretch reflex seen, for example, in the reflex knee jerk (Sherrington, 1906). More complex reactions are represented in spinal mechanisms that have been described as "central pattern generators" (Grillner, 2003; Grillner et al., 2007; Rossignol and Frigon, 2011). Their action can be demonstrated in a spinal cord that has been disconnected from all higher centers. They are capable of producing walking or running movements that will adjust to the speed of a walkway, adjusting the speed and the pattern of the movements in a species-appropriate manner. Within the brain stem, there are other pattern generators that are concerned with actions such as breathing, chewing, and swallowing (figure 1.3).

The brain stem and spinal cord contain a variety of lower motor centers, not all of which can be readily defined as central pattern generators. The activity of all of these lower motor mechanisms is controlled from two major sources: one consists of the afferents that enter from the peripheral nerves bringing messages from the many different types of sensory receptors, and the other comes from efferents of higher centers. Each plays an important role in the control of movements. There are five major higher control centers in mammals that have to be included in any consideration of the way in which thalamus and cortex relate to behavior. These are the cerebral cortex itself, the basal ganglia, the cerebellum, the superior colliculus, and cell groups in the reticular formation of the midbrain and pons (figure 1.3). The last four are phylogenetically old centers present in nonmammalian vertebrate brains (Grillner et al., 2007; Stephenson-Jones et al., 2012). The cerebral cortex has a phylogenetically old olfactory portion (the paleocortex) and a hippocampal portion (also known as archicortex) as well as a newer neocortex, which is

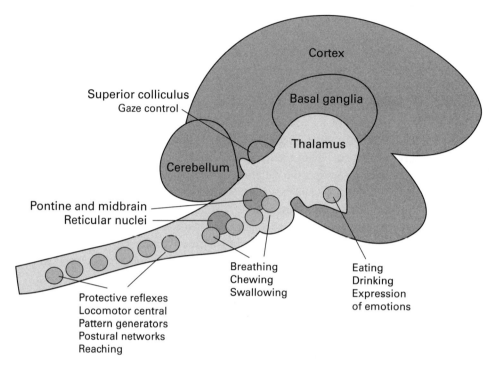

Figure 1.3
Schematic representation of central pattern generators and lower motor centers shown as small pale-gray circles, which receive inputs from several different higher motor areas shown in a darker shade of gray. Based on details in Grillner (2003). Further details in the text.

the part of the cortex that is the major recipient of thalamic inputs.[2] We shall refer to neocortex throughout the book simply as cortex, and all parts of the cortex connect either to the other four control centers or to the lower motor centers themselves. The pathways by means of which the mammalian higher control centers interact with each other and with the lower motor centers are currently not very well understood. Anatomically, these higher control centers are richly interconnected. The cortex sends major outputs to the other four, and these four have outputs that go back through the thalamus to the cortex. However, the influence of any one of these higher centers upon activity in any of the others is not currently well defined. The thalamus sends information about the world, the body, and lower centers of the brain to the cortex, and each area of the cortex sends its outputs for action to one or more of the

2. In nonmammalian vertebrates, neocortical structures are absent, although homologs have been suggested (Rowell et al., 2010; Butler et al., 2011; Dugas-Ford et al., 2012).

higher control centers or directly to the lower motor centers. The cortex can only produce action through these subcortical centers, and all cortical areas have links to several of these subcortical centers. This last point is often left out of consideration when the functions of any one cortical area are being considered.

One key question that we will be asking about thalamus and cortex concerns how they relate to the control of behavior, and an answer will necessarily depend on how these four higher control centers relate to each other and to the lower motor centers themselves. We cannot study cortex on its own, either as a whole or one cortical area at a time, and hope to understand the cortical role in the control of behavior without learning more about the functional organization of these links. Nor can we reasonably expect to understand the role of any one cortical area by looking at its connections with other cortical areas and hoping to assess its role in behavior by tracing its links through other cortical areas to the motor cortex alone (see figure 1.1), ignoring the many interactions that are relevant to understanding the final direct output that any one cortical area sends to the other motor control centers. We will argue that all cortical areas connect more or less directly with one or more of the other four higher control centers or with the lower motor centers and that these connectional patterns need to be understood for any cortical area if we wish to understand its particular role in behavior. Our book is concerned with drawing attention to the missing information. We will be raising questions about unknowns as much as introducing some new knowns. As in our earlier books on thalamus, we will include a short list of major questions at the end of chapters 3–8, not to serve as a quiz for students but as a focus on major areas where answers are not currently available and are badly needed.

1.2 Thalamocortical Connections

We have focused the book heavily on the thalamic nuclei and their connections with structurally and functionally distinct cortical areas, and in figure 1.4 we show a schematic and simplified view of the major relationships between thalamic nuclei and cortical areas. Although, as we shall see in chapter 7, there are many differences in the literature about the details, and although many of the relationships will appear more complicated than this figure shows, the basic relationships illustrated are widely understood and provide a useful basis for our account. Essentially each major thalamic cell group or nucleus relates to a specific cortical area, relaying messages from lower centers or from other cortical areas. The differences illustrated in figures 1.1 and 1.2 relate to the pathways that connect thalamus and cortex to each other and also to the rest

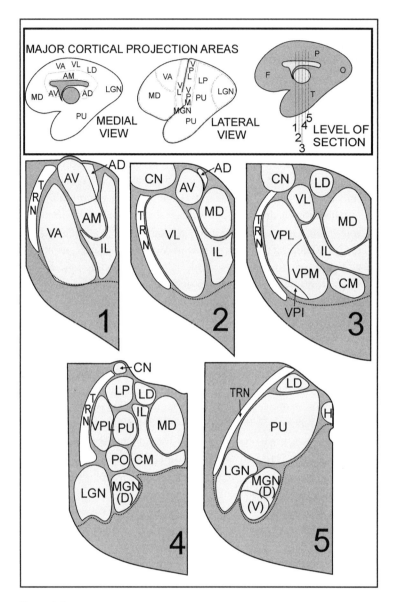

Figure 1.4
Schematic view of the major thalamic nuclei identifiable on five coronal sections (numbered 1 to 5) through the mammalian thalamus. The major cortical projection areas of the thalamic nuclei are shown at the top of the figure together with a view of the section levels illustrated in the main part of the figure. AD, anterior dorsal nucleus; AM, anterior medial nucleus; AV, anterior ventral nucleus; CM, center medial nucleus; CN, caudate nucleus; IL, intralaminar nuclei; H, habenular nuclei; LD, lateral dorsal nucleus; LGN, (dorsal) lateral geniculate nucleus; MD, medial dorsal nucleus; MGN (D and V), dorsal and ventral medial geniculate nucleus; MGN(v), ventral medial geniculate nucleus; LD, lateral dorsal nucleus; LP, lateral posterior nucleus; PO, posterior nucleus; PU, pulvinar; TRN, thalamic reticular nucleus; VA, ventral anterior nucleus; VPI, ventral posterior inferior nucleus; VPL, ventral posterior medial nucleus; VPM, ventral posterior medial nucleus.

of the brain, and, as indicated, only some of these differences are illustrated in figures 1.1 and 1.2.

Figure 1.2 shows some of the basic connection patterns that our account is based on. They are considered in detail in chapters 3, 5, 6, and 7. Figure 1.2 shows that we distinguish first order thalamic relays from higher order relays on the following basis: inputs to first order thalamic relays come from either ascending sensory pathways or subcortical brain centers (first order, because they are sending to cortex reports about events in the world or in subcortical structures that have not previously been reported to cortex), whereas inputs to higher order relays come from cortex (higher order, because here information about events already in cortex is now being relayed though these thalamic nuclei from one cortical area to another). First order and higher order relays are reintroduced in chapter 2 and discussed in more detail in chapter 5, and in that chapter we present evidence that the transthalamic corticocortical connection has functional capacities that differ significantly from those of the direct corticocortical connections shown in both figures 1.1 and 1.2. Figure 1.2 also shows that instead of a single exit point for information processing from areas of motor cortex to the lower centers, all of the cortical areas send a message down to subcortical motor centers. It also shows that all of the inputs to the thalamus, the ascending ones entering the first order relays as well as the corticothalamic ones entering higher order relays, come from branching axons whose other branch(es) innervate(s) one or more motor centers.

Because the branch innervating the thalamus (both first order and higher order) carries the same message as that innervating motor structures, these messages bring information to cortex about ongoing instructions for upcoming motor actions. They inform cortex about probable future motor actions. For example, a message about an instruction for an upcoming eye movement would serve to distinguish apparent motion in the visual world (caused by the eye movement) from real motion in the world. This overall monitoring of motor outputs provides information about actions that we have generated ourselves. When the actions occur, they are expected on the basis of information about the prior brain activity that made them happen. They are distinct from events produced by changes in our environment and contribute to the distinction between self and non-self. This capacity for the thalamocortical pathways to distinguish changes in the environment from those produced by the organism itself has in the past been widely recognized. Such a distinction has been considered to be dependent on "efference copies" or "corollary discharges" (Sperry, 1950; von Holst and Mittelstaedt, 1950; von Holst, 1954; Wurtz, 2008), which are messages about instructions for motor actions fed back into the central circuits, where they can serve to correct motor actions and allow for evaluation of the significance of sensory changes. We will show that essentially all of the messages that the thalamus sends to the cortex contain such an efference copy (see chapter 6).

Two other features where our account differs from ones more widely accepted at present are not illustrated in the figures and concern the nature of the relay in thalamus and in cortex. The thalamus has for long been described as the "gateway" to the cortex. This was usually interpreted as a way into the cortex through which messages from the lower centers and the outside world must pass before they can reach the cortex. As pointed out by Crick (1984), this is an unusual gateway because messages that go to the cortex pass through it, whereas messages that leave the cortex take a different route. The view of the thalamus as an entrance to a city or a castle is of limited use, but we will find it is helpful to recognize that there are two respects in which the thalamus in fact acts like a gate that can be open or shut or partly open. One is that only a minority of the synapses in the thalamus (less than 10% and in some nuclei much less) transmit messages that are relayed to the cortex; that is, less than 10% pass through the gateway. As indicated above, we earlier called these *drivers* and called all of the others *modulators* (Sherman and Guillery, 1998) because although the modulators do not carry messages that are passed on to the cortex, their activity can change the way that the driver message is passed to cortex and even whether it is passed to cortex at all. That is, the thalamic "gate" functions in two ways. One is to relay information to cortex from a very limited proportion of its inputs (a gate that separates the sheep from the goats or lets through only a portion of the sheep), and the other is to modify the way in which that relay is functioning on the basis of ongoing activity in the modulators; in this sense, it is more of a filter than a gate. The drivers that come from the cortex to higher order nuclei arise from cells in cortical layer 5. Such drivers do not go to first order relays (by definition). However, corticothalamic modulators, which arise in cortical layer 6, innervate all thalamic nuclei and form numerically a much greater input than the drivers.

In general, it is important to distinguish drivers from modulators in order to understand how any neural connections relate to behavior or cognition. Modulators include a variety of different inputs, including GABAergic, cholinergic, serotonergic, and others, whereas the known drivers, as well as some modulators, are glutamatergic, and in practice we need to know which glutamatergic inputs are modulators and which are drivers; most of the further discussions concerning the distinction between drivers and modulators will be limited to glutamatergic inputs (see chapter 5).

A summary of the points that we have made in this section shows that a minority of the inputs to the thalamus carries messages that are relayed to the cortex. These messages are copies of motor instructions that allow higher areas of the nervous system to be kept informed about motor instructions that are being issued by lower centers. The major part of the input to thalamus plays a role in determining whether at any one time the message will be sent to cortex or not, and if it is sent whether it will be passed as a close copy of the input or in some other form. The details of this are presented in chapter 5.

1.3   Tracing the Message

A significant part of the book is concerned with the problems involved in tracing information flow through the brain and in defining the nature of the information. Ideally, we need to understand the nature of the messages that are relayed from the thalamus to the cortex. The fact that we know more about information flow in the sensory pathways than in other parts of the brain is in part due to the fact that the anatomy of the pathways has been defined in considerable detail but owes more to the fact that the nature of the information appears to be relatively easy to define in terms of receptive field properties that are transmitted from one cell group to another and modified on the way. Defining the nature of the information in many thalamocortical pathways where we know nothing about receptive fields is a difficult question and one that is distinct from knowing the pathway, although knowing where a pathway comes from or goes to is often a useful clue as to the sort of information that may be involved.

For these reasons, the best place to start a consideration of thalamocortical relationships is with the thalamic relays for the major sensory pathways, the visual, auditory, or somatosensory pathways. Of these thalamic relay nuclei, the lateral geniculate,[3] ventral medial geniculate, and ventral posterior nuclei respectively (figure 1.4), each receives its driver inputs from a single group of afferent pathways and sends the great majority of its thalamocortical axons to one related cortical area (figure 1.4). These thalamic relay nuclei have the same basic structural characteristics when seen in electron micrographs, with the driver inputs clearly identifiable by morphological criteria, and have generally the same populations of modulators, with some exceptions (see chapter 5). Functionally, the thalamic cells in each of these nuclei show receptive field properties that are remarkably similar to those of their driver inputs. That is, the thalamic relays appear not to change significantly the nature of

3. This should strictly be the dorsal lateral geniculate nucleus, but common usage generally omits "dorsal" unless there is a possibility for confusion with the ventral lateral geniculate nucleus, which is not part of dorsal thalamus and does not project to cortex.

the message that is transmitted. Other thalamic nuclei in general show the same basic structural features, although as we progress we will note some exceptions.

Because it was possible to identify the messages that are being processed in each of these sensory thalamocortical pathways and to trace the messages as they pass from thalamus to cortex and then through further cortical relays [for example, for the visual pathways (Hubel and Wiesel, 1962; Hubel and Wiesel, 1965)], the sensory nuclei and their cortical targets have served as an important model for understanding the basic functional organization of thalamic nuclei. If we ask what is known about the nature of the messages that are processed, there are many thalamic nuclei and associated cortical areas, other than the sensory relays, for which we have no answer. We do not know about the nature of the activity in their driver afferents and often do not know which of the afferents are their drivers. We know that the smallest of the three anterior thalamic nuclei (the anterior dorsal nucleus) relays messages about head position (Taube, 2007), and it turns out that this is a sensory (vestibular) relay, but we know almost nothing about the nature of the messages that many of the other thalamic nuclei are transmitting to the cortex. Where little or nothing is known about the nature of the messages carried by the afferents to a particular nucleus, it will take a clever insight based on known connectivity patterns or a piece of luck during a recording session to define the nature of the message that is being passed from thalamus to cortex. This is a piece of information that is badly needed for many thalamic nuclei, and it may be very difficult to get. Even in the visual cortex, defining the nature of the stimuli to which cortical cells would respond initially depended on a piece of luck described by Hubel and Wiesel (2005), who found that a cortical cell responded strongly to the edge of a slide that carried the visual stimuli to which none of the cortical cells had been responding.

It may prove to be a serious mistake to think that the responses recorded from a nerve cell are interpreted by the brain in terms of exactly the same meanings as they are by the investigator. We will see examples in chapter 6 where the functional significance of a message can turn out to have more than one interpretation, because the same message is carried by different branches of one axon, and the interpretation of that message as, for example, visual sensory, depends on the different targets of the axon branches.

These thoughts about the nature of the message that any one thalamic nucleus sends to the cortex are important, because they stress another large area of current ignorance. For many thalamic nuclei, we interpret their functional role in terms of their connections. Thus, we can describe the ventral lateral nucleus as receiving inputs from the cerebellum and sending outputs to the motor cortex, the pulvinar as receiving inputs from visual cortex[4] and also from the superior colliculus and in turn sending outputs to higher areas of cortex, the medial dorsal nucleus as receiving inputs from the amygdala and from the prefrontal cortex and sending outputs back to prefrontal cortex, and the anterior medial nucleus as receiving inputs from the mamillary bodies and sending outputs to the anterior limbic cortex. These admittedly oversimplified descriptors provide some clues about possible functions, expressed in terms of what is known about the functions of the thalamic afferents or the cortical areas concerned, but they tell us little or nothing about the nature of the messages that are being relayed from the thalamus to the cortex. Another problem of basing a function of a thalamic relay entirely on the structures that are connected to that relay is that this interpretation assumes that the input to the thalamus is a driver. Two examples are described in chapter 4 for the ventral anterior nucleus, which receives input from the globus pallidus, commonly treated as input for relay to motor cortex, and for the dorsal medial geniculate nucleus, which receives input from the inferior colliculus, commonly treated as input for relay to auditory cortex; in each case, there is now good evidence that neither of these inputs to the thalamus acts as a driver (Smith and Sherman, 2002; Lee and Sherman, 2010). The available anatomical evidence on its own did not suffice to tell us that these are driving inputs. Where clues about the functions of the cortical areas are available, they may be helpful in discovering the nature of the messages that are relayed through the thalamus, but examples are rare.

Although we are mainly interested in how information about the world, the body, or other brain parts reaches each cortical area and what actions each area has on lower centers, we also need to learn about the way in which the message is processed within the cortex. And here, as in the thalamus, one important step is to distinguish drivers from modulators (or other types of input if they exist). There are two ways in which that distinction can be made. One that is direct is to show that the incoming and outgoing messages are closely related. Another that is more indirect is to show that the actions of the inputs on membrane properties and postsynaptic receptors have properties that favor the transmission of a message. Recent studies have produced some useful results of this sort for the glutamatergic thalamocortical and corticocortical

---

4. In the primate, the anterior pulvinar has connections with somatosensory cortex, but in other species (for example, rodents and carnivores), the combination of pulvinar and lateral posterior nucleus appears to be essentially limited to vision. Whether there are parts of this complex in other animals that are somatosensory and have not yet been charted or whether the nomenclature "anterior pulvinar" for the primate reflects a mislabeling based on incorrect homologies is not clear.

pathways, and these are described in chapter 4. For reasons discussed there, it is generally not possible actually to identify a driver function in terms of the message that is transmitted, but it is possible to define properties that are either necessary for driver functions or that would make a driver function unlikely or impossible. This more indirect evidence about driver or modulator functions has led to the suggestion (Covic and Sherman, 2011; Viaene et al., 2011b, c; Petrof and Sherman, 2009) that it is better to speak of glutamatergic inputs as Class 1 (drivers) or Class 2 (modulators) where direct evidence of the transmission of the message is not available, and in the rest of the book we adopt that terminology when the actual function, driver, or modulator of the input remains unclear, using the driver/modulator terminology only where it is justified and adds to clarity. That is, the Class1 and Class 2 terminology describes the *properties* of the input, and in the few cases where the *function* of the input is also known, we can refer to Class 1 as driver and Class 2 as modulator; however for most, we extrapolate from these examples to offer the hypothesis that Class 1 inputs are drivers, and Class 2, modulators.

These studies of the Class 1 and Class 2 inputs in cortex indicate the degree to which it will be important to learn a great deal more about how the inputs to any one cortical area are processed to produce the outputs. We regard the details of the relevant cortical circuits as a current work in progress in many laboratories, with details that are (necessarily) beyond the aims and the scope of this book. Thus, this is not a book about the details of the structure of the cortex in itself, nor is it a book about many of the details of the brain beyond the thalamus in relation to the cortex.

1.4   An Overall Summary of Some of the Major Points

We view the cortex as several functionally distinct areas each receiving distinct inputs from the thalamus and each having outputs to subcortical centers. The subcortical motor centers themselves have motor actions that are often complex and efficient and may be able to function with little or no cortical control. The following list includes a brief summary of some of the major points that are raised in the following chapters.

1. Whereas some thalamic cells relay messages that come from ascending pathways to the thalamus, many thalamic cells in the so-called association nuclei relay messages that come from the cortex itself. On this basis, we have distinguished first order thalamic relays, which receive messages from subcortical centers, from higher order relays, which receive messages from the cortex and that we no longer refer to as association nuclei. For information about the

world or activity in subcortical neural centers that has already been processed once in one cortical area, higher order relays provide a transthalamic input from that cortical area to another, higher, cortical area.

2. We shall show that in thalamus and cortex, the glutamatergic inputs in cortical and thalamic circuitry are either Class 1 or Class 2. We suggest that in both thalamus and cortex, Class 1 inputs are the drivers, bringing information to the postsynaptic neurons for further processing, and that Class 2 inputs are the modulators, modulating the action of the driver inputs.

3. There are two functionally and morphologically distinct types of cortico-thalamic axons. Those coming from cells in layer 6 innervate all thalamic nuclei and are Class 2 (modulators). Those coming from layer 5 only innervate higher order thalamic relays (by definition), and all the evidence currently available indicates that they are Class 1 (drivers). Overall in the thalamus, the modulators greatly outnumber the drivers.

4. We show that the great majority of driver inputs to the thalamus, possibly all, are branches of axons that innervate lower motor centers. These branched inputs provide the thalamocortical pathways with information that is not pure sensory, including also information about forthcoming actions.

5. All cortical areas appear to receive thalamic inputs and to send outputs to subcortical motor centers. It will be important to define which of the thalamo-cortical axons act as drivers in the cortex, to define the extent to which any one cortical area depends on its thalamic inputs, and to define the extent to which an area may be dependent on its cortical inputs for the information that is processed.

6. The contribution that one cortical area makes to the functional properties of any other cortical area will depend on whether the corticocortical connections are Class 1 or Class 2. They will also depend on the extent to which the corticofugal pathways to lower motor centers interact with each other and on the interactions between the direct and transthalamic pathways that connect cortical areas. This is all largely terra incognita at present.

Although a large part of the material we present serves to focus on issues about which we currently need more information, a significant part presents new views of thalamocortical relationships, and these stress the importance of following the message from receptors through central relays to the motor outputs and indicates that essentially everywhere in the brain, these messages are neither pure sensory nor pure motor. They are messages that have sensory and motor components and that reveal us as organisms actively involved in the world, not as passive observers of a world projected onto some internal cerebral screen.

2 Cell and Synaptic Properties

2.1 Intrinsic Cell Properties

Thalamic neurons, like many other neurons in the brain, receive synaptic inputs from many sources, mostly onto their dendrites, and they produce an output via a single axon.[1] How these synaptic inputs are integrated to affect the axonal output is an important basic issue. This integration is strongly influenced by intrinsic properties of the cell.

2.1.1 Cable Properties

A starting point for determining how synaptic inputs combine to affect the axonal output of a cell is to use a simple, linear model of integration in which the dendritic arbor acts like a passive, branched cable to transfer currents, generated by synaptic inputs, to a region where the axon joins the cell body [the axon hillock, and the adjacent unmyelinated initial segment of the axon; here, the level of depolarization reached determines when the axon fires action potentials and at what rate (Jack et al., 1975; Rall, 1977)]. This mathematical process is known as "cable modeling." Before developing this point, it is important to recognize the many assumptions such cable modeling requires. Chief among them is the requirement that the dendrites be passive, meaning that they do not support active processes in the form of voltage or other gated conductances. Also, various parameters, which are often impractical to measure in neurons, such as electrical resistance of the cytoplasm and electrical resistance and capacitance of the membrane, can be estimated from measurements made in other cell types but have not been explicitly tested for neurons, and

---

1. Among the exceptions to this, and described in chapter 3, are the thalamic interneurons, which have multiple dendritic outputs as well as the single axonal output.

these parameters are typically, but not always, assumed to be uniform spatially and temporally (Jack et al., 1975).

Imagine that a synaptic input at some specified dendritic site locally changes an ionic conductance in the postsynaptic membrane, allowing current in the form of charged ions to flow into (or out of) the cell at that site. How does this event change the membrane potential at the soma or axon initial segment? This will depend on the flow of current that is initiated and on how it spreads through the cell. Figure 2.1 schematically illustrates this. If the membranes were not permeable to ions (that is, had infinite electrical resistance), any current injected into the dendritic process by a synaptic event would flow within the cell, although it would divide at the entry point to flow in both directions, toward and away from the soma (figure 2.1A). However, neuronal membranes are leaky or permeable to ions, and thus some of the current will flow out of all parts of the dendrites and the cell into the extracellular spaces (figure 2.1B). Furthermore, the actual pattern of current flow is mostly related to the complex, three-dimensional architecture of dendritic arbors (figure 2.1C). As noted, not all the current will flow directly to the soma because some will flow in the opposite direction toward more distal dendrites. As each dendritic branch point is passed in either direction, the current flow will divide from the active branch into adjacent branches. Thus, if we start with the current flowing toward the soma, once a branch point is reached, the current will divide, some heading down the parent branch toward the soma, and some heading out along other branches away from the soma (figure 2.1C). This process is repeated as other branch points are reached, regardless of the direction of current flow toward or away from the soma, until either the soma or a dendritic ending is reached. Some of the current eventually reaches the soma and axon initial segment, depolarizing or hyperpolarizing these structures. If the depolarization is sufficient, one or more action potentials will be generated, and if the hyperpolarization is sufficient, firing will be reduced, cease, or be prevented.

For example, activation of an excitatory synapse causes positive current to flow into the cell. Often this is in the form of opening Na^+ or Ca^{2+} channels, allowing these positive ions to flow into the cell. An inhibitory synapse, by opening other channels, will do the opposite, causing positive current to flow out of the cell. For instance, the opening of Cl^- channels causes negative Cl^- to flow into the cell, which is formally equivalent to positive current flowing out of the cell, and the opening of K^+ channels causes these positive ions to flow out of the cell. The effect of an inhibitory synapse on current flow inside the cell can be imagined by reversing the direction of the arrows in figure 2.1, and this direction of current flow within the cytoplasm will eventually cause

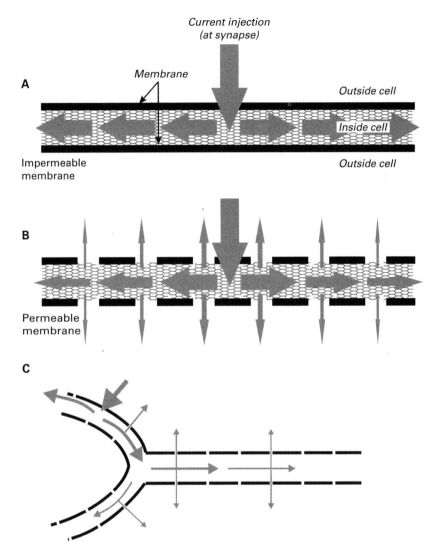

Figure 2.1
Schematic view of current flow in dendrites as a result of current injected into the process (for example, by synaptic activation). The gray arrows indicate the direction of current flow, and their width indicates current amplitude. (A) Example of an impermeable membrane. All of the current flows within the process, although it divides to flow in both directions from the point of current injection. (B) Example with partly permeable membrane. As current flows inside the dendrite, some of it leaks across the membrane, so that the farther from the injection site, the smaller the amplitude of the remaining intracellular current flow. (C) Effect of dendritic branching. As the current flow reaches a dendritic branch point, the current divides to enter all of the resultant branches.

hyperpolarization at the axon initial segment, which, as noted, will reduce the frequency of action potential firing or block it.

Estimating how much current flow reaches the axon initial segment and thus how much of a voltage change is induced can be done by cable modeling, which remains the best means of relating these parameters to the morphology of the neuron (Jack et al., 1975; Rall, 1977).

The cross-sectional area of a dendritic segment is proportional to the square of its diameter, but the surface membrane area is only linearly proportional to the diameter. Thus, the ratio of its cross-sectional area to its membrane area for a dendrite increases with increasing thickness. Because the relative resistance of the cytoplasm is related to its cross-sectional area and that of the membrane to its total area, the larger the ratio of cytoplasmic cross-sectional area to membrane area, the lower the resistance of the cytoplasm relative to that of the membrane. As a result, a thicker dendrite allows more current from synaptic activation to flow down the path of least resistance through the cytoplasm to the soma, and less of this current will leak out through the membrane. A major reason that the thalamic relay cells are electrotonically compact is that their dendrites are relatively thick through most of their extent (figure 2.2A).

2.1.1.1 Cable Properties of Interneurons and Reticular Cells Interneurons of the cat's lateral geniculate nucleus appear to be much larger electrotonically than are relay cells, meaning that if the dendrites act as passive cables, voltage changes created at distal dendrites would be likely to be much more attenuated at the soma. Two examples based on modeling from morphological and electrophysiological data from the interneurons are shown in figure 2.2C and D, based on the same methods as for the relay cells of figure 2.2A and B to compute measures of membrane voltage expected at the soma and at various dendritic loci when a current is injected at one distal dendritic locus (Bloomfield et al., 1987; Bloomfield and Sherman, 1989). A major reason for the electronically extensive dendritic arbor of interneurons appears again to be related to dendritic diameter: although dendrites of interneurons are roughly as long as are those of relay cells, those of interneurons are much thinner, and this results in more current from synaptic activation leaking across the membrane en route to the soma.[2] However, cable modeling of thalamic interneurons has so far been limited to a few examples from the cat's lateral geniculate nucleus (Bloomfield and Sherman, 1989), and given evidence of other classes of thalamic interneuron (Bickford et al., 1999; Carden and

2. See chapter 3 and figure 3.9 for further discussion of properties of interneurons.

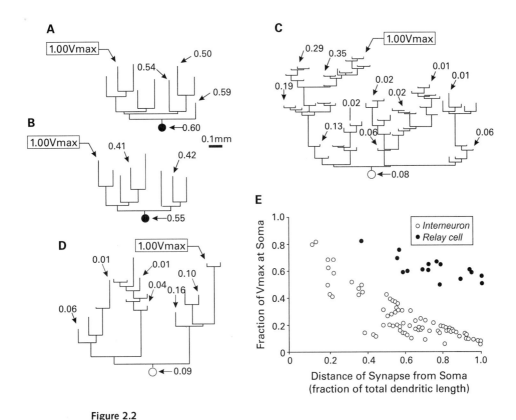

Figure 2.2
Cable modeling of the voltage attenuation that occurs within the dendritic arbors of two relay cells (A, B) and of two interneurons (C, D) after the theoretical activation of a single synapse (that is, of a single current injection leading to a relative depolarization as indicated). The cells from the cat's lateral geniculate nucleus were labeled by intracellular injection of a dye in vivo, and the stick figures represent a schematic view of one primary dendrite from each cell with all of its progeny branches. Each branch length is proportional to its calculated electrotonic length (see text for details). The site of a theoretical voltage injection is indicated by the boxed value labeled 1.00 V_{max} (maximum voltage). Computed voltage attenuation at various dendritic endings within the arbor and soma is indicated by arrows and given as fractions of V_{max}. (E) Attenuation at the soma of a single theoretical voltage injection placed at different dendritic endings within the dendritic arbor as a function of anatomical distance of the voltage injection from the soma. Each voltage injection mimics the activation of a single synapse. The abscissa represents relative anatomical distances normalized to the greatest extent of each arbor, and the plotted points represent values from the four cells shown in (A–D). Redrawn from Bloomfield and Sherman (1989).

Bickford, 2002) (see chapter 3), it is possible that other patterns of cable properties are present in other interneurons. More information is needed about the general properties of interneurons and their variation across thalamic nuclei and species.

Cable modeling of neurons of the thalamic reticular nucleus in rats suggests that they, too, are electronically extensive (Destexhe et al., 1996), although these data for reticular cells so far are limited to data from the rat.

2.1.1.2 Cable Properties of Cortical Cells   Although there has been some attempt to use passive cable modeling to describe current flow in cortical pyramidal cells (for example, Stafstrom et al., 1984), it now seems clear that membranes of these cells are not passive and that active conductances strongly affect current flow in general and postsynaptic flow of synaptic currents more specifically. For instance, one tenet of cable modeling is that the further a synapse is from the cell body, the less effective will be its postsynaptic effect due to loss of current across the membrane as it flows to the cell body, and this tenet has been challenged with evidence that active membrane properties can compensate for synaptic location to make synapses more or less equally effective regardless of their position (Magee and Cook, 2000). Although this approach has not been much if at all applied to stellate cells, it nonetheless seems clear that understanding of synaptic integration in cortical cells requires an appreciation of active dendritic properties, which is considered further in section 2.1.4, and this renders cable modeling of limited utility in describing the behavior of these neurons.

2.1.2 Passive Membrane Conductances

The membranes of neurons are not completely impermeable to ions, and thus there is a constant leak of ions across the membranes. The ionic conductances underlying these leaks generally do not depend on membrane voltage. These leaks are offset by the action of various ionic pumps that move specific ions across the membranes in either direction. The result of the leaks and pumps is a stable, equilibrium concentration gradient for each ion, and this leads to the typical resting potential. This, for a typical neuron, is usually between -65 and -75 mV. Of the ions that dominate this process, namely K^+, Na^+, and Cl^-, the passive leakage is greatest for K^+. This is the so-called K^+ leak conductance. However, if only K^+ leaked across the membrane, the resting potential would be at the reversal potential for K^+, or about -100 mV. The smaller leakage of Na^+ and Cl^-, which are driven by more positive reversal potentials, combine with the K⁺ leak conductance to create the resting membrane potentials observed, typically between −65 and −75 mV. As we shall see in section 2.2.1, one important effect of certain synaptic actions is a change in this K⁺ leak conductance.

2.1.3   Active Membrane Conductances: Thalamic Neurons

Cable modeling, while a useful first approximation, assumes passive membrane properties but does not allow for active or changing conductances in the cell membranes. However, such active conductances are quite common in all neurons, including thalamic neurons, and this greatly complicates our understanding of how the synaptic inputs in these cells are integrated to affect the axonal firing. As noted above, a K⁺ conductance is a common source of current leak across the membrane. If this were increased by opening more K⁺ channels, more current generated from a synaptic input would leak across the membrane en route to the soma and axon initial segment, further attenuating the synaptic effect; the converse occurs when many K⁺ channels are closed. Synaptic processes themselves produce change, usually an increase, in membrane conductance that can affect cable properties. Finally, active conductances, because they involve ion flow, also tend to affect membrane voltage in much the same way a synaptic event will, and these can also affect the axon initial segment apart from any effect these processes have on cable properties. Thus several different processes, including synaptic events and the turning on or off of active conductances, can lead to voltage changes at the axon initial segment and thus affect cell firing.

Most known active conductances are voltage dependent because they are turned on or off by changes in membrane voltage. However, they also have a time dependency so that a minimum duration of membrane voltage change is a prerequisite for a change in the conductance, and usually the greater the voltage change, the faster the conductance change. Most readers will be familiar with the Na⁺ and K⁺ conductances underlying the action potential (Hodgkin and Huxley, 1952), and because the conductance changes underlying action potentials share many features with other conductances found in thalamic neurons, it is worth briefly reviewing this process.

2.1.3.1   The Action Potential   As shown in figure 2.3A, the action potential has a very rapid rise or upstroke from rest and a somewhat slower fall or downstroke that slightly overshoots rest to create a period of relative hyperpolarization, known as the afterhyperpolarization. Figure 2.3B shows the

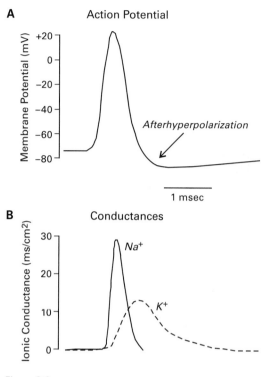

A   Action Potential

B   Conductances

Figure 2.3
Typical action potential. (A) Voltage trace. Note that just following the action potential there is a period of modest hyperpolarization, known as the *afterhyperpolarization*. (B) Voltage-dependent Na$^+$ and K$^+$ conductances underlying the voltage changes of the action potential.

underlying conductances that lead to the voltage changes in figure 2.3A. These consist of an inward (depolarizing) Na$^+$ conductance and an outward (hyperpolarizing) K$^+$ conductance, and each conductance occurs due to opening of the appropriate Na$^+$ or K$^+$ ion channels.

The Na$^+$ channel has two voltage-sensitive gates, an *activation gate*, opened when the conductance is activated, and an *inactivation gate*, closed when the conductance is inactivated; generally, depolarization opens the activation gate and closes the inactivation gate, whereas hyperpolarization closes the activation gate and opens the inactivation gate. Thus, the two gates of the Na$^+$ channel have opposite voltage dependencies, and both respond relatively quickly to voltage changes. Both must be open for Na$^+$ to flow into the cell, which leads to an inward Na$^+$ current (I_{Na}). At the resting membrane potential (for example, –65 mV), the inactivation gate is open (so I_{Na} is *de-inactivated*), but the activation gate is closed (so I_{Na} is *deactivated*), and thus there is no inward flow of Na$^+$ (figure 2.4A). When the membrane is depolarized beyond the activation threshold for I$_{Na}$ (figure 2.4B), the activation gate rapidly opens (so I$_{Na}$ is both *activated* and still de-inactivated), and Na$^+$ flows into the cell, producing the depolarizing upswing of the action potential. After a suitable period of 1 ms or so, this resultant depolarization leads to closing of the inactivation gate (so I$_{Na}$ is *inactivated*), blocking further flow of Na$^+$ ions (figure 2.4C). This, plus the opening of various slower K$^+$ channels, which do not inactivate because they have only an activation gate, produces a hyperpolarizing outward flow of K$^+$ (I$_K$). This repolarizes the membrane to near its starting position (figure 2.4D). I$_{Na}$ remains briefly inactivated because it takes roughly 1 ms of this hyperpolarization to open the inactivation gate, restoring the initial conditions of figure 2.4A. Finally, note that the roughly 1 ms of hyperpolarization needed to de-inactivate the Na$^+$ channel (figure 2.4C and D) provides a refractory period limiting firing rates for the action potential to roughly 1000 Hz.

For the action potential to propagate as an all-or-none wave, it is essential that the density of Na$^+$ channels be sufficiently high so that once threshold is reached, the further depolarization caused by the first channels opening causes a self-regenerating, explosive opening of nearby Na$^+$ channels. If the Na$^+$ channel density were too low, the initial channels opening would lead only to a local depolarization that would decay exponentially. The Na$^+$ and K$^+$ channels involved in the action potential are located in the soma (and also typically in dendrites) and in the axon hillock and initial segment, where they are involved with the initiation of action potentials, and are also present all along the axon, where they produce the conduction of the nerve impulse. The refractory period indicated in figure 2.5A dominates the wake of the action potential as it sweeps down the axon and prevents a backward propagating spike. It is important to realize that under normal conditions, a spike is usually generated in the initial segment of the axon, nearest the cell body, and this property means that the spike will travel only away from the cell body toward targets elsewhere and will not start at some point to reverse direction toward the cell body. However, under experimental conditions, it is possible to initiate an action potential at any location along the axon, in which case the spike will propagate in both directions away from the initiation site. The point here is that the refractory period means that, once initiated, a propagating spike cannot reverse direction.

A special case occurs for myelinated axons, where the voltage-sensitive Na$^+$ and K$^+$ channels are concentrated at the nodes of Ranvier, so instead of a smooth, wave-like propagation, the action potential jumps from node of Ranvier to node of Ranvier down the axon in a saltatory fashion (figure 2.5B).

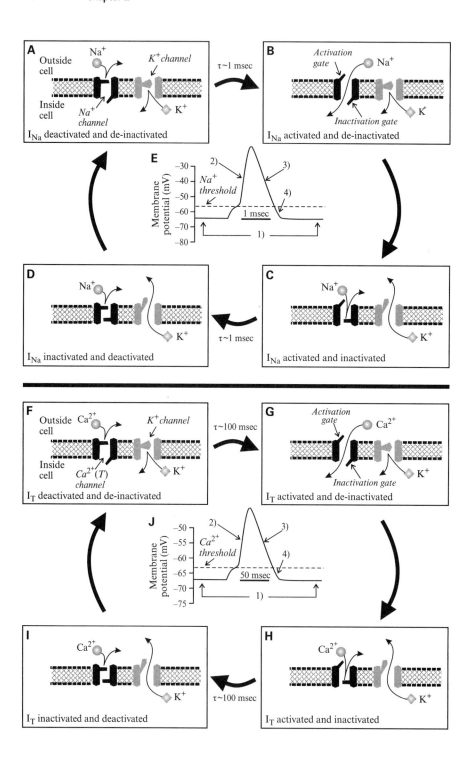

◀ **Figure 2.4**
Schematic representation of voltage-dependent ion channels underlying the conventional action potential (A–E) and the low-threshold Ca^{2+} spike (F–J). For the action potential, (A)–(D) show the channel events, and (E) shows the effects on membrane potential. The Na^+ channel has two voltage-dependent gates: an *activation gate* that opens at depolarized levels and closes at hyperpolarized levels, and an *inactivation gate* with the opposite voltage dependency. Both must be open for the inward, depolarizing Na^+ current (I_{Na}) to flow. The K^+ channel (actually an imaginary combination of several different K^+ channels) has a single activation gate, and when it opens at depolarized levels, an outward, hyperpolarizing K^+ current is activated. (A) At a resting membrane potential (roughly –60 to –65 mV), the activation gate of the Na^+ channel is closed, and so it is deactivated, but the inactivation gate is open, and so it is de-inactivated. The single gate for the K^+ channel is closed, and so the K^+ channel is also deactivated. (B) With sufficient depolarization to reach its threshold, the activation gate of the Na^+ channel opens, allowing Na^+ to flow into the cell. This depolarizes the cell, leading to the upswing of the action potential. (C) The inactivation gate of the Na^+ channel closes after the depolarization is sustained for roughly 1 ms ("roughly," because inactivation is a complex function of time and voltage), and the slower K^+ channel also opens. These combined channel actions lead to the repolarization of the cell. While the inactivation gate of the Na^+ channel is closed, the channel is said to be inactivated. (D) Even though the initial resting potential is reached, the Na^+ channel remains inactivated because it takes roughly 1 ms ("roughly" having the same meaning as above) of hyperpolarization for de-inactivation. (E) Membrane voltage changes showing action potential corresponding to the events in (A)–(D). For the representation of actions of voltage-dependent T (Ca^{2+}) and K^+ channels underlying the low-threshold Ca^{2+} spike, the conventions are as in (A)–(E); (F)–(I) show the channel events, and (J) shows the effects on membrane potential. Note the strong qualitative similarity between the behavior of the T-type Ca^{2+} channel here and the Na^+ channel shown in (A)–(E), including the presence of both activation and inactivation gates with similar relative voltage dependencies. (F) At a relatively hyperpolarized resting membrane potential (roughly –70 mV), the activation gate of the T-type Ca^{2+} channel is closed, but the inactivation gate is open, and so the T-type Ca^{2+} channel is deactivated and de-inactivated. The K^+ channel is also deactivated. (G) With sufficient depolarization to reach its threshold, the activation gate of the T-type Ca^{2+} channel opens, allowing Ca^{2+} to flow into the cell. This depolarizes the cell, providing the upswing of the low-threshold spike. (H) The inactivation gate of the T-type Ca^{2+} channel closes after roughly 100 ms ("roughly," because, as for the Na^+ channel in figure 2.2, closing of the channel is a complex function of time and voltage), inactivating the T-type Ca^{2+} channel, and the K^+ channel also opens. (I) These combined actions repolarize the cell. Redrawn from Sherman and Guillery (2006).

In general, myelinated or not, thicker axons conduct action potentials to their targets with a faster conduction velocity. However, this saltatory conduction in myelinated axons produces a much faster conduction velocity than would be expected in unmyelinated axons of the same diameter. In any case, whatever the speed of thalamocortical transmission, the transmission itself depends critically on the presence of a sufficiently dense distribution of Na^+ and K^+ channels in the axons.

2.1.3.2 I_T and Burst Firing Perhaps the most important voltage-dependent conductance besides those involved in the action potential is that involving T-type (for transient) Ca^{2+} channels (Jahnsen and Llinás, 1984; Stanley et al., 1999; Sherman, 2001; Swadlow et al., 2002; Lesica and Stanley, 2004). When the channel opens, an inward, depolarizing current, I_T, is generated. The presence or absence of this current largely determines which of two firing modes,

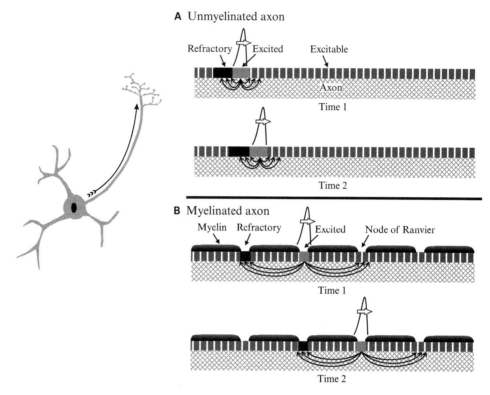

Figure 2.5
Schematic view of propagation of the action potential along an axon. (A) Unmyelinated axon. The upswing of the action potential leads to a large current injection (due to inward flow of Na⁺ as depicted in figure 2.2), and this spreads in both directions inside the axon, depolarizing neighboring patches of axon membrane. In the direction the spike is traveling, the newly depolarized membrane patch will lead to opening of its Na⁺ channels and a renewal of the action potential. This continues in this direction, and so the spike sweeps down the action. However, the current flow in the reverse direction depolarizes a patch of membrane that recently produced an action potential, and so it is in a refractory period and thus cannot regenerate another one. As a result, the action potential will not reverse direction. Normally, an action potential is generated at the initial segment next to the cell body and travels down the axon to its terminals. However, it is possible experimentally to initiate an action potential anywhere in the axon by injecting current into the axon at that site. If this happens away from the cell body, the spike will travel in both directions from the initiation site, so two spikes actually are created moving away from each other, but, as noted, neither reverses. (B) Myelinated axon. Myelin acts like an effective insulator, preventing ion flow across it, so current can only flow into and out of the cell at gaps in the myelin, where Na⁺ and K⁺ channels are especially concentrated, known as nodes of Ranvier. Thus, instead of a smooth flow of the action potential, as in (A), it jumps from node to node. This is saltatory conduction and is much faster than the smooth conduction in (A). As in (A), the refractory period ensures that the action potential flows in only one direction.

tonic or burst, describes the relay cell response, and as discussed below, the firing mode plays an important role in how information is relayed through the thalamus (Stanley et al., 1999; Sherman, 2001; Swadlow et al., 2002).

As shown in figure 2.4F–J, the behavior of this Ca^{2+} channel is qualitatively the same as that of the Na^+ channel, with the same two types of voltage-dependent activation and inactivation gates. At rest (figure 2.4F), the activation gate is closed, but sufficient depolarization opens it (figure 2.4G), allowing the inward I_T that further depolarizes the cell. This produces a self-regenerating, propagating spike, much like the Na^+/K^+ action potential. This depolarization eventually closes the inactivation gate (figure 2.4H), which, along with the activation of K^+ channels, repolarizes the cell (figure 2.4I). This repolarization eventually leads to de-inactivation of I_T and restoration of the rest condition (figure 2.4F). Just as there is a refractory period for action potentials of about 1 ms, representing the time of repolarization needed to open the inactivation gate (figure 2.4D), there is also a refractory period for I_T. However, the kinetics of the inactivation gate for this Ca^{2+} channel are much slower than for the Na^+ channel—roughly 100 ms rather than roughly 1 ms—and so the refractory period for I_T lasts for roughly 100 ms.

As for Na^+ channels, if a sufficiently high density of T-type Ca^{2+} channels exists, the threshold opening of the initial T-type Ca^{2+} channels leads to an explosive all-or-none spike. This is the case for thalamic cells, and the result is a spike-like depolarization of 25 to 50 mV that propagates throughout the dendrites and soma (Jahnsen and Llinás, 1984; Huguenard and McCormick, 1994; Zhan et al., 1999; Sherman, 2001). However, because the refractory period for I_T is roughly 100 ms, I_T spiking cannot occur faster than about 10 Hz. T-type Ca^{2+} channels are quite common in neurons throughout the central nervous system, but only in rare cells is the density high enough to support all-or-none Ca^{2+} spikes. Thus, this property of all-or-none spiking based on T-type Ca^{2+} channels is unusual, but in the thalamus nearly every relay cell of every nucleus in every mammalian species so far tested shows this property (Sherman and Guillery, 2006), and it is also common in interneurons and reticular cells (Huguenard and Prince, 1992; Pape and McCormick, 1995).

Figure 2.4 illustrates the *qualitative* similarity between the Na^+ and T-type Ca^{2+} channels, but there are important quantitative differences. One is that the inactivation kinetics are much faster for the Na^+ channel. Also, the T-type channel operates in a more hyperpolarized regime (compare figure 2.4E and J); the resulting depolarization, which in thalamic cells is an all-or-none Ca^{2+} spike, is also known as the "low threshold spike." A final important difference not shown in figure 2.4 is the distribution of these T-type channels, which are

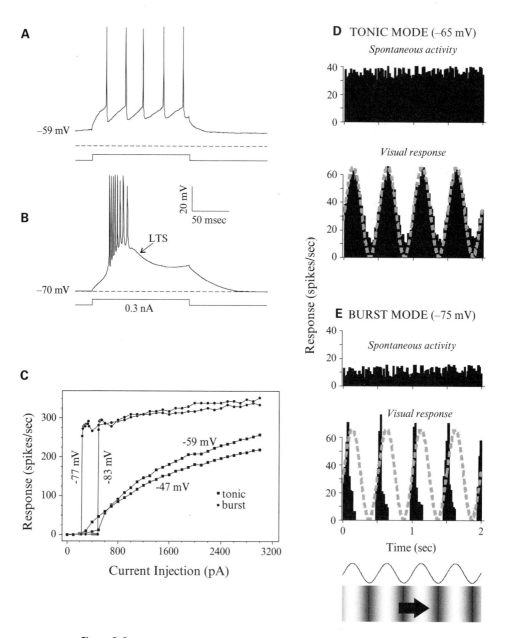

Figure 2.6
Properties of I_T and the low-threshold Ca^{2+} spike. All examples are from relay cells of the cat's lateral geniculate nucleus recorded intracellularly, either in an in vitro slice preparation (A–C) or in an anesthetized in vivo preparation (D, E). (A and B) Voltage dependency of the low-threshold spike. Responses are shown to the *same* depolarizing current pulse delivered intracellularly but from two different initial holding potentials. When the cell is relatively depolarized (A), I_T is inactivated, and the cell responds in tonic mode, which is a stream of unitary action potentials to a suprathreshold stimulus. When the cell is relatively hyperpolarized (B), I_T is de-inactivated, and

effectively limited to the soma and dendrites, whereas Na$^+$ channels, although often found there as well, are notable for their distribution along the axon. This allows action potentials to travel from the soma to a target far away, such as cortical targets for thalamic relay cells. Whereas T-type Ca^{2+} spikes propagate in the dendrites and soma, they do not travel along the axon and thus do not reach cortex. Thus, the significance of these Ca^{2+} spikes for what arrives in the cortex ultimately rests with their effect on conventional action potentials as described in the following paragraphs. This effect is dramatic and important.

Figure 2.6 illustrates the relationship between I$_T$ and tonic or burst firing. If the cell is initially depolarized for at least 100 ms by as little as 5 mV or so, I$_T$ is inactivated (as indicated in figure 2.4H) and plays no role in the ensuing firing pattern; the cell now responds in tonic mode so that a suprathreshold depolarization evokes a string of unitary action potentials (figure 2.6A). If, instead, the cell is hyperpolarized by ≥5 mV for ≥100 ms, inactivation of I$_T$ is removed (as in figure 2.4H and I) so that the next sufficiently large depolarization will activate an all-or-none I$_T$ (as in figure 2.4F). This, in turn, evokes the low threshold spike that propagates through the dendritic tree and soma, reaching the initial segment, and riding its crest is a burst of two to nine high-frequency action potentials (figure 2.6B). This is the burst mode of firing. Figure 2.6A and B thus show that the exact same depolarizing stimulus evokes a very different response in the cell depending on its recent voltage history; that is, the same EPSP can evoke either a tonic or a burst response depending on this history.

the cell responds in burst mode, which involves activation of a low-threshold Ca^{2+} spike (LTS) with multiple action potentials (eight in this example) riding its crest. (C) Input–output relationship for another cell. The abscissa is the amplitude of the depolarizing current pulse, and the ordinate is the firing frequency of the cell for the first six action potentials of the response, as this cell usually exhibited six action potentials per burst in this experiment. The initial holding potentials are shown, and –47 mV and –59 mV produces tonic mode, whereas –77 mV and –83 mV produces the burst mode. (D and E) Responses of a representative relay cell in the lateral geniculate nucleus of a lightly anesthetized cat to a sinusoidal grating drifted through the cell's receptive field. The trace and grating at the bottom reflect the sinusoidal changes in luminance contrast with time. Current was injected into the cell through the recording electrode to alter the membrane potential. Thus in (D), the current injection was adjusted so that the membrane potential without visual stimulation averaged –65 mV, promoting tonic firing, because I$_T$ is mostly inactivated at this membrane potential; in (E), the current injection was adjusted to the more hyperpolarized level of –75 mV, permitting de-inactivation of I$_T$ and promoting burst firing. Shown are average response histograms to the visual stimulus (bottom histograms in D and E) and during spontaneous activity with no visual stimulus (top histograms), plotting the mean firing rate as a function of time averaged over many epochs of that time. The sinusoidal changes in contrast as the grating moves across the receptive field are also shown as a dashed, gray curve superimposed on the responses in the lower histograms. Note that the response profile during the visual response in tonic mode looks like a sine wave, but the companion response during burst mode does not. Note also that the spontaneous activity is higher during tonic than during burst firing. Redrawn from Sherman and Guillery (2006).

One big difference between these two firing modes is the linearity of transmission through the thalamus. In tonic mode, with no contribution from I_T, the larger the afferent suprathreshold EPSP, the greater the postsynaptic firing rate until some saturation is reached. This is shown in figure 2.6C. However, the burst mode leads to a strikingly nonlinear transmission. This is because the evoked action potentials are not directly created by a variable-sized EPSP, but rather by a low-threshold Ca^{2+} spike that is itself all-or-none. That is, a larger EPSP will not evoke a larger Ca^{2+} spike, and because the size of the Ca^{2+} spike determines the number of action potentials in a burst, there is no longer a linear relationship between the size of the afferent EPSP and the response in terms of either firing frequency or number of action potentials (figure 2.6C).

Although the Ca^{2+} spike is all-or-none, its amplitude can vary in the following way. At any time, the number of T-type Ca^{2+} channels that are de-inactivated and thus available to produce I_T depends stochastically on the extent of hyperpolarization experienced by the cell: the more hyperpolarized the cell, the more of these channels are available, and the greater the amplitude of the Ca^{2+} spike (Zhan et al., 1999). Thus for any given amount of hyperpolarization, the Ca^{2+} spike is all-or-none, but different durations and amplitudes of hyperpolarization will produce different amplitudes of the Ca^{2+} spike and thus different numbers of action potentials in the burst.

Figure 2.6D and E further show differences between burst and tonic firing, including the linearity difference. In this example from an in vivo experiment in a cat in which the activity of a geniculate cell was recorded intracellularly, the firing modes could be controlled by injecting a constant current into the cell to maintain different "resting" potentials: a relatively depolarized one for tonic firing (figure 2.6D) and a hyperpolarized one for burst firing (figure 2.6E). The spontaneous activity is shown in the upper histograms, and responses to a sinusoidal grating drifting through the receptive field are shown in the lower histograms. Four differences between firing modes are clear (Sherman, 2001; Sherman and Guillery, 2006):

1. Spontaneous activity is lower during burst mode, almost certainly due to the fact that the cell is being held at a more hyperpolarized level for burst mode.

2. The visual stimulus applied provides a sinusoidal change in contrast level with time as the grating drifts across the receptive field. Thus, a linear response would be one that varies sinusoidally with the stimulus. It is clear that the response during tonic mode is quite linear but that in burst mode it is highly nonlinear. This is a simple consequence of the difference shown in figure 2.6C. Even the difference in spontaneous activity contributes to the greater linearity

of tonic firing, because the higher spontaneous level allows the inhibitory components of the visual response to appear as a reduction in activity instead of "bottoming out" and producing a rectification of the response, which is a clear nonlinearity. Note also that the burst response shows a small phase advance (it occurs earlier with respect to the visual stimulus), and this appears to be a consequence of the fact that the burst response starts with the low-threshold Ca^{2+} spike, which is initiated earlier from a more hyperpolarized level than are the action potentials of the tonic response.

3. The peak firing rates of both visual responses are quite high. If the visual response represents the signal relayed to cortex, then the spontaneous activity represents noise, because, by definition, the spontaneous firing bears no relationship to any visual stimulus. In this regard, then, the ratio of the signal (visual response) to noise (spontaneous activity) is greater during burst firing. Thus, whatever the message that is transmitted during burst firing, it should be more readily perceived as an important signal.

4. A final difference between firing modes has to do with certain synaptic properties of thalamocortical axons, which are considered in more detail in section 2.2.2. The vast majority of thalamocortical synapses show the property of paired-pulse depression, meaning that two action potentials occurring with an interspike interval shorter than roughly 100 ms will produce EPSPs of different amplitudes, the first being larger than the second, and so in a train of EPSPs evoked at a high enough rate, the EPSPs depress sequentially in amplitude. The shorter the interval between action potentials, the greater is the depression. To avoid depression and evoke a maximum EPSP, an action potential must follow a silent period ≥ 100 ms or so. Typically during tonic mode, the relay cell fires at rates exceeding 10 Hz, so that the evoked EPSPs in cortex are continuously depressed. However, the generation of a low-threshold Ca^{2+} spike requires ≥ 100 ms of sustained hyperpolarization to remove inactivation of the T-type Ca^{2+} channels (figure 2.4), and this, in turn, means that there can be no action potentials generated during that period. Thus, the first spike of a subsequent burst occurs with no paired-pulse depression, and this first spike would then evoke the maximum EPSP so that EPSPs generated by the first spike in a burst should, on average, exceed those generated by tonic firing, and this has, in fact, been demonstrated (Swadlow and Gusev, 2001; Swadlow et al., 2002).

From the properties of tonic and burst firing just described, we can imagine certain benefits of each leading to the following hypothesis (Sherman, 2001; Sherman and Guillery, 2006). The advantage of tonic firing seems obvious: Its more linear transmission to cortex minimizes nonlinear distortion of information and thus supports a higher fidelity of information transfer. Burst firing, with its greater signal-to-noise ratio and stronger activation of cortex, could be useful to alert cortex to qualitative changes in the environment as a sort of "wake-up call," particularly useful if attention needs to be directed at other environmental features. This could promote an attention shift to the novel stimulus, perhaps with a shift to tonic firing for a more faithful analysis of the new stimulus. Chapter 4 explores how thalamic circuitry is arranged to control the firing mode and switch between tonic and burst firing.

Although this hypothesis seems plausible, there are others. For instance, some suggest that burst firing is relevant only during slow-wave sleep and that its presence during active behavior is too rare to be relevant (for example, Llinás and Steriade, 2006). Other suggestions include the notion that burst and tonic firing are intermingled due to considerable stochastic synaptic noise causing the T channels to move frequently between inactivation and de-inactivation and that this mixture helps to linearize responsiveness to incoming signals (Wolfart et al., 2005). Overall, a consensus has emerged that relay cell burst firing plays an important role in behaving animals and does produce strong activation of cortex (for example, Swadlow et al., 2002; Lesica and Stanley, 2004; Alitto et al., 2005; Bezdudnaya et al., 2006).

2.1.3.3 Other Active Membrane Conductances Relay cells, like other neurons in the brain, exhibit a variety of other active membrane conductances. A complete description of these can be found elsewhere (Hille, 1992; Huguenard and McCormick, 1994; Kandel et al., 2000; Sherman and Guillery, 2006), and they will be only briefly described here. Various K^+ conductances exist, and a prominent one leads to an outward (hyperpolarizing) current known as I_A, which occurs when K^+ channels open. These channels have a voltage dependence similar to that of the T-type Ca^{2+} channels, with the main difference that they lead to an outward rather than an inward current. Thus, at depolarized levels, I_A is inactivated and plays no role. However, if the cell has been hyperpolarized for ≥ 100 ms or so, the inactivation is removed, and the next suitable depolarization will activate I_A. Instead of leading to a depolarizing event seen with I_T, this produces a hyperpolarization that will work against the depolarizing event (for example, an EPSP) that evoked it. The result is that I_A will act to slow down and diminish EPSPs, delaying resultant action potentials and reducing their rate of firing. Thus, it has been suggested that I_A extends the dynamic range of a neuronal input–output relationship by reducing its slope (Connor and Stevens, 1971).

Another series of subtly different K^+ conductances are triggered by Ca^{2+} entry into the cell, and these also lead to outward currents. These commonly occur as the result of activation of I_T or action potentials, or both, which activates various high-threshold Ca^{2+} channels described in the next paragraph, and the increased Ca^{2+} concentration that results can activate Ca^{2+}-dependent K^+ conductances. Some of these activate rapidly and will help to repolarize the cell as in figure 2.4C and H. Others that are slower to activate will build up with more and more action potentials, leading to spike frequency adaptation (Adams et al., 1982; Powers et al., 1999). This phenomenon results in the slow reduction of the firing frequency of a neuron to a constant stimulus, and the higher the frequency of initial firing, the more adaptation occurs. Spike frequency adaptation has been demonstrated for thalamic cells (Smith et al., 2001).

In addition to the Ca^{2+} conductance underlying I_T, there are two or more much higher threshold Ca^{2+} conductances that are located in the dendrites and synaptic terminals of virtually all thalamic and cortical cells (Llinás, 1988; Johnston et al., 1996; Zhou et al., 1997). One involves L-type Ca^{2+} channels ("L" for "long-lasting," because it slowly inactivates) and the other N-type channels ("N," wryly, for "neither," being neither T nor L type; it inactivates more rapidly than the L-type channel). Other types of high-threshold Ca^{2+} channels also exist (Wu et al., 1998). The higher threshold than for T channels means that a much larger depolarization (to about -20 mV) is needed to activate these Ca^{2+} conductances, and they allow the Ca^{2+} entry into the cell as a result of an action potential. In synaptic terminals, these conductances represent a key link between the action potential and neurotransmitter release, because the action potential will activate these channels, and the resultant Ca^{2+} entry is needed for neurotransmitter release. Less is known about these Ca^{2+} conductances in dendrites, but by providing a regenerative spike that can travel between the site of an activating EPSP and the soma, they may help ensure that distal dendritic inputs that are strong enough to activate this conductance will significantly influence the soma and axon hillock. These channels may also play a role in the backpropagation of action potentials that depends on Na^+ channels through the dendritic arbor (see section 2.2.3).

Another conductance is activated by membrane hyperpolarization and deactivated by depolarization. This *hyperpolarization-activated cation conductance*, leads, via influx of cations, to a depolarizing current, which is called I_h (McCormick and Pape, 1990). It is sometimes called the "sag current" because it is activated by hyperpolarization and causes the membrane potential to drift back, or sag, toward the initially more depolarized level. Activation of I_h is slow, with a time constant of >200 ms.

2.1.3.4 Active Processes in Dendrites The observation that dendrites of many neurons are not simply passive, acting like cables, but rather have ion channels that can be controlled by membrane voltage or other processes has important implications for synaptic integration within neurons. Whereas the Na^+ and K^+ channels underlying the action potential are ubiquitous in axons, their presence in dendrites of certain cells, like cortical or hippocampal pyramidal cells, has only recently been appreciated (Johnston et al., 1999; Magee and Johnston, 2005; Gasparini et al., 2007; Larkum et al., 2007; Spruston, 2008). Recent studies suggest that they are also present in dendrites of thalamic relay cells (Williams and Stuart, 2000), interneurons (AcunaGoycolea et al., 2008), and cells of the thalamic reticular nucleus (Crandall et al., 2010). The presence of these conductances, if of sufficiently high density, means that action potentials may be generated initially in the dendrites and that those generated in the soma or axon hillock can back-propagate up the dendritic tree. This backpropagation can evoke high-threshold Ca^{2+} conductances (see below) and affect integration of synaptic inputs.

Likewise, T-type Ca^{2+} channels are distributed throughout the dendrites of relay cells, interneurons, and reticular cells. This means that I_T can be evoked in dendrites by synaptic inputs, and this, too, will have important consequences for synaptic integration.

Note that these active conductances—Na^+, K^+, Ca^{2+}, or other—can be dynamically controlled. Membrane potential offers one control: for instance, if the dendritic arbor is relatively depolarized, I_T will be inactivated and not contribute to synaptic integration. Another control seen for cortical and hippocampal neurons (but not yet reported for thalamic neurons) is the ability of various neuromodulators, like acetylcholine, to affect the functioning of these channels (Hoffman and Johnston, 1999; Magee and Johnston, 2005; Gasparini et al., 2007). This creates considerable flexibility in how dendrites act to integrate synaptic input, behaving more like simple cables when their active conductances are down-modulated, and providing a more complicated and nonlinear integration when these channels are up-modulated.

2.1.4   Active Membrane Conductances: Cortical Neurons

Active processes in cortical cells produce four main firing patterns that have been associated with the morphological type of the cell (Connors and Gutnick, 1990; Nowak et al., 2003; Llano and Sherman, 2009).

1. Regular spiking neurons respond to suprathreshold depolarization with a train of action potentials that slowly decrease in frequency during the train; this is a property known as "spike-frequency adaptation" and is typically caused by buildup of K^+ during the afterhyperpolarization (see figure 2.3A),

which hyperpolarizes the cells and thereby reduces firing frequency. These neurons are found in layers 2–6 and include both spiny stellate cells and pyramidal cells. Both cell types are thought to be glutamatergic and excitatory and project strictly within cortex.

2. Another pyramidal cell type, also glutamatergic and excitatory, often responds to depolarization with an initial burst of action potentials followed by single spike firing. These are pyramidal cells located mostly in layer 5 and include those that project to various subcortical sites.

3. Fast spiking neurons respond at very high firing rates without spike-frequency adaptation. These are aspinous stellate cells, thought to be GABA-ergic and inhibitory and are found in layers 2 and 3.

4. More recently, "chattering cells" were described as firing frequent short-duration bursts (Gray and McCormick, 1996). These are both spiny stellate and pyramidal cells, thought to be glutamatergic and found mostly in layer 3.

2.2 Synaptic Properties

Virtually all of the known synaptic inputs in thalamus and cortex operate by conventional chemical transmission,[3] and so an appreciation of how such synapses operate is required to understand how thalamic and cortical circuits function. Among the important synaptic properties considered here are the different types of postsynaptic receptor involved in synaptic transmission and the temporal properties that characterize various synapses.

2.2.1 Ionotropic and Metabotropic Receptors

Clearly, postsynaptic receptors play a large role in the functioning of synapses. These receptors form two different functional classes known as *ionotropic* and *metabotropic*. Details of differences between these receptors can be found elsewhere: for a general review of metabotropic receptors, see Nicoll et al. (1990); for metabotropic glutamate receptors, see Pin and Bockaert (1995), Pin and Duvoisin (1995), and Conn and Pin (1997); and for GABA$_B$ receptors, see Mott and Lewis (1994) and Padgett and Slesinger (2010). Key features are briefly summarized here.

---

3. In both cortex and thalamus, some GABAergic neurons, in addition to receiving synaptic inputs using chemical transmission, also form gap junctions between one another to create another means of cell to cell communication. This includes cells of the thalamic reticular nucleus (Liu and Jones, 2003; Landisman and Connors, 2005; Lam et al., 2006; Haas et al., 2011) and certain GABAergic cortical cells (Kandler and Katz, 1998; Liu and Jones, 2003; Caputi et al., 2009; Ma et al., 2011).

Receptors have their effects on cells typically by causing certain ion channels to open or close, allowing flow of ions across the membrane, which can depolarize or hyperpolarize the cell. For instance, depolarization results when Na^+ or Ca^{2+} channels open, allowing these positive ions to flow into the cell, or when K^+ channels close, retarding the leak of K^+ from the inside of the cell; hyperpolarization occurs when K^+ or Cl^- channels open, the former because it allows a positive ion to flow out of the cell, and the latter because it allows a negative ion to flow in. One of the main differences between ionotropic and metabotropic receptors is the nature of their relationship to ion channels that may be opened or closed by their activation.

The ion channels that are here treated as gates that are either open or closed are now recognized as proteins with four alpha-helical transmembrane segments that can act as voltage sensors and move through the membrane in response to voltage changes across the membrane. As the helical structure of the protein moves through the membrane in response to voltage changes, the control of ion fluxes through the membrane is changed. Differences in the amino acid sequence of the proteins produce differences in the kinetics of the channels (see Catterall, 2010).

Ionotropic receptors are generally complex protein chains that wrap back and forth to span the membrane several times and comprise several subunits that may combine in different ways to affect receptor functioning subtly (for details of receptor structure and differences among them, see Kandel et al., 2000). When a neurotransmitter binds to an ionotropic receptor, it acts in a fairly direct fashion through a conformational change in the receptor to open a specific ion channel, which is actually embedded in and thus part of the receptor (figure 2.7). Because of the direct linkage between receptor activation and opening of the ion channel, the potentials evoked from activation of ionotropic receptors are typically fast: they have a short latency, a fast rise to peak, and generally last only 10 ms or so.

Metabotropic receptors, per se, are simpler structures, typically composed of a single polypeptide, but they form part of a more complex chain of events leading to postsynaptic responses. When the neurotransmitter binds to a metabotropic receptor, a much more complicated series of events is triggered. The conformational change in the receptor ends in the release of a G-protein, which in turn leads to a cascade of biochemical reactions in the membrane and/or cytoplasm of the postsynaptic cell. This process is known as a *second messenger pathway*, because the postsynaptic effects of the neurotransmitter are carried indirectly through second messengers by these processes. Several reaction chains ensue, and one of these eventually causes specific ion channels to open or close. In the thalamus and cortex, the main effect is on K^+ channels,

Ionotropic Receptor (AMPA, GABA_A, Nicotinic)

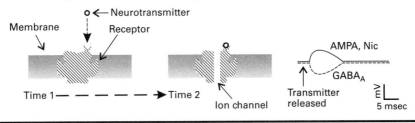

Metabotropic Receptor (mGluR, GABA_B, muscarinic)

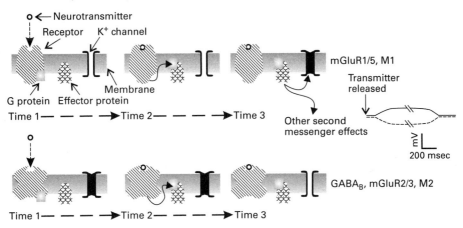

Figure 2.7

Schematic depiction of ionotropic and metabotropic receptors, each type shown repeatedly at different times (*time 1* and *time 2* for the ionotropic receptor, and *time 1, time 2,* and *time 3* for the metabotropic receptor), and the evoked postsynaptic potentials are shown on the right. The actual receptor protein complexes are shown with hatching. For the ionotropic example, time 1 represents the period before binding to the transmitter, and time 2 is the period after binding. The binding causes a conformational change that opens the ion channel, which forms the central core of the receptor complex. For the metabotropic receptors, time 1 is the period before transmitter binding, and just after binding (time 2) a G-protein is released, which reacts with an effector protein to produce a cascade of biochemical reactions eventually resulting in opening or closing of an ion channel, usually a K+ channel (time 3). Not shown is the possibility that for some receptors, the G-protein can directly affect the ion channel. Note the much longer time course for the metabotropic than ionotropic receptor examples. Further details are provided in the text.

although Ca^{2+} channels can also be affected. Postsynaptic potentials produced in this way by metabotropic receptors are quite slow. They begin with a long and somewhat variable latency, usually >10 ms, take tens of milliseconds to reach their peak, and they remain present for a long time, typically hundreds of milliseconds to several seconds or even longer (see figure 2.7) (Govindaiah and Cox, 2004). They thus affect the postsynaptic cell on a timescale that is significantly prolonged compared to the effects of ionotropic receptor activation.

For glutamatergic inputs, the ionotropic subtypes are AMPA, kainate, and NMDA receptors; and metabotropic glutamate receptors include eight subtypes in three groups (Conn and Pin, 1997). In thalamus and cortex, these receptors are mainly group I, which causes closing of K^+ channels and thus an EPSP, and group II, which causes opening of K^+ channels and thus an IPSP. For GABA, the ionotropic version is the $GABA_A$ receptor, and the metabotropic version is the $GABA_B$ receptor. For acetylcholine, the ionotropic receptors are nicotinic receptors, and the metabotropic receptors are several subtypes of muscarinic receptor. Five subtypes are known (M1 to M5), but two, M1 and M2, are most commonly found in thalamus and cortex. The M1 receptors, when activated, depolarize the cell, like the group I metabotropic glutamate receptors; M2 receptors, when activated, hyperpolarize the cell, like the group II metabotropic glutamate receptors. Other neurotransmitter systems relevant to thalamus and cortex, such as noradrenergic, serotonergic, and dopaminergic, appear to employ only metabotropic receptors.

Other differences exist between these ionotropic and metabotropic receptor classes. One is that activation of metabotropic glutamate receptors usually requires higher levels of presynaptic firing. This is thought to be due to the location of these receptors, which have been shown to be farther from the synapse center than are their ionotropic counterparts (Nusser et al., 1994; Lujan et al., 1996). Thus, at lower afferent firing rates, only ionotropic receptors are activated, and as these firing rates increase, metabotropic receptors also become activated. Another difference has to do with the fact that metabotropic glutamate receptor activation triggers intracellular second messenger pathways, and these can affect many neuronal functions other than the state of ion channels, such as synaptic plasticity and gene expression (Jia et al., 2001; Bellone et al., 2008; Kullmann and Lamsa, 2008; Mao et al., 2008; Anwyl, 2009; Tsanov and ManahanVaughan, 2009). However, these other potential functions of activation of metabotropic glutamate receptors have not yet been much tested for thalamus or cortex.

The above treatment refers to receptors that are postsynaptic. However, some synaptic terminals themselves have functional receptors, and these are known as presynaptic receptors (Miller, 1998). These presynaptic receptors with rare exceptions are located on synaptic terminals that are not postsynaptic to other structures. This means that these presynaptic receptors must be activated by neurotransmitters that have not been cleared by various uptake mechanisms and whose source is unknown. Usually, but not always, these presynaptic receptors are metabotropic, and their activation typically acts to affect presynaptic neurotransmitter release. For example, many and perhaps all thalamocortical terminals have group II metabotropic glutamate receptors on them, which when activated by glutamate extrasynaptically reduces neurotransmitter release and thus the size of the evoked EPSP (Mateo and Porter, 2007). Retinal terminals in the lateral geniculate nucleus also have a variety of presynaptic receptors associated with them, including $GABA_B$, metabotropic glutamate, and serotonergic receptors, and when activated, these serve to reduce retinogeniculate EPSP amplitudes (Chen and Regehr, 2003; Govindaiah et al., 2012).

Not all presynaptic receptors involve traditional neurotransmitter systems. An intriguing example described now for several brain areas involves endogenous endocannabinoids that act across the synapse as a retrograde messenger to affect cannabinoid receptors on presynaptic terminals and thus affect synaptic strength in numerous brain areas, including the cerebellum (Brown et al., 2003), hippocampus (Takahashi and Castillo, 2006), auditory brain stem (Tzounopoulos et al., 2007), cortex (Huang et al., 2008; Marinelli et al., 2008; Bacci et al., 2005), and thalamus (Sun et al., 2011). A synaptic response that leads to significant Ca^{2+} entry into the postsynaptic cell, typically via activation of metabotropic glutamate receptors, triggers the release of endocannabinoids from the postsynaptic cell, and these travel to the presynaptic terminals to activate cannabinoid receptors there.

2.2.2 Short-Term Plasticity

Many synapses in the brain, including thalamic and cortical synapses, behave in a frequency-dependent manner (Thomson and Deuchars, 1994, 1997; Lisman, 1997; Chung et al., 2002). This is because the presynaptic interspike interval can strongly influence the size of the evoked postsynaptic potential. This is often explored by comparing the sizes of postsynaptic potentials evoked by a first action potential to one evoked by the next action potential as a function of the intervening time interval; because this represents responses evoked by pairs of stimuli, this property is often called a *paired-pulse effect*. Different paired-pulse effects are shown in figure 2.8. *Paired-pulse depression* (figure 2.8A) occurs when the second postsynaptic potential is smaller than the

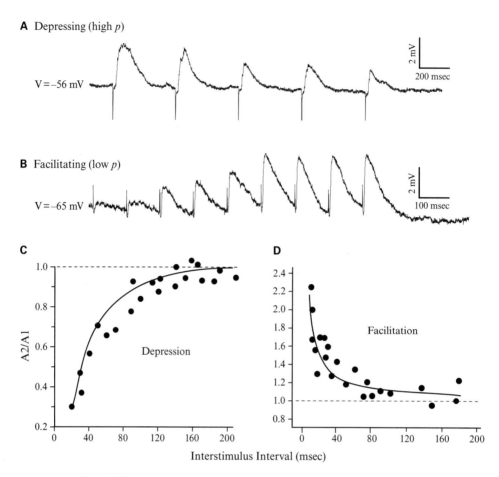

Figure 2.8
Examples of paired-pulse effects. Recordings from single cells in slices of mouse lateral geniculate nucleus in response to electrical stimulation of optic tract (retinal) or layer 6 cortical inputs. (A) Paired-pulse depression in response to retinal stimulation at 10 Hz. Note that during the train, EPSPs become smaller. Such a response is associated with a high probability of transmitter release for each action potential invading a terminal. (B) Paired-pulse facilitation in response to cortico-geniculate stimulation at 10 Hz. Note that during the train, EPSPs become larger. Such a response is associated with a low probability (p) of transmitter release. (C) Time course of paired-pulse depression showing the ratio of amplitudes of the second EPSP divided by the first (A2/A1) as a function of interstimulus interval. (D) Time course of paired-pulse facilitation showing the ratio of amplitudes of the second EPSP divided by the first (A2/A1) as a function of interstimulus interval. Data from (C) and (D) are from different examples than shown in (A) and (B).

first for interspike intervals less than a certain value. *Paired-pulse facilitation* (figure 2.8B) is the opposite: the second evoked postsynaptic potential is larger for a range of interspike intervals. The effective interspike intervals for both depression and facilitation are similar, with time constants of several tens of milliseconds that vary among synapses; with longer intervals, there is no facilitation or depression (figure 2.8C and D).

There may be many different cellular mechanisms for these phenomena, and there is still considerable debate about this. One idea is that the mechanisms are largely presynaptic and that they relate to a stochastic process of transmitter release: namely that, for a given presynaptic terminal, an invading action potential will cause transmitter release with a probability between zero and one (Dobrunz and Stevens, 1997). It is possible that the probability for any given synapse can change over time. Note that most axons contribute multiple synapses, sometimes hundreds, to one or more postsynaptic target cells, and that a probability <1 does not mean that an action potential often has no postsynaptic effect. That is, if the probability for release for each synapse, on average, is 0.5, and an axon with several thousand terminals contacting many cells contacts a given cell with 50 synapses, then an action potential will cause an average of 25 synapses on that target to release transmitter and produce a postsynaptic potential. The point is that for any one pair of synaptically linked cells, the larger the probability of release, the larger the resultant postsynaptic potential will be.

Depression is least likely to occur in a synapse if the preceding action potential fails to elicit transmitter release because, as noted, this release is always a stochastic occurrence. Such failure naturally will occur most commonly in synapses with low probability of release. This probability may be closely related to the Ca^{2+} concentration inside the synaptic terminal, because the probability of release is monotonically related to internal Ca^{2+} concentration (Dunlap et al., 1995; Matthews, 1996; Reuter, 1996). Synaptic terminals contain high-threshold Ca^{2+} channels (these have been described in section 2.1.3.3; see Dunlap et al., 1995; Matthews, 1996; Reuter, 1996) that differ from the T-type channels involved in the low-threshold Ca^{2+} spike because they have a much higher (that is, depolarized) threshold for activation. An invading action potential depolarizes the terminal sufficiently to activate these channels, leading to Ca^{2+} entry, and, as the internal Ca^{2+} concentration increases, so does the probability of transmitter release. However, when a single action potential and the subsequent Ca^{2+} influx fail to promote transmitter release, the Ca^{2+} concentration will remain elevated for several milliseconds. If a second action potential follows the first while the internal Ca^{2+} concentration remains elevated, it will cause a second wave of Ca^{2+} entry that will sum with what remains from the first, much like temporal summation in postsynaptic potentials. Because transmitter release increases with internal Ca^{2+} concentration nonlinearly as a power function with a power of 3 to 4 (Landò and Zucker, 1994), the result will be a higher probability of transmitter release for the second action potential. A typical axon innervates any target with many synaptic terminals, as long as the average probability of release for all these synapses is very low (in many connections within cortex this has been computed as <0.1), then most synapses will fail to release transmitter in response to the first action potential. However, the probability of release from each terminal may be significantly enhanced for the second action potential if it follows the first within tens of milliseconds or so.[4] This can result in more transmitter release and thus a larger postsynaptic potential for the second action potential, thus producing paired-pulse facilitation.

Depression occurs because once transmitter is released, it takes time (often hundreds of milliseconds) before the probability of release to another action potential returns to baseline levels. This may be partly due to depletion of transmitter stores or to other effects. The result is that for some time after transmitter release, the probability of release is reduced. This will result in paired-pulse depression in afferents having an average probability of release for their synapses high enough that many or most release transmitter to the first action potential.

It follows from the above that if an axon contacts a postsynaptic cell with synapses having a low probability of release (that is, "low p" synapses), these synapses are more likely to show paired-pulse facilitation (Dobrunz and Stevens, 1997; Markram et al., 1998; Varela et al., 1999; Thomson, 2000; Frick et al., 2007). This is because most of the synapses will not release transmitter but will instead show an increased probability for some time because of an increased internal Ca^{2+} concentration. Conversely, if the contacts are made

---

4. Just how low this probability must be for this to occur is a complex function of many factors. Consider an afferent axon with 20 synapses on one of its target cells. If the average baseline probability of release is 0.1, only two synapses will release transmitter, creating a postsynaptic potential of proportional amplitude. The 18 synapses failing to release transmitter will have a greater probability of release to a second action potential if it arrives soon enough, and the two synapses that released transmitter will, in turn, have a much lower probability (near zero) of release. For the second action potential to create a larger EPSP than the first (that is, to result in paired-pulse facilitation) would require that >2 synapses released transmitter, and this would happen if the average probability for the remaining 18 were ≥2/18. Thus, the phenomenon of paired-pulse facilitation depends on the initial probability of release as well as the increase seen in those terminals failing to release initially. The phenomenon of paired-pulse depression described below has a similar dependence on these variables.

with synapses having a high probability of release (that is, "high p" synapses), they are more likely to show paired-pulse depression because more synapses will release transmitter and show relative refractoriness until their transmitter pools are restored (Dobrunz and Stevens, 1997; Markram et al., 1998; Varela et al., 1999; Thomson, 2000; Frick et al., 2007).

Note that this explanation for paired-pulse effects is based on the assumption that they are related to *presynaptic* factors involving probability of release. However, it is possible that postsynaptic factors also play a role, perhaps even a dominant role. For example, if the probability of release were unchanged by paired-pulse effects, then one could simply consider the probability that an evoked EPSP sufficiently depolarized the postsynaptic cell to fire an action potential. Paired-pulse effects would then be related to the nature of temporal summation of the EPSP and activation of NMDA receptors; often, a single EPSP does not depolarize the postsynaptic cell sufficiently to overcome the Mg^{2+} block of the NMDA receptor, but two summed EPSPs could do so. Whatever the explanation for paired-pulse facilitation or depression, the fact that effectively all synapses onto relay cells and cortical cells show one or the other behavior (as described more fully in chapter 4) underscores the importance of these phenomena in circuit functioning.

Perhaps the most important point about these paired-pulse effects is that they play a key role in the relationship between firing patterns of an afferent and the efficacy of synaptic transmission. For a synapse showing paired-pulse facilitation, low firing rates in the afferent would be relatively ineffective in influencing the postsynaptic cell because such low rates would not lead to facilitation of the synapse. The result would be no or a very small PSP evoked from most afferent action potentials. Such a synapse would be most effective when the afferent fired at rates high enough to elicit paired-pulse facilitation. For a synapse showing paired-pulse depression, a different pattern of firing evokes the greatest PSP. That is, if the afferent fired at high rates, the synapse would be persistently depressed. The largest PSP would result for an afferent action potential that followed a silent period long enough to ameliorate synaptic depression. Thus, for such a synapse, very low firing rates actually evoke the largest individual PSPs.

One interesting suggestion for paired-pulse depression is that it provides a gain control mechanism for synaptic processing (Chung et al., 2002). That is, the depression reduces EPSP amplitudes monotonically with firing rates. The lower amplitudes at higher rates help to avoid saturation of the postsynaptic response, thereby extending the dynamic range of afferent frequency over which the synapse can function.

2.2.3 Synaptic Integration

Neurons each commonly receive thousands of synapses from many different sources, and an obvious challenge is understanding how combinations of synaptic inputs interact to affect the postsynaptic cell. It should be clear from the above sections that understanding the effects of even a single input is quite complicated, given the issues of cable properties, active dendrites, and paired-pulse effects. To add to this, we now know that the summation of multiple inputs is far from linear (Gulledge et al., 2005; Gasparini et al., 2007; Spruston, 2008).

A particularly interesting example of this involves a phenomenon related to synaptic integration seen in cortical layer 5 pyramidal cells (Larkum et al., 1999). These cells have apical dendrites that extend into layer 1, where they terminate in a small apical tuft, and they appear to receive inputs there from axons traveling in layer 1. Some of these respond strongly to direct electrical stimulation of layer 1 only if paired with an action potential evoked by current injection at the soma, thus operating as a sort of AND gate or coincidence detector. That is, stimulation of layer 1 alone produces little or no measurable response; moderate somatic current injection itself produces a single action potential; but conjoint stimulation of layer 1 and somatic current injection properly timed produce a strong burst of action potentials related to backpropagation of the somatically evoked action potential, which, when properly timed with layer 1 stimulation, activates a Ca^{2+} conductance underlying the burst.

2.2.4 Summary and Conclusions

2.2.4.1 Thalamic Cells It is important to understand how a neuron's cellular properties—or more specifically, the passive and active properties of its dendritic (and somatic) membranes—affect its responses to synaptic input. Indeed, the interplay between a relay cell's cellular properties and the nature of its synaptic inputs is at the heart of understanding the functioning of thalamic relays.

Computed cable properties of thalamic relay cells, as suggested by consideration of their dendritic arbors, indicate relatively little electrotonic attenuation along the dendrites, so that synaptic inputs even on the most distal dendritic locations will have significant impact at the soma and axon hillock. However, it must be remembered that attenuation along a cable is frequency dependent, meaning that faster events, or faster postsynaptic potentials, will attenuate more during conduction to the soma than will slower ones. The cable properties of interneurons, in contrast, may be quite different. These cells have two outputs: a conventional one via the axon and another via terminals from peripheral dendrites. Finally, the active dendritic processes of cortical neurons (a factor that may also apply to thalamic cells, but this is not yet entirely clear; see chapter 4) appear to override issues regarding cable properties.

However, thalamic neurons have numerous voltage-dependent conductances in their dendritic membranes, and such active conductances can override issues related to cable properties. The state of these channels, especially their inactivation or activation by membrane voltage or neuromodulators, can strongly affect how the relay cell responds to synaptic input and thus how it relays information to cortex. Understanding how these channels are controlled by various synaptic inputs and how they might interact with one another is an ongoing challenge for students of the thalamus. Two other factors have complicated our understanding of how these cells respond to their synaptic inputs. First, paired-pulse effects mean that the rate of afferent firing plays a role in the magnitude of the postsynaptic response, and second, the idea that synaptic integration from multiple inputs is often nonlinear. These factors add to the complexity of understanding how these neurons respond to their afferent inputs.

2.2.4.2 Cortical Cells   The same points as noted for thalamic neurons apply to cortical cells, with the added point that cortical circuitry is much more complex than thalamic circuitry, and the responses of cortical cells to synaptic inputs are even more dominated by active membrane properties than is the case for their thalamic counterparts.

3 The Basic Organization of Cortex and Thalamus

3.1 The Cortex

3.1.1 The Cortical Areas

Thalamus and cortex[1] are two closely interconnected structures and are a characteristic development of the mammalian brain. They form the focus of this book. The cortex is widely regarded as having six layers (Brodmann, 1909), although different interpretations of the laminar structure are possible (see figure 3.1). Thus there are cortical areas where several subdivisions of layer 4 are clearly recognized, or where layer 6 is also separable into two distinct layers. In spite of these local differences, the complex multilaminar structure distinguishes the neocortex from the phylogenetically older paleo-cortex and archicortex, which have a simpler organization with fewer distinguishable layers.

Several different cell types are distinguishable in the cortex. The Nissl method (figure 3.1 left panel) allows the distinction between the pyramidal cells with an ascending dendrite and the smaller, rounded stellate cells, but the Golgi method reveals a greater variety of cell types (see figures 3.1 and 3.2) and shows the distribution of the dendrites and axons more clearly. Thus, figure 3.2 shows occasional stellate cells that have a slender ascending dendrite not unlike that of the pyramidal cells (figure 3.2, "E") and also shows a few of the many other cell types that characterize the neocortex. However, the details of these are beyond the needs of this book. From the point of view of the material in this book, a crucial point to record is that the stellate cells have their dendritic receptive surfaces limited to essentially one or two layers, whereas the pyramidal cells have ascending apical dendrites that can receive inputs from axons terminating in any of the more superficial layers.

---

1. As indicated earlier (see chapter 1), cortex here refers to neocortex.

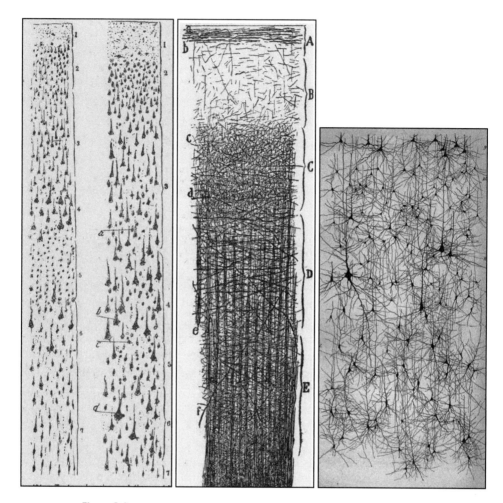

Figure 3.1
The appearance of the cortical layers as demonstrated by three different methods of staining cells and fibers. The left panel (from Cajal, 1911) shows two sections stained by the Nissl method, from the postcentral sensory area on the left and from the precentral motor area on the right. The central panel (also from Cajal, 1911) shows a section stained by the Weigert method for myelinated axons from the precentral motor area. The right panel (from Kölliker, 1896) shows a Golgi stain of only the lower layers of the precentral motor area, showing mainly the distribution of the pyramidal cells. Notice the variation in the laminar structures, with the granular layer (labeled as 5 in the left panel) absent in the motor cortical areas. Further details in the text.

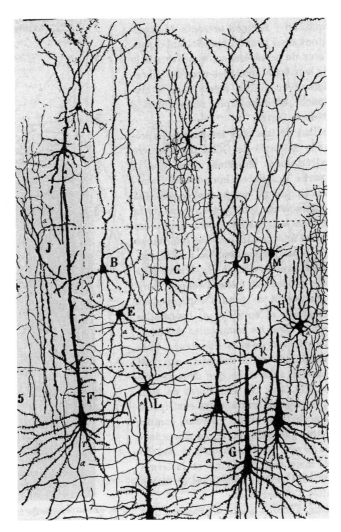

Figure 3.2
Cortical cells from the temporal cortex of a kitten to show some of the different cell types in the cortex. Golgi method. A, B, F, G, pyramidal cells; C,D, granule cells with ascending axons; E, stellate cell with apical dendrite; H, I, J, bitufted cells; K, cell with long descending axon; L, stellate cell; M, neuron with an ascending axon; a, axons. From Cajal (1911).

The mammalian cortex and its thalamic input is not only a particular mammalian feature of the brain but is highly developed in primates where it must be regarded as providing much of the neural circuitry that produces primate and, more particularly, human cognitive and behavioral skills. We stress, however, and will continue to stress in later chapters that a large repertoire of such skills is available to nonmammalian vertebrates, who are able to do a great deal with their subcortical centers, centers that survive in mammals. Frogs can catch flies. Cortex depends on these subcortical centers for the execution of all of its functions.[2]

In all mammals, it is possible to recognize many separate cortical areas. These areas are structurally distinguishable, and, functionally, many of them relate to distinct sensory modalities or motor functions (figure 3.3). There are relatively few such areas in some small mammals, particularly in the small, smooth cerebral hemispheres of species such as the opossum, tenrec from Madagascar, or the European hedgehog, which are thought to represent a phylogenetically early stage of mammalian cerebral evolution (Kaas, 2005, 2011). These simple brains of early mammals lack most of the great many higher cortical areas that characterize primates, but nonetheless they generally have more than just a single cortical representation for any one of the major sensory systems (auditory, somatosensory, or visual: see Kaas, 1995; Karlen and Krubitzer, 2007; Ashwell et al., 2008) with not very much beyond that. In some marsupial brains, the primary somatosensory and motor cortical areas, which lie adjacent to each other in eutherian mammals, are represented by a fused sensorimotor area or by such a fused area with an extra motor area adjacent anteriorly and a second sensory area posteriorly. The opossum brain illustrated in figure 3.3 appears to lack a distinct motor cortex. The doubling of the sensory areas and what appears to be the absence of a distinct motor cortex in these simple mammalian brains indicates that sensory cortical areas evolved earlier than motor areas and that a duplication of the sensory areas appears to be a very basic feature of early cortical evolution. These features raise two questions about the phylogenetic history of cortical subdivisions. One concerns the necessary motor functions of a sensory cortex where no motor cortex is identifiable, and the other relates to the possibility that an essential feature of sensory cortical areas in general is the duplication of these

---

2. Elliot Smith (1910) has written: "In its most primitive form the brain of the vertebrate animal might be compared to a close confederation of States, none of which is absolutely dominant. The cerebral hemisphere is little more than an instrument . . . whereby . . . impressions may bring their influence to bear on the nervous mechanisms which regulate movements, and so contribute their quota to the forces which control the behavior of the animal and its reactions to its environment."

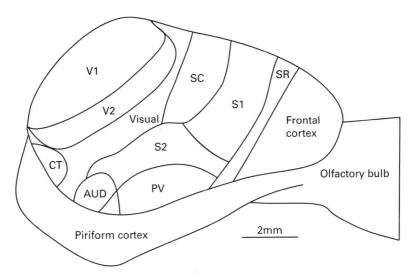

Figure 3.3
A dorsolateral view of an opossum brain showing some of the major cortical areas. AUD, auditory cortex (which may include more than one field); CT, caudotemporal cortex (this may be a visual field with an extra possible visual field rostral to V2); PV, a posteroventral somatosensory field; S1 and S2, two other somatosensory fields; SC and SR, additional somatosensory fields; V1 and V2, two visual fields. From Kaas (1995), with permission from S. Karger AG Basel.

areas or, at later stages of evolution, their multiplication, allowing for complex interactions in the cortex, interactions that would not be possible with just one such area. We will show that all areas of cortex have outputs to lower motor centers and that one important characteristic of cortical organization is that the cortex monitors ongoing motor outputs not only from the phylogenetically older subcortical parts of the nervous system but also plays a significant role in monitoring itself, with one cortical area receiving information through the thalamus about the motor outputs of a related cortical area (discussed further in chapters 6 and 8). That is, early in the evolution of the cortex, the areas that receive sensory inputs also had motor outputs,[3] and it may well be that the development of double or multiple cortical areas for any one modality, one monitoring the outputs of the other, preceded the separate development of motor cortex.

Here, we need to stress that the thalamus relays messages to each cortical area from the world, the body, or from other parts of the brain (including cortex) with specific thalamic nuclei linked to specific cortical areas (see

3. It is an important part of our account that they still have such motor outputs, even in higher primates (see chapter 8).

chapters 1 and 7). For the sensory cortical areas, the visual pathways have a relay in the lateral geniculate nucleus, the auditory pathways in the ventral medial geniculate nucleus, and the somatosensory pathways in the ventral posterior nuclei (see figure 1.4 of chapter 1). The development of higher order thalamic relays like those of the medial dorsal, lateral dorsal, and pulvinar nuclei and their extensive cortical fields in frontal, parietal, occipital, and temporal cortex[4] is reasonably regarded as a late evolutionary development, particularly striking in primates where well over 50 distinct areas have been recognized, and the number of areas tends to increase as we learn more (see Campbell, 1905; Von Economo and Koskinas, 1929; Van Essen et al., 1992). These higher areas all receive inputs from higher order thalamic nuclei; that is, from thalamic nuclei that are relaying messages from another area of cortex.

3.1.2   Localization of Function in the Cortex

The distinct functional characteristics of some individual cortical areas were originally defined on the basis of clinically identified losses and also on the basis of the positive outcomes that could be produced by direct electrical stimulation of the cortex. This dual approach to functional localization was pioneered by Hughlings Jackson (Hughlings Jackson, 1884), who used it to define the motor cortex on the basis of clinical observations, showing that some lesions in this cortical area (see figure 1.4) produced a loss of movement (that is, paralysis), whereas other lesions in the same area produced an excitatory discharge that was seen in the patient as grand mal epilepsy. Hughlings Jackson not only showed that there is an area of cortex that is concerned with the production of movements but also showed that there is an orderly representation of the body in the motor cortex, with the head controlled from ventral and lateral areas of the motor cortex, the feet from dorsal and medial areas, and the rest of the body in between. At much the same time, Fritsch and Hitzig (1870), using electrical stimulation of the cortex of dogs, identified an area of motor cortex, again, with evidence for a topographical representation of the body parts. Demonstrations of cortical areas concerned with somatosensory inputs came later. Their identification depended mainly on stimulation of cortex in conscious patients (figure 3.4) (Cushing, 1909; Penfield and Boldrey, 1937) and in part on studies of sensory losses produced by lesions. The motor cortex lies just anterior to the central sulcus in primates with the somatosensory cortex just posterior to the sulcus with a parallel representation of the body (figure 3.4).

---

4. The lobes of the brain are named after the bones they lie next to.

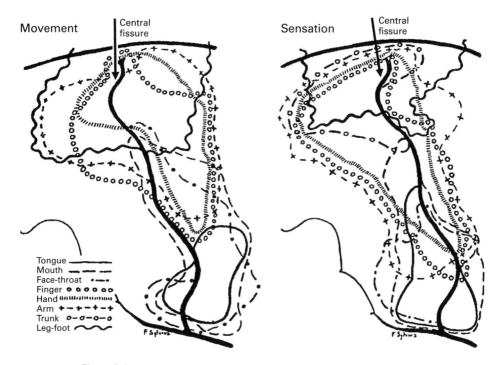

Figure 3.4
Summary of observations made in several patients of the responses to cortical stimulation in the areas outlined. Modified from Penfield and Boldrey (1937).

Studies of patients with lesions that produced a loss of vision played a large part in defining the area of the primary visual cortex (Brodmann's area 17, now often called V1 or striate cortex) and in showing the topographical representation of the visual field within this area (figure 3.5). Generally, lesions seen clinically do not allow sufficiently accurate localization of the injured area for detailed conclusions such as those shown in figure 3.5, and several earlier studies of lesions in the occipital and temporal cortex had led to significant disagreements, which are well summarized by Polyak (1957). However, advancing military technology during the 19th century produced high-speed bullets that could pass straight through the human head leaving clear entry and exit wounds, on the basis of which it was possible to make a precise estimate of the bullet's course and of the resulting cerebral damage. By comparing the locus of the cortical injury with the loss of vision in limited parts of the visual field, Inouye (1909), studying veterans of the Russo-Japanese War (1904–1905), and Holmes (Holmes, 1918), with veterans of the First World War (1914–1918), defined the map of the visual hemifield

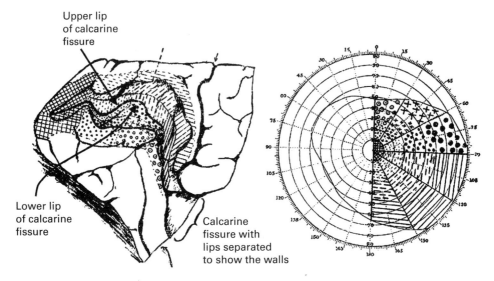

Upper lip
of calcarine
fissure

Lower lip
of calcarine
fissure

Calcarine
fissure with
lips separated
to show the walls

Figure 3.5
The medial surface of the left occipital lobe is shown on the left, and a chart of the right visual
hemifield is shown on the right. The sectors of the visual field are identified to correspond to the
parts of the visual cortex (area 17) on either side of the calcarine fissure. Note that a significant
part of the cortex is hidden in the depths of the calcarine fissure. Modified from Holmes (1918).

that is represented in the occipital cortex of each hemisphere and shown in
figure 3.5.

These studies of motor cortex on the one hand and of somatosensory and
visual cortex on the other, with auditory cortex defined later (Walzl and
Woolsey, 1946), led to a widely accepted view of the cortex. It was regarded
as organized so that the major sensory inputs, all of which except olfaction
were known to be relayed to cortex through the thalamus, provided a way into
the cortex from the body and the world. That is, the way into the cortex was
through the thalamus, and the way out back to the body was through the motor
cortex to the lower centers of the brain for control of behavior. The cortical
areas in between were naturally treated as a route between the input and the
output and for many years were regarded as "association cortex," implying an
ill-defined function aimed at associating information from the various input
sources and preparing for the necessary outputs on the way to the motor cortex
that eventually would lead to movement control or would lead along a different
route to memory.

More recently, many of these intermediate nonprimary areas have become
closely linked functionally to one or another of the primary sensory areas, and

a view of cortical processing has developed that has a single entry point for each modality and a limited number of motor exits. Multiple cortical areas for auditory, somatosensory, or visual functions began to appear in the late 1940s and the 1950s (see chapter 7), and then in the late 1960s and the 1970s, physiological recordings from cortical regions near the primary visual area (V1) demonstrated further multiple maps of the contralateral visual hemifield in each hemisphere (for example, Allman and Kaas, 1971; Allman and Kaas, 1975) and showed that some of these areas had distinct visual functions (for example, Zeki, 1969; Desimone et al., 1984; Maunsell and Newsome, 1987). Today, functional mapping combined with studies of response properties and anatomical connections have provided more specific functions for much of cortex relating to somatosensory, auditory, visual, and motor activities. For vision, more than 30 discrete areas have been mapped as specific visual areas in the macaque cortex, and these areas occupy all of occipital and much of parietal and temporal cortex (Allman and Kaas, 1975; Felleman and Van Essen, 1991). Comparably, a number of distinct auditory and somatosensory cortical areas have been defined in the temporal and parietal lobes of the brain, respectively (Kaas and Collins, 2001). Many of these functionally distinct cortical areas, but by no means all, have been identified as structurally distinguishable entities on the basis of architectonic studies.

3.1.3 Structurally Distinct Cortical Areas and Their Functions

There are several ways of distinguishing the architectonic structure of cortical areas (figure 3.1). Myelo-architectonic studies are based on the distribution of myelinated nerve fibers and show that some cortical areas have particularly thick bundles of myelinated fibers entering or leaving through white matter, have particularly marked layers of myelinated fibers relating to one or more of Brodmann's six cellular layers, or show characteristic patterns of incoming or outgoing nerve fibers as indicated in figure 3.1. Cyto-architectonic studies have mostly been based on the Nissl stain, which shows the distribution of the nerve cell bodies and glial cells but none of the axons or dendrites except for the most proximal parts of the largest apical dendrites of pyramidal cells. This stain reveals primarily the ribosomes and their distribution in the nerve cells. These represent the machinery for making the many different proteins manufactured by each distinct cortical nerve cell type and thus give each of the many different neuron types in the cortex a more or less characteristic appearance in the Nissl sections. In this way, the Nissl preparation provides a summary overview of how cortical areas differ in terms of the overall distribution of the major functionally distinct cell types.[5] Some investigators, particularly Cajal (1911), have used the Golgi method to demonstrate the characteristic distribution of different nerve cell types in the different cortical areas (figure 3.1). Because this method reveals the cell body as well as the dendrites and often the axon of individual cells as well, it allows another approach to defining the differences between cell types in different cortical areas. There have been several different maps of cortical areas proposed for different species, and although there is nothing like a clear agreement as to any one definite cortical "map" of all the areas across species or even for any one species, there is general agreement on several of the major areas, particularly the primary sensory (somatosensory, auditory, visual) and motor cortical areas and a few others.

For some sensory cortex beyond the primary areas, it has been possible to assign specific functions to particular areas. This has been most successful for the visual areas of primates, where, for example, in an architectonically distinct area in the mid-part of the temporal lobe, area MT (for "middle temporal"),[6] most of the neurons respond preferentially to moving stimuli with each cell having a preferred direction of motion (Dubner and Zeki, 1971; Maunsell and Van Essen, 1983; Movshon et al., 1985), whereas in other areas most of the cells appear to respond best to features such as color, stereopsis, optic flow, or faces. Exactly what this cortical division of labor achieves for the mammalian brain is not entirely clear. The capacity to recognize and respond to one or another of these features is present in animals that have no visual cortex. Frogs can catch flies and fish can respond to colors, although we know nothing about their capacity to recognize faces. The phylogenetically older, subcortical circuits can do at least some of the things now generally ascribed to cortex [see also Sprague (1966), who showed that cats lacking cortex can orient to visual stimuli].

We know very little about how these many functionally distinct separate areas of the primate visual cortex, each with its own more or less complete representation of the visual hemifield, its own structural specializations, and its own boundaries and connections to thalamus and to other cortical areas, evolved from the relatively simple brains of the earliest mammals (see figure

---

5. Note that many investigators recognized that the Nissl method showed the distinctive structures of functionally distinct nerve cell types long before the functions of the ribosomes had been defined, at a time when other investigators were still asking whether the inclusions stained by the Nissl method might not represent some artifact of the method, quite unrelated to the functions of the nerve cell. Today there are still many who question that the Nissl method, which stains almost no parts of the neural processes, can be useful in distinguishing functionally distinct nerve cells and groups of nerve cells.

6. Sometimes called "V5."

3.3), nor do we know exactly what special skills their presence adds to a primate's visual capacities. A reasonable conclusion would be that these are perceptual skills that the midbrain visual centers lack, but a detailed comparison of exactly what each group of cortical visual centers can do that the lower centers cannot do has not been attempted to our knowledge.

A comparable evolutionary development of multiple areas characterizes the auditory, somatosensory, and motor pathways and raises the same questions: how did the many areas evolve from a few, and how do the many interact to produce what appears to be one world?

For the visual pathways, Van Essen and colleagues (Felleman and Van Essen, 1991; Van Essen et al., 1992) have proposed complex hierarchical schemes of more than 30 separate cortical areas and their interconnections, which they distinguish as feedforward, feedback, or lateral, based on the laminar origins and terminations of the connections. They integrated earlier connectional studies from many laboratories into schemes that showed each of the cortical areas with between two and more than 20 connections to other cortical areas. Their figures show several parallel paths proposed for visual information processing in a pattern that includes a number of hierarchical levels and can in some of the figures be traced to motor outputs or memory. It should be emphasized that these figures rely almost entirely on neuroanatomical data with practically no information regarding the nature of the messages transmitted or the synaptic properties of the corticocortical connections as likely drivers or modulators,[7] a point we shall revisit in chapter 5.

Given a mammalian cortex that receives sensory inputs through the thalamus and that is considered to send its motor instructions for behavioral control out through a few areas of motor cortex only, the crucial neural computations that relate the many distinct cortical areas to each other would all have to be done in the cortex as proposed by Van Essen and colleagues. The possibility that corticocortical interactions may be possible for all cortical areas through corticofugal axons to many subcortical centers, which in turn can supply afferents through the thalamus back to cortex, or that corticocortical links may pass directly through the thalamus, is not considered by these authors; the interactions that they illustrate imply that once information reaches a primary sensory cortical area from the thalamus, it is processed entirely within cortex with no contribution from any subcortical source and with no outputs to any lower centers before the major cortical motor areas have been reached.

This essentially corticocentric view of cortical processing confines the cortex to a constrained set of functional interactions that has the cortical

---

7. Defined in chapter 1.

computations going on without concurrent reference to the rest of the nervous system, or to the body and the world. It is not a view limited to considerations of sensory processing alone. For example, Passingham et al. (2002), in their account of cortical areas concerned with motor functions, have argued that the functions of any cortical area can be determined by its intrinsic and extrinsic connections, but they limited their review of the extrinsic connections to the cerebral cortex; details of the connections with subcortical centers were not considered. Similarly, Milner and Goodale (2008), in considering that two distinct cortical streams may be concerned with action and perception, state that the functional differences between the two streams are best understood in terms of their outputs, but the outputs that they consider are those going to other cortical areas, not those that are going to subcortical centers relevant for the behavioral capacities crucial for their theory. We will show in chapter 8 that, for any one cortical area, there are links with the body and the world that do not go through other cortical areas, and we will suggest that all of these links play a close and essential role in the cognitive life of the individual organism.

There are other key issues that arise from contemporary views of cortico-cortical communication and that are relevant to our treatment of cortex. One is that understanding cortical processing will depend on distinguishing connections that can be understood to have driver functions from those that have modulator functions, and this is considered in chapter 4. A second is that, as we show in section 3.2.1.2, cortical areas communicate with each other through the thalamus, not only through direct corticocortical connections; and a third is that, as indicated above and considered in more detail in chapter 8, all cortical areas have their own outputs to lower centers through which their role in perceptual and cognitive processes can be directly linked through the phylogenetically older lower motor centers to the body and the world and also back to other cortical areas.

3.2 The Thalamus

The thalamus is a part of the diencephalon, and we are particularly concerned with two parts of the diencephalon that have a separate developmental origin and consequently a distinct structural organization. The first and most important is the dorsal thalamus, which is the main part of the diencephalon and includes the relay nuclei that project to cortex; this is generally referred to simply as the "the thalamus," and we will do the same in the following text. The other is the ventral thalamus, including the thalamic reticular nucleus, which does not project to cortex. We are concerned with the reticular nucleus and with one other region that is sometimes included in the ventral thalamus, the zona incerta.[8] Both of these structures, ventral thalamus and zona incerta, provide GABAergic inputs to thalamic relays, but the reticular nucleus innervates all thalamic nuclei, whereas the zona incerta targets mostly the higher order relays (see chapter 5; reviewed in Sherman and Guillery, 2006; Jones, 2007).

The independent developmental origin of the thalamus on the one hand and the reticular nucleus on the other represents an important key to the way in which we treat them. This key, which sees a common developmental origin of any one major brain part as producing a common structural plan for all subdivisions of that part, is one that has been widely used for all parts of the nervous system in the past. Thus, we expect the spinal cord to demonstrate a common plan even though that common plan will be modified where the limbs are innervated or the sympathetic outflow originates, and we expect the cortex to have a common plan, even though we recognize major differences between the cortical areas considered in the previous section. We expect the common plan to be preserved to a significant extent across species but also recognize that some differences are almost certain to arise in species that have particular specializations: we do not expect snakes to have marked limb specializations of the spinal cord and are not surprised that the arrangement of cortical cells that is a neural replica of the mystacial vibrissae, the characteristically organized "barrel cortex" or first somatosensory cortex of some rodents (Woolsey and Van der Loos, 1970), is not found in primates. This approach represents one basis on which we see all parts of the thalamus as having a common plan and recognizes that there are specializations in some parts of the thalamus and variations between species. Similarly, the reticular nucleus has an overall common structure with some, but few, variations that are clearly defined at present.

3.2.1   The Thalamic Nuclei and Their Inputs

The thalamus is divided into a number of distinct nuclei, which are shown schematically in figure 1.4 and are briefly introduced in chapter 1. For the thalamus, as for the cortex, the major divisions are most readily based on myelo-architectonic or cyto-architectonic studies. The names of the nuclei are broadly based on their position within the thalamus, and their functional

---

8. The extent to which the zona incerta and the reticular nucleus actually share a developmental origin is uncertain. Certainly, they differ significantly in the organizational plan of their cell types and connections.

identity depends on the source of the driver inputs or on the neural center that receives their outputs (for a further discussion, see chapter 7).

3.2.1.1 The Major Thalamic Inputs   Four main types of input, identifiable by their appearance in the electron microscope (Sherman and Guillery, 2006), account for roughly 95% of the identified inputs to the thalamic relay cells. The RL terminals (for round vesicle, large profile) are the driver inputs (RL in figure 3.6). They come from different sources depending on the nucleus. These are the largest terminals in the nucleus, they are glutamatergic, and each forms about 10 asymmetric synapses[9] onto its targets. The round vesicles and asymmetric synapses are marks of excitatory synapses. The RS terminals (for round vesicle, small profile) (RS in figure 3.6) can be either from cortical layer 6 glutamatergic axons or from brainstem cholinergic axons. These rarely form more than one asymmetric synapse and are also excitatory. There are two types of F terminal (for flattened or pleomorphic vesicles; F in figure 3.6), both of which form symmetric synapses, and they are GABAergic and inhibitory. One type (F1) is formed by terminals of interneuronal axons, reticular cells, or of other GABAergic inputs to thalamus (for example, from the zona incerta, substantia nigra, or pretectal region), and these are always presynaptic, never postsynaptic. The other type (F2) arises from interneuronal dendrites and represents the axoniform dendritic appendages of interneurons discussed in section 3.2.2.2; these are both presynaptic and postsynaptic and their postsynaptic targets are relay cell dendrites (which also receive synaptic inputs from RL, RS, or F1 terminals). Further details of their interconnections are also presented in section 3.2.3.

Groups of these synaptic processes are often seen to be closely packed in a structure described as a glomerulus (Szentágothai, 1963). The main features of a glomerulus include a large relay cell dendrite contacted by RL, and F profiles in a region almost entirely free of astrocytic processes (figure 3.6), a feature that clearly distinguishes them from nearby neural tissue. Triadic synaptic arrangements that involve an RL profile presynaptic to the dendrite and to an F2 profile (details in section 3.2.2.1), with the F2 profile also presynaptic to the dendrite, are common within glomeruli.

3.2.1.2 The First Order Nuclei   In chapter 1, we defined the first order nuclei as nuclei that relay messages to cortex from sensory pathways or subcortical centers. The auditory pathways bring messages to the medial

---

9. An asymmetric synapse is one that has a pronounced postsynaptic thickening, whereas a symmetric synapse is one that lacks a pronounced postsynaptic thickening.

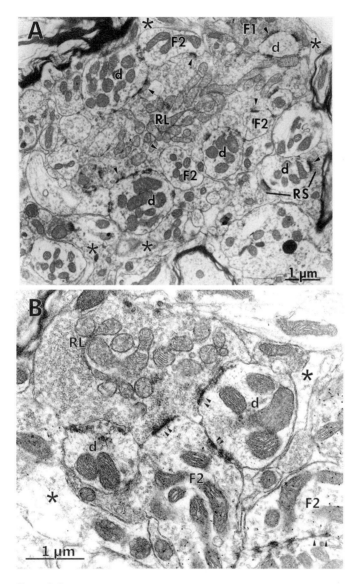

Figure 3.6
Electron micrographs of two glomeruli from the lateral geniculate nucleus of a cat. (A) A large RL terminal making many synaptic contacts with dendrites and F2 terminals; a (darker) F1 terminal, top right, makes contact with a relay cell dendrite (d). Arrowheads indicate the presynaptic aspect of synaptic junctions. Two RS terminals are shown on the lower right. Asterisks show astrocytic processes that lie outside the glomerulus. Further details in the text. From Sherman and Guillery (2001). (B) A section through a part of a glomerulus that contains a triad with a large RL terminal presynaptic to a dendrite (d) and an F2 terminal; the F2 terminal, in turn, is presynaptic to the same dendrite. Another F2 terminal is shown in the lower right. In this figure, the GABAergic F2 terminals are labeled with a post-embedding gold procedure. The double arrowheads indicate the postsynaptic aspect of the synaptic junctions. From Erişir et al. (1998).

geniculate nucleus from the inferior colliculus. The inputs from the retina identify the lateral geniculate nucleus as the major thalamic visual relay. The ventral posterior nucleus receives inputs from the medial lemniscus, bringing messages about touch and kinesthesis from the limbs and the trunk, and also from the spinothalamic tracts, which carry messages about pain and temperature from the trunk and limbs. The ventral posterior medial nucleus receives corresponding somatosensory messages from the face from two pathways that come from the trigeminal nuclei in the brainstem. These first order nuclei are the main nuclear groups that play a central role in our presentation; they relay messages to the auditory cortex, the visual cortex, and the somatosensory cortex, respectively (see figure 1.4). They are important because they link the thalamus to the outside world, and through the thalamus they link the world to the cortex. In addition to these, there are ascending gustatory and vestibular inputs to parts of the thalamus that we consider briefly in chapter 6.[10] Two other major groups of ascending pathways enter the thalamus for relay to cortex (that is, as drivers). One comes from the deep cerebellar nuclei and is recognized as an important input destined for relay to the motor cortex. It innervates a region of the thalamus that is shown as the ventral lateral nucleus (VL) in figure 1.4.[11] The other comes from the mammillary bodies and innervates the three anterior thalamic nuclei (anterior medial, anterior ventral, and anterior dorsal) for relay to the limbic cortex, which lies on the medial aspect of the hemisphere above the corpus callosum (figure 1.4).

These ascending pathways to the thalamus are all glutamatergic and all form RL terminals in the thalamus. They are reasonably regarded as driver inputs that carry messages for relay to cortex because they are carrying sensory messages for relay to areas of sensory cortex, or cerebellar messages for relay to the motor cortex, or mamillary messages for relay through the anterior thalamic nuclei to the limbic cortex. As described in detail in section 3.2.3 and

10. The olfactory pathways reach the paleocortex without passing through the thalamus, and for this reason we have excluded them here. The pyriform cortex in turn sends two types of axon to the medial dorsal nucleus (Kuroda et al. 1992b), which correspond to the drivers and modulators innervating other thalamic nuclei and are discussed later in this section.

11. This is a region that is sometimes referred to as a part of VA/VL (that is, the ventral anterior and ventral lateral nuclei) and that has been subdivided in various ways. For the sake of simplicity and in accordance with other results that show in rat and monkey two mostly non-overlapping territories in the VA/VL region, one innervated by axons from the deep cerebellar nuclei and the other from the globus pallidus (Schell and Strick, 1984; Sakai et al., 1996; Kuramoto et al., 2011), we will refer to the region of cerebellar input as the ventral lateral nucleus and the region of the pallidal input as the ventral anterior nucleus; the latter is innervated by a layer 5 input from frontal cortex (McFarland and Haber, 2002; Kultas-Ilinsky et al., 2003).

illustrated in figure 3.6, these afferents all have a shared characteristic appearance. They often terminate in glomeruli where they tend to show the same arrangement of synaptic contacts, although we will note some differences between thalamic nuclei relating to the species being studied or the identification of the nucleus as first or higher order. It is on the basis of these driver inputs that bring messages from regions of the brain other than cortex or thalamus for relay to cortex that these thalamic nuclei are recognized as "first order." That is, these nuclei transmit messages to the cortex that do not come from the cortex. The other major thalamic nuclei, particularly the medial dorsal nucleus, the pulvinar, the lateral posterior nuclei, the intralaminar nuclei, and the ventral medial nucleus, receive a significant part of their driver inputs from axons that arise in the cortex itself and that form the characteristic RL terminals in the thalamus. For this and other reasons considered in the next section, they are regarded as higher order nuclei. The input from the globus pallidus to the ventral anterior nucleus (see footnote 11) is GABAergic and is not regarded as carrying messages for relay to cortex (see chapter 4; also, Smith and Sherman, 2002), and we include this nucleus with the higher order nuclei for reasons explained in the next section.

3.2.1.3 The Higher Order Nuclei   The general organization of the higher order nuclei does not differ greatly from that of the first order nuclei, but they are distinguished because many, possibly all, of their RL terminals (that is, the terminals that in first order nuclei carry messages for relay to cortex) can be shown to come from the cortex, not from ascending pathways (see also chapter 5). We treat a thalamic nucleus as higher order if at least a significant component of its RL (driver) inputs come from the cortex, and we distinguish a higher order nucleus from a higher order relay, which is defined on the basis of individual relay cells. That is, just as functionally distinct circuits (such as the X and Y relays described in section 3.2.2) are mixed in some parts of the cat's lateral geniculate nucleus, so may first and higher order relays be mixed in a higher order nucleus.[12] To date, all evidence of first order nuclei indicates that they do not contain higher order circuits. However, whereas in first order nuclei there is no evidence, to date, of integration (that is, convergence and interaction) of driver inputs from different sources carrying different information in

12. A possible example is the pulvinar, which we regard as a higher order nucleus because it receives substantial driver input from cortex, but it also receives input from the superior colliculus, which may also prove to be a driver (for example, Kelly et al., 2003). Whether higher order nuclei such as the pulvinar, medial dorsal nucleus, posterior medial nucleus, and dorsal medial geniculate nucleus also contain first order relays remains an unanswered question.

single relay cells, this issue has not been sufficiently studied in higher order nuclei.[13]

Not only is the appearance of the RL profiles that come from corticothalamic axons and the details of their synaptic relationships essentially the same as seen in the first order nuclei, but, where their functional role has been defined (in the lateral posterior nucleus of the cat and rat, the pulvinar of the monkey, the posterior medial nucleus of the mouse, and the dorsal medial geniculate nucleus of the cat and mouse), they also represent the axons that carry messages for relay to the cortex.

The first indications of such a cortical input to thalamus for relay to cortex was the observation that axon terminals resembling the RL terminals, having the same basic structure and the same patterns of synaptic relationships, often

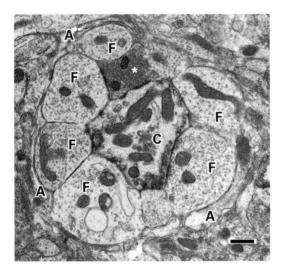

Figure 3.7
Glomerular synapses from the dorsal medial geniculate nucleus of a cat. The portion of the axon terminal marked with an asterisk is presynaptic to the central dendrite, C, and corresponds to what we have called an RL terminal; it is also presynaptic to the upper of the vesicle containing profiles marked F to the left of C. The A's show astrocytic processes. Scale bar: 500 nm. With permission from Ojima and Murakami (2011) with some modifications in the labeling. We thank Professor Ojima for providing us with an original image for this figure.

13. By integration, we mean that a relay cell's responses reflect input from drivers representing different information, such as that seen in layer 4 cells of visual cortex (Alonso et al., 2001), which receive driver inputs from a number of geniculate relay cells differing in receptive field location and properties (for example, on-center versus off-center). Whereas geniculate relay cells often receive several convergent retinal inputs, these virtually always have the same receptive field location and properties (Cleland et al., 1971; Usrey et al., 1999), and so this is not an example of integration.

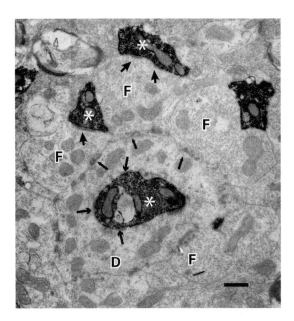

Figure 3.8
Glomerular synapses from the dorsal medial geniculate nucleus of a cat. The dark profiles (three of which are identified by an asterisk) have been labeled by phaseolus vulgaris leucoagglutinin that was injected into the auditory cortex. These represent sections of corticothalamic axon terminal(s) that innervate this higher order thalamic nucleus and also correspond to what we have called an RL terminal. D identifies a central dendrite of the glomerulus, and the F's indicate the F terminals in the glomerulus. The short arrows with large heads pointing upward show synapses that the corticothalamic terminals make upon the F terminals; the three arrows with small heads pointing upward show synapses that the F terminals make upon the central dendrite; the other three intermediate-size arrows within the central dendrite show synaptic junctions that the labeled corticothalamic terminal makes upon the central dendrite. Scale bar: 500 nm. With permission from Ojima and Murakami (2011) with some modifications in the labeling. We thank Professor Ojima for providing us with an original image for this figure.

within glomeruli, could be shown to degenerate in some thalamic nuclei after cortical lesions. Thus, after lesions in visual cortex of monkeys, such degenerative changes occur in the pulvinar (Mathers, 1972; Ogren and Hendrickson, 1979), and after lesions in the frontal cortex they appear in the medial dorsal nucleus (Schwartz et al., 1991). These lesions also produced degenerative changes in corticofugal axons that formed the smaller RS terminals making relatively simple synaptic junctions on more peripheral dendrites and not in the glomeruli. These are not regarded as carrying messages for relay to cortex (discussed in section 3.2.3). The large RL terminals were subsequently shown to come from thick axons that arise in layer 5 of cortex (figures 3.7 and 3.8), whereas the smaller extraglomerular corticothalamic RS terminals came from thinner axons that arise in cells in cortical layer 6 (Ojima, 1994; Ojima et al., 1996; Rockland, 1996; Rouiller et al., 1998; Cappe et al., 2007).

Injections of retrogradely transported tracers into thalamic nuclei have shown that layer 6 cortical cells are labeled by injections into any thalamic nucleus, but layer 5 cells are only labeled after injections into higher order thalamic nuclei (Gilbert and Kelly, 1975; Abramson and Chalupa, 1985; Llano and Sherman, 2008). That is, all thalamic nuclei receive layer 6 inputs but only some receive layer 5 inputs, and we have called the latter higher order nuclei. Figures 3.7 and 3.8 show sections through a large corticothalamic RL axon terminal typical of layer 5 cell terminals in the dorsal medial geniculate nucleus, a higher order nucleus. The profiles of these RL terminals lie in glomeruli and establish synaptic pattern comparable to the patterns made by ascending driver afferents to the first order thalamic nuclei (figure 3.6).

For the visual pathways, elimination of the layer 6 (modulatory) input to the lateral geniculate nucleus has only subtle effects on receptive fields of geniculate neurons (Kalil and Chase, 1970; Baker and Malpeli, 1977; Schmielau and Singer, 1977; Geisert et al., 1981). However, the receptive fields of some pulvinar cells resemble those of cells in layer 5 of the visual cortex, and these receptive field properties are lost after lesions of striate cortex (Bender, 1983; Chalupa, 1991), thus supporting the morphological evidence and showing that these corticothalamic axons from layer 5 are the ones that carry messages for relay to cortex. A similar pattern has been reported for the somatosensory pathways, where we know that layer 5 cells in the first somatosensory cortex, S1, project to the posterior medial thalamic nucleus (Deschênes et al., 1994) and where it has been shown that elimination of S1 cortex has little effect on responses in the ventral posterior medial nucleus, which receives a layer 6 input from that area of cortex, but virtually silences responses in the posterior medial nucleus [which receives a layer 5 input from this cortical area (Diamond et al., 1992) as well as a layer 6 input].

For the auditory pathways, we know that the primary auditory cortex projects to the dorsal medial geniculate nucleus (figures 3.7 and 3.8) (Ojima, 1994; Ojima and Murakami, 2011), but we lack the information about the characteristics of the receptive fields or about the losses that occur in the higher order thalamic nuclei after destruction or silencing of the relevant cortical inputs.

Many layer 5 cells have axons that do not go to the thalamus but go to other centers. The important point to stress here is that the layer 5 cells that do project to the thalamus all appear to have long descending branches that innervate one or another of the lower motor centers (striatum, superior colliculus, pontine nuclei, pontine reticular formation, or spinal cord; see chapter 6 for details). The contrast with layer 6 corticothalamic axons is important. The axons from layer 6 cells form a large part of the input to all thalamic nuclei. They send a modulatory signal as a feedback message to the thalamus. They send branches to the thalamic reticular nucleus but send no branches beyond the diencephalon. They are introduced here because they are often confused with the layer 5 corticothalamic axons. Whereas the layer 6 cells have receptive field properties that are not seen in the thalamus and are not relayed to cortex, the layer 5 cells, where they have been studied, have receptive field properties that are seen in the thalamus and are relayed to cortex on a presumed feedforward corticothalamocortical pathway from one cortical area to another (probably higher) cortical area (Van Horn and Sherman, 2004; Llano and Sherman, 2008). The layer 6 cells have branches that innervate the thalamic reticular nucleus and lack long descending branches to brainstem motor centers, whereas the layer 5 cells do not have branches that innervate the thalamic reticular nucleus and have long descending branches to brainstem motor centers.

3.2.2   The Cells of the Thalamus

3.2.2.1   The Relay Cells   The relay cells represent the majority of the thalamic cells, making up 75% to 80% of the cells in many thalamic nuclei of cats and monkeys and up to 99% in some of the thalamic nuclei of rats and mice (Spreafico et al., 1994; Arcelli et al., 1997). As we have seen, the relay cells serve to transmit a message about activity in another part of the nervous system to a particular, specialized part of the cerebral cortex. This transfer of information has to be regarded as the main function of the thalamus, so that the relay cells are key structures in thalamic function. The basic pattern of thalamocortical connections links each thalamic nucleus to one or more particular specialized cortical areas, and essentially all of the information that the cortex receives about activity in other parts of the central nervous system or, through these, about the outside world, comes from the relay cells of the thalamus.

So far as we know at present from observations of some of the first order thalamic nuclei, the message passed to cortex is a reasonably close copy of the message that the relay cell receives. This is known to be true for the primary sensory relays (visual, auditory, somatosensory) and for a small part of the mamillothalamic relay, the anterior dorsal nucleus, where information about head position is relayed from the small lateral mamillary nucleus to the retrosplenial cortex (see chapter 6; also, Taube, 2007). Indirect evidence supports a similar relationship between visual cortex and the higher order pulvinar, where pulvinar receptive fields, which can be quite complex, are similar to those recorded in cortex (for example, Dumbrava et al., 2001); the missing link is evidence that individual layer 5 inputs to each of these thalamic cells have the same response properties.

There are typically subtle changes in the message for relay cells where the receptive fields have been defined, as in the size of the receptive field or the center/surround relationships of a receptive field, but the major features of the message are not changed in these thalamic relays. For the visual system, this is in marked contrast with the other relays in the path from the retinal receptors to the higher cortical areas, and within visual cortical areas, where at every stage the relay appears to produce significant changes in the characteristics of the receptive field. That is, the synapses from the main Class 1 driver input to the relay cells are typically unique within the sensory pathways in not participating in significant receptive field elaboration. We regard thalamic circuitry as providing a significant but different function that is treated in detail in chapter 4. In other words, there is no evidence to date that any real integration of inputs takes place in thalamus, as it does in cortex. However, specific evidence for this feature of thalamus is limited to a few first order relays, and the possibility that integration exists in some thalamic regions, especially higher order relays, needs to be explored.

In spite of what might appear to be a relatively uniform function of thalamic relay cells in general, the relay cells vary significantly in size, and they vary in the details of their dendritic arbors. It has long been recognized that the dendritic arbors of thalamic relay cells can be distinguished as either radiate or bushy (Kölliker, 1896; Guillery, 1966; LeVay and Ferster, 1977). The bushy arbors have several branches originating together from the ends of the short primary stems of dendrites, and the arbor can have a bipolar structure, whereas the radiate cells have arbors whose primary dendrites give off single branches progressively as they pass from the cell body, and the arbor often has a roughly spherical structure. In cats, ferrets, monkeys, and bushbabies there are three parallel visual pathways with relays in the lateral geniculate nucleus. These are the X, Y, and W pathways in the cat and the parvocellular, magnocellular, and koniocellular pathways in the monkey (Fukuda and Stone, 1974; Leventhal et al., 1981; Sherman and Spear, 1982; Stone, 1983; Leventhal et al., 1985; Casagrande and Norton, 1991). The W and konio relay cells are the smallest, and the Y and magno relay cells are the largest, with evidence from cat that the Y cells tend to be radiate and the X cells bushy (Friedlander et al., 1981; Stanford et al., 1983). However, exactly how these morphological differences relate to the way in which the message passes through the thalamus is at present not clear, nor do we know how such morphological differences seen in other thalamic nuclei relate to the relay functions. Bartlett and Smith (1999) reported both bushy (tufted) and radiating dendrites in the rat's ventral medial geniculate nucleus (first order) but found only radiate dendrites in the dorsal medial geniculate nucleus (higher order), and Yen et al. (1985) reported the same two types of relay cell in the ventral posterior nucleus (first order) of the cat.

At present, there is no clear evidence to indicate the functional significance of the two different types of dendritic arbor; that is, to relate the structure of the radiate and tufted dendrites to the computational requirements of these two types of relay cell. The point is raised as one of the many puzzles about the thalamus that still merit attention. If the thalamic relay cells are not significantly changing the message that is passed to the cortex, then one may wonder why the cell type varies in its appearance. This line of thinking suggests that the differences may relate more to the role of the modulators (see chapter 4) or certain details of circuitry not clearly revealed by the structure of dendritic arbors, such as glomerular and triadic circuitry seen more commonly for X than Y cells (see section 3.2.3 and figure 3.9A), than to the role of the drivers, but the puzzle of the functional significance of the branching patterns remains.

As we pointed out in chapter 1, the nature of the messages that are sent to the cortex through the relay cells has been defined for all of the major sensory pathways that pass through the thalamus but is essentially unknown for the great majority of the thalamic relay cells. This is still a major unanswered question for much of the thalamus.

When it first became possible to record individual action potentials from axons or nerve cells, the temporal pattern of the action potentials could be read as a code representing the message that was being passed by the axon or sent by the nerve cells to the thalamus. Adrian (1928) early showed how to interpret this code for the sensory pathways where sensory receptors respond to changes in the environment, and he could relate those changes in terms of the temporal patterns of action potentials because the axons were responding to sensory receptors whose activation could be experimentally manipulated. Once we move away from the sensory receptors toward thalamic afferents that come from central relays like the deep cerebellar nuclei or other cortical areas, matching the incoming action potentials to activity of the centers is much more complicated, because the action potentials in any one axon or nerve cell represent a small part of the total input pattern that can be related to specific actions or perceptions. The messages are then much harder to interpret. Identifying the nature of the message that passes through the thalamus will not be easy, but we cannot expect to understand the functional role of the inputs to any one cortical area if we do not know anything about the nature of the

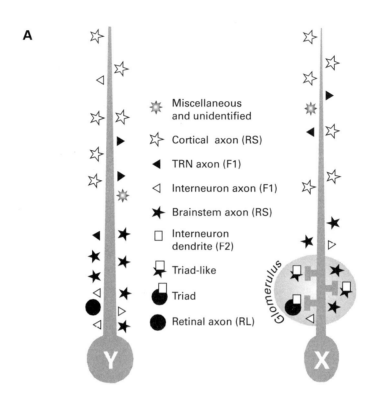

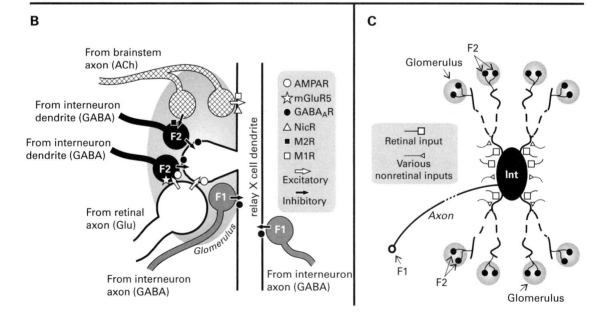

Figure 3.9
Schematic views of features of connectivity in the A layers of the cat's lateral geniculate nucleus. (A) Synaptic inputs onto an X and a Y cell. For simplicity, only one, unbranched dendrite is shown. Synaptic types are shown in relative numbers. Redrawn from Sherman and Guillery (2006). (B) Synaptic inputs in and near a glomerulus. Shown are the various synaptic contacts (arrows), whether they are inhibitory or excitatory, and the related postsynaptic receptors. The conventional triad includes the lower F2 terminal and involves three synapses (from the retinal terminal to the F2 terminal, from the retinal terminal to an appendage of the X cell dendrite, and from the F2 terminal to the same appendage). Another type of triad includes the upper F2 terminal and also involves three synapses: a branched (cholinergic) brainstem axon produces one synaptic terminal onto an X cell relay dendrite and another onto the F2 terminal, and a third synapse is formed from the F2 terminal onto the same dendrite. For simplicity, the NMDA receptor on the relay cell postsynaptic to the retinal input has been left off. Abbreviations: ACh, acetylcholine; AMPAR, AMPA receptor; F1 and F2, two types of synaptic terminal; $GABA_A R$, type A receptor for GABA; Glu, glutamate; M1R and M2R, two types of muscarinic receptor; mGluR5, type 5 metabotropic glutamate receptor; NicR, nicotinic receptor. Redrawn from Sherman (2004). (C) Schematic model for functioning of the interneuron (Int). The F1 axon outputs are controlled by inputs onto the cell body and proximal dendrites. The peripheral dendritic outputs (F2) are controlled locally by direct inputs, mostly either from retinal or the cholinergic brainstem afferents (see text for details), and these F2 circuits are within glomeruli. The dashed lines connecting proximal dendrites to the F2 circuits represent 5–10 levels of dendritic branching. Redrawn from Sherman (2004).

messages that are being sent. There is an important difference between knowing when a cortical area is active as a whole, as in a functional magnetic resonance imaging study, and understanding how that activity is generated and relates to specific, interpretable patterns of activity in the relevant neural pathways. It is this latter relationship that is basic to our approach to the neural connections of thalamus and cortex. In addition to the problem of identifying the message that is transmitted by the relay cell, we will find that there is a different question, considered further in chapter 6, concerning the extent to which the message reported for any one relay cell by an experimentalist is the same as the message that is received by the relay cell and transmitted to the cortex. We raise the issue here to signal that understanding the message raises some novel and generally unexplored issues. They are considered more fully in chapters 6 and 8.

Thalamic relay cells have also been categorized as "core" or "matrix" first on the basis of the calcium binding proteins that they contain in primates (parvalbumin or calbindin; Jones, 1998b) and then on the basis of the pattern of their cortical terminations (Jones, 2001). However, based on what is currently known about the functions of the calcium binding proteins, the functional significance of this categorization of relay cells is unclear, and, because the calcium binding proteins do not provide clear markers in nonprimate species, the function of these two classes of relay cell is more readily considered in terms of the cortical terminations of their axons, and these are discussed in chapter 8.4.

3.2.2.2 The Interneurons The interneurons and relay cells lie intermingled with each other in the main part of the thalamus and have had to be distinguished on grounds other than their position. Whereas the axons of relay neurons, which are glutamatergic, have an excitatory action on their postsynaptic targets, the interneurons, which are GABAergic, have an inhibitory action. The interneurons are smaller, but that is an uncertain criterion, and today the simplest and most reliable method for distinguishing the two cell types is the presence of GABA in the interneurons but not in the relay cells. An immunohistochemical stain for GABA or glutamic acid decarboxylase (the GABA synthesizing enzyme) readily identifies the interneurons, leaving the glutamatergic relay cells unstained, and as indicated in section 3.2.2.1, these stains have demonstrated that the presence of interneurons, expressed as a ratio of cell numbers relative to relay neurons, varies greatly. They represent 20% to 25% of the cells in thalamic nuclei in cats and monkeys but only 1% in most of the thalamic nuclei of some rodents (Spreafico et al., 1994; Arcelli et al., 1997). Strangely, the lateral geniculate nucleus in these rodents has an interneuronal population not differing greatly from that of thalamic nuclei generally in cats and monkeys, but the other nuclei all appear to have very few interneurons.

The relative frequency of interneurons may vary even more if accounts that their relative number in the human anterior thalamus may be as high as 42% (Dixon and Harper, 2001) can be confirmed and shown to be independent of the conditions necessarily present for human postmortem material. The range of variation of interneuronal numbers between nuclei and between species raises some important questions about the role that these cells play in, for example, the rat ventral posterior nucleus (1%) or the anterior thalamus in the rat (very few), on the one hand, and the rat lateral geniculate nucleus (15% to 25%), the thalamic nuclei of cat or monkey (mostly about 20%), or the human anterior thalamus (possibly 42%), on the other. It is possible that the GABAergic axons of the reticular nucleus, zona incerta, or other extrathalamic inhibitory pathways (section 3.2.3.) take over some or all of the functions of the interneurons in nuclei that have few interneurons, but important questions remain, and they reveal how little we understand about inhibitory actions in thalamic circuits.

Given that a single interneuron can produce a great many inhibitory synapses (figure 3.10; see also the discussion of interneuronal processes in section 3.2.3), it is possible that there is still a significant interneuronal inhibitory activity to be reckoned with in all thalamic nuclei, even those that have very few interneurons, but the virtual absence of interneurons in some thalamic nuclei is striking and raises a serious question about what role interneurons

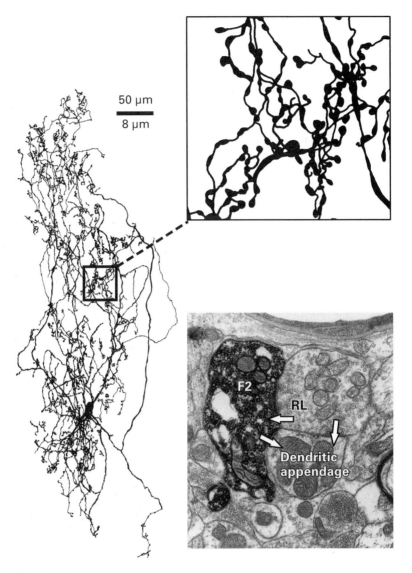

Figure 3.10
Interneuron from the lateral geniculate nucleus of a cat. Preparation by intracellular injection with horseradish peroxidase. Redrawn from Hamos et al. (1985). The inset at the top shows an enlarged view of axoniform dendritic terminals. The electron micrograph shows a triad in one section, with the labeled F2 terminal from the interneuron indicated. Arrows lie across the three synapses of the triad: from the F2 terminal to the dendritic appendage of an X relay cell, from the retinal terminal (RL) to the F2 terminal, and from the retinal terminal to the dendritic appendage.

may be playing and to what extent the thalamus of some rodents, other than possibly the lateral geniculate nucleus, may be a useful model for understanding the human or, more generally, the mammalian thalamus, and why should the lateral geniculate nucleus be different?

Golgi methods had shown that the interneurons are not only smaller, but also generally can be seen to have a local axon[14] that distributes its terminals in the close vicinity of its dendritic arbor with no evidence of any branches reaching beyond the thalamus or, more generally, beyond the thalamic nucleus in which the axon originates (Guillery, 1966; Tömböl, 1967; Tömböl, 1969; Ralston, 1971). Further, these smaller neurons have an unusual dendritic morphology with dendritic appendages (figure 3.10) that were described as "axoniform" (Guillery, 1966) because they resemble the terminal parts of axons more than other dendritic appendages that are more commonly found on many other dendrites. These interneuronal dendritic appendages, the F2 profiles when seen in electron micrographs, further resemble axons because they lie at the ends of slender stalks that can be up to 10 μm in length, contain synaptic vesicles, and are presynaptic to other dendrites (see section 3.2.3). They are only seen in the astrocyte free glomeruli participating in the synaptic triads (section 3.2.1.1 and figure 3.6). These axoniform dendrites of the interneurons can form an extremely dense local arbor (figure 3.10), bearing many hundreds of axoniform appendages, and these are distributed in the same general region as axonal terminals (that is, some of the F1 profiles described in the next section) of the same interneuron.

3.2.3   Thalamic Connectivity Patterns

The several inputs to the first and higher order relays of the thalamus have been studied in several different species and several different nuclei, where they form terminal structures and relationships that are similar in their electron microscopic appearance and in the synaptic relationships that they establish for all thalamic nuclei, both first and higher order (Sherman and Guillery, 2006; Jones, 2007). The synaptic organization of the lateral geniculate nucleus of the cat will here serve as a model for that of thalamus more generally, and we shall describe this in some detail. Figure 3.9A shows the various terminal types and the synaptic contacts they form in the lateral geniculate nucleus in relation to the X and Y relay cells of a cat.

---

14. Where an interneuron appears to lack an axon, it is probable that this reflects a failure of the method rather than a demonstrably real absence of an axon. The possibility that there are axonless interneurons remains open but will be hard to demonstrate.

The fraction of synapses onto relay cells formed by each identified input are about 5% to 10% for RL (retinal) inputs, 60% to 65% for RS terminals, divided roughly equally between cortical and brainstem inputs, and 30% to 35% for F terminals, which, with rare exceptions, are all F1 terminals on Y cells and roughly equally divided between F1 and F2 on X cells (Wilson et al., 1984; Van Horn et al., 2000). Figure 3.9A also emphasizes several other key points. The dendritic arbor can be divided into a retinorecipient zone on proximal dendrites and a corticorecipient zone on distal dendrites, with effectively no overlap between them. Interneuronal (both F1 and F2) and cholinergic brainstem inputs are found in the retinorecipient zone, whereas reticular inputs are found predominantly in the corticorecipient zone (Wilson et al., 1984; Cucchiaro et al., 1991; Erişir et al., 1997a; Wang et al., 2001). Also, all triad and triad-like synaptic complexes (shown in figure 3.9B) are limited to glomeruli, which are found on X cells but rarely on Y cells, and virtually all retinal input to X cells is located within glomeruli.

Figure 3.9B schematically illustrates two distinct types of triad. One, which we refer to as a "true" triad, involves a retinal (RL) terminal that contacts both a relay cell dendrite and an F2 terminal, with the F2 terminal also contacting the same relay cell dendrite. Thus, three synapses are involved, with a single RL synapse contacting two different targets, F2 and dendrite (see figures 3.6A and 3.9B). The other, which we call "triad-like," also involves three synapses, but here two terminals from the same cholinergic brainstem axon are involved: one of these contacts an F2 terminal that itself is presynaptic to the same dendritic process as the one targeted by the other cholinergic terminal (see figure 3.9B).

Figure 3.9B also shows the different postsynaptic receptors involved in these triadic arrangements (Cox and Sherman, 2000). For synapses involving an RL terminal, the synapse onto the F2 terminal activates both ionotropic and metabotropic glutamate receptors and these lead to EPSPs, so that activation of the retinal input will increase GABA release from the F2 terminal onto the relay cell dendrite. This leads to disynaptic inhibition of the relay cell following the monosynaptic excitation of the retinal input to the dendrite. However, the fact that there is a metabotropic receptor on the F2 terminal and not on the relay cell dendrite means that the action leading to GABA release is prolonged and can outlast activity in the retinal afferent (Cox and Sherman, 2000; Govindaiah and Cox, 2004).

One suggested consequence of this is as follows (Sherman, 2004). Because firing rate in retinal axons is monotonically related to contrast of the visual stimulus, a high contrast stimulus will produce a high firing rate and activation of metabotropic receptors on the F2 terminals, increasing the inhibitory outputs of these terminals. If the visual contrast is suddenly reduced and the firing of the retinal axon reduces in step, the extra inhibition from the activated F2 terminals will continue for several seconds and thus reduce the gain of the retinogeniculate transmission. This would lead to a reduction in contrast sensitivity lasting for several seconds, a well-known psychophysical phenomenon referred to as contrast adaptation or contrast gain control. Contrast gain control is an important property of the visual system that, like other forms of adaptation (for example, to brightness or motion), helps to adjust the sensitivity of visual neurons to ambient levels of stimulation. The point here is that triadic circuitry in this way provides one neuronal mechanism for this perceptual process, although it is thought that such mechanisms also exist in retina and cortex (Sclar et al., 1989; Carandini and Ferster, 1997; Sanchez-Vives et al., 2000; Demb, 2002; Solomon et al., 2004).

The triad-like complex involving cholinergic terminals is associated with M1 (excitatory) and nicotinic receptors on the relay cell dendrite and M2 (inhibitory) receptors on the F2 terminal (Cox and Sherman, 2000), and thus activation of the cholinergic input leads to clear excitation of the relay cell via both direct excitation and disinhibition, the latter because the M2 receptors on the F2 terminal lead to a reduction of GABA release there.

We have noted that the interneuron has two different types of synaptic terminal: F1 terminals (found both within and outside of glomeruli) come from the axon, and F2 terminals (found strictly within glomeruli) come from the dendrites. The use of cable modeling as described in chapter 2 suggests an interesting hypothesis for the functioning of the interneurons as regards the processing that involves the F1 and F2 outputs (Sherman, 2004). This is shown schematically in figure 3.9C. Because inputs onto more distal dendrites are distant electrotonically from the cell body and spike-generating region (see figure 2.2), perhaps only inputs onto proximal dendrites significantly influence the generation of action potentials in the axon, and so the axonal outputs represent conventional integration that is limited to the more proximal inputs to the interneuron. Inputs to the F2 terminals will affect their release of GABA, but these postsynaptic influences are very local within segments of the distal dendritic arbor and are electrotonically isolated from one another as well as from the cell body. In this way, the interneuron may multiplex in terms of its input–output functions, with one input–output algorithm for the axon, reflecting integration of proximal inputs to the dendrites and cell body, and many others for electrotonically isolated groups of F2 terminals, thereby allowing the interneuron independently to provide multiple input–output channels to its target cells. Recent evidence for this model of interneuron function has been reported (Crandall and Cox, 2012).

3.2.4   An Overall View of Thalamic Inputs

In addition to the ascending driver axons with their characteristic large thalamic terminals that carry messages for relay to cortex and are described for
first and higher order nuclei in section 3.2.1.1, there are many other types of
axon in any thalamic nucleus, and these other inputs form 90% or more of the
synapses in a thalamic nucleus but do not carry a message for relay to cortex.
Their function is modulatory.

A major proportion of incoming axons are those from glutamatergic cells
in layer 6 of the cortex that are organized mainly as a feedback pathway.
These are relatively thin axons that distribute small RS terminals to peripheral
dendrites of relay cells and also to interneurons. These terminals are significantly smaller than the RL terminals and contain fewer mitochondria, suggesting that they are less active, making lower energetic demands. In addition
to this, other axons that terminate as RS terminals are the cholinergic terminals considered in the previous section, and these arise from cells in the
parabrachial region (also called the pedunculopontine nucleus or the lateral
dorsal tegmental nucleus). These distribute to proximal dendritic segments of
relay cells and also contribute to some triads with F2 (interneuronal) terminals
(see figure 3.9).

Several other modulatory synapses that come from serotonergic cells in the
raphe nuclei of the brainstem, from noradrenergic cells in the parabrachial
region, from histaminergic cells in the tuberomamillary nucleus of the hypothalamus, and from dopaminergic cells likely located in the ventral tegmental
area (reviewed in Sherman and Guillery, 2006; Jones, 2007) contribute to the
relay cells and interneurons of most if not all thalamic nuclei. These are all
able to affect the membrane potential of the relay cells; also, because all of
these inputs can activate metabotropic receptors, which evoke second messenger pathways in the postsynaptic cell (see chapters 2 and 4), more complex
and longer-lasting changes might be affected by these inputs.

Finally, GABAergic inputs provide modulatory input to thalamic neurons.
These include local interneurons, reticular cells, and various sources of
extrathalamic inputs, such as the pretectum, zona incerta, and globus pallidus.
As noted above, interneurons supply both F2 (dendritic) and F1 (axonal)
inputs, and all of the others provide purely F1 axonal inputs. The F1 terminals
from reticular cells are the dominant form of F1 terminal found on peripheral
dendrites of relay cells (Cucchiaro et al., 1991; Wang et al., 2001), implying
that the other F1 inputs target these cells proximally.

Overall, these modulatory inputs all play a role, discussed further in chapter
4, in the way in which the incoming messages are relayed to the cortex.

3.3   The Thalamic Reticular Nucleus

The thalamic reticular nucleus forms a narrow band of GABAergic cells that lies adjacent to the dorsal and lateral aspects of the thalamus (see figure 1.4) and forms a curved shell or cover that lies in the path of all the axons that connect the cortex and the thalamus in both directions. That is, the reticular cells can be recognized on the basis of where they are. Any cell that lies within the borders of the thalamic reticular nucleus is a reticular cell, no matter what it looks like. In general, the reticular cells are treated as a homogeneous group, having slender dendrites that stretch out in the plane of the reticular nucleus and a cell body that is also often elongated in this plane (figure 3.11). Although there are some rostrocaudal differences in the appearance of the cells, their

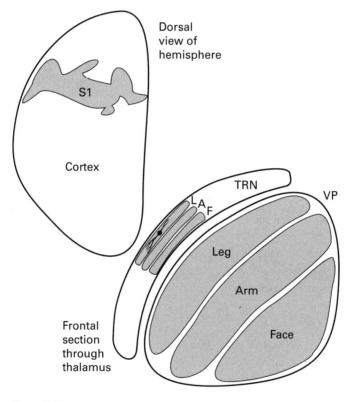

Figure 3.11
Schema to illustrate how the parts of the body are represented in the thalamus (VP), the thalamic reticular nucleus (TRN) and the cortex of a rabbit. Only one reticular cell is shown to illustrate the orientation of the reticular dendrites. A, arm; L, leg; T, trunk. Based on data from Crabtree (1992b).

functional properties and connections are not known to vary significantly in different parts of the nucleus, and there is at present no reason for recognizing distinct groups or subtypes of reticular cells.

The main inputs to the reticular nucleus come from branches of the layer 6 corticothalamic axons and from branches of thalamocortical axons. Both are glutamatergic. The ramifications of these two sets of inputs to the reticular nucleus define the topographical relationships between the thalamus, its cortical target, and the reticular nucleus. The reticulothalamic connections have a large reciprocal component with some evidence for patterns of branching that are not reciprocal (Crabtree et al., 1998; Crabtree and Isaac, 2002). The axons of the reticular cells pass into the main nuclei of the thalamus forming well-localized terminal arbors (Uhlrich et al., 1991; Pinault et al., 1995a; Pinault et al., 1995b; Deschênes et al., 2005).

In the past, the thalamic reticular nucleus was considered to have "diffuse" connections with the thalamus and through the thalamus with other neural centers and the cortex (Scheibel and Scheibel, 1966; Jones, 1985). However, there is now significant evidence in favor of well-organized topographical relationships between separate sectors of the reticular nucleus with each sector connecting to a separate thalamic nucleus or a part of a nucleus (Crabtree and Killackey, 1989; Crabtree, 1992a, b; Crabtree and Isaac, 2002; Lam and Sherman, 2007) linking each nucleus to a separate sector of the reticular nucleus, and the terminals of these reticulothalamic axons make contact with the distal portions of the relay cell dendrites and to a lesser extent with interneurons in the thalamic zone from which they receive their inputs (Cucchiaro et al., 1991; Wang et al., 2001). They also form connections with one another, involving both conventional chemical synapses and gap junctions (Landisman et al., 2002; Lam et al., 2006). The long dendrites of the reticular cells that stretch out in the plane of the nucleus have been used as an argument against a well-organized topography, with the peripheral parts of the dendrites being regarded as necessarily stretching beyond the mapped representations (Jones, 1985). However, this does not take into account the unexpected observation that the topographical maps in the reticular nucleus show the body parts or the sensory surfaces stretched out in the plane of nucleus, following the orientation of the dendrites (figure 3.11). The representations of the body parts do not lie perpendicular to the plane of the nucleus as might be expected from the direction of the main thalamocortical and corticothalamic axons as they pass between the thalamus and the cortex. That is, if one views the sheet of the reticular nucleus as a miniature version of the cortical sheet, then the topographical representations of the body parts or sensory surfaces in the reticular sheet are perpendicular to those in the cortical sheet.

This is a surprising relationship, the functional or developmental significance of which remains as a puzzle. It appears that the axons passing through the reticular nucleus between thalamus and cortex give off branches that turn at right angles to follow the dendrites through the nucleus.

There is evidence that both first order and higher order thalamic nuclei are represented in the reticular nucleus, with suggestions that they occupy different parts of the reticular nucleus for some systems (Conley and Diamond, 1990) and share sectors in other systems, with branched axons going to the ventral posterior nucleus (first order) and the posterior medial nucleus (higher order) in the cat (Crabtree, 1992a; Crabtree, 1996). The evidence available for cat and rabbit suggests that the first order inputs to the nucleus form a stronger terminal plexus than do the higher order inputs, occupying more of the reticular nucleus.

3.4   Outstanding Questions

1. What is the functional significance and the evolutionary origin of multiple sensory cortical areas? Is this primarily a division of labor, with each field analyzing different aspects of the sensory input, or can the interactions between different cortical fields play another role? (This is discussed further in chapter 8.)

2. Is there any evidence for significant integrative activity in any thalamic relay? That is, do driver inputs representing different forms of information (for example, from subcortical and layer 5 sources) converge onto single relay cells and produce a novel output on the basis of the interaction of the separate inputs? At present, we know of no evidence, but not enough is known about higher order relays to rule this out.

3. What are the functionally important differences between thalamic nuclei or relays that include significant relationships with GABAergic interneurons and those that lack such relationships?

4. What can one deduce or demonstrate about the functional differences between radiate and bushy thalamic relay cells?

5. What are messages that thalamic nuclei other than those primarily concerned with sensory pathways are transmitting to cortex?

6. What are the special functional relationships that are established within the astrocyte-free regions of the thalamic glomeruli?

Classification of Afferents in Thalamus and Cortex

It is clear that classification of different inputs to a cell or cell group provides a crucial step toward a better understanding of circuits in the brain. This is particularly true of glutamatergic inputs, because these are the dominant ones for information processing. We next consider such a classification, starting with glutamatergic inputs, and explore what insights this provides concerning circuits in thalamus and cortex.

We began such a classification with a distinction between drivers and modulators; this was originally based on thalamic relays for which the properties of the synapses and their function in the relay of information were known (reviewed in Sherman and Guillery, 1998, 2006, 2011). The drivers represented the inputs carrying a message for relay to cortex, and the modulators could affect how driver input was relayed; such modulation can take many forms, including affecting overall excitability of relay cells, controlling the inactivation state of many voltage-gated conductances in relay cells, or affecting the gain of transmission of the driver input. It is important to note the point made in the previous chapter: thalamic relay cells receive, in addition to the glutamatergic drivers, a variety of modulatory inputs discussed there, and these include the feedback projection to the thalamus from layer 6, which is also glutamatergic (see section 4.1.3 below for further details). That is, of the glutamatergic inputs to thalamus, some are drivers and others, modulators.

A key distinction between drivers and modulators depended heavily on the identification of the receptive field properties that are relayed to cortex with relatively little modification. On this basis, the driver inputs to the medial and lateral geniculate nuclei, the ventral posterior nucleus, and the anterior dorsal nuclei among first order nuclei could be identified, and all showed the same morphological and, where studied, functional characteristics of the input synapses. By extension, there were other inputs to first order and also to higher order thalamic nuclei, all of which showed the same morphological synaptic characteristics and were regarded as probably also drivers; although for most of these the nature of the driving message was not identified, and only the source of the input was known.

The drivers were distinguished from the modulators, which in the sensory relays were not carrying a message for relay to cortex. It is important to stress the role that the sensory relays played in these classifications. In these relays, the characteristics of the receptive fields could be traced through the thalamus, demonstrating clearly that only the driver inputs carried the message for relay to cortex. For instance, in the lateral geniculate nucleus, the center/surround receptive fields of relay cells clearly derived from retinal inputs with essentially the same receptive fields, and not from other inputs, such as cortical or reticular, which have quite different receptive fields, or brainstem, which have no clear ones at all.

When the functional characteristics of the drivers were compared with those of the modulators, based initially on observations of the lateral geniculate nucleus (Sherman and Guillery, 1998), it became clear that these characteristics related closely to the functional capacities of the drivers or modulators as inputs. That is, as explained in some detail in section 4.1.3, the synaptic properties of drivers, which have fast, large responses and evidence for a high probability of transmitter release in response to afferent firing, are well suited for passing on basic information; modulators, with weaker responses associated with evidence for low probability of transmitter release, and activation of metabotropic postsynaptic receptors, involving long-lasting responses, are poorly suited for carrying basic information but are well suited for various forms of modulation. On the basis of these functional characteristics, it became possible to classify inputs as resembling either the drivers on the one hand or the modulators on the other, without involving any information about the message that was being transmitted.

A number of such studies of synaptic inputs in thalamus and cortex have recently been undertaken (Reichova and Sherman, 2004; Lee and Sherman, 2008, 2009b, 2010; Petrof and Sherman, 2009; Theyel et al., 2010b; Covic and Sherman, 2011; DePasquale and Sherman, 2011; Viaene et al., 2011a, 2011b, 2011c), all of them based on slice preparations where questions about the nature of the message being transmitted cannot be asked.[1] These studies now provide a powerful extension of the driver/modulator classification. However, it is important to stress that the extension of this classification is no longer based explicitly on following information flow, partly because this classifica-

---

1. It should be noted that these studies are based on activation of pathways involving multiple axons, and some issues raised regarding possible variability in properties of single axons are noted in section 4.1.2.8.

tion has been extended to the more complex circuitry of cortex, where following information flow is difficult, and partly because it has depended heavily on slice preparations where information content of an input cannot readily be determined. For this extension, we have thus adopted less evocative terminology: we refer to glutamatergic inputs with the synaptic properties of drivers in thalamus as Class 1 and those with modulator properties as Class 2. We then offer the hypothesis that Class 1 inputs generally are in fact drivers (that is, carrying the main information), and Class 2, modulators. This is elaborated in the following sections.

4.1 Classifying Glutamatergic Afferents

4.1.1 Terminology: Class 1 and Class 2 Glutamatergic Inputs

"Class 1" and "Class 2" refer to the structural and functional parameters that are listed in table 4.1, whereas "driver" and "modulator" refer to the actual role of the input. However, there is another important distinction to be made. Currently, all known driver inputs (for example, to thalamus or certain thalamic inputs to layer 4 of cortex) are Class 1, but it is possible that other types of driver inputs, that is, main glutamatergic information-bearing afferents, will be discovered with a completely different set of synaptic properties: they might be a "class 3." Conversely, it is possible, but has so far not been encountered, that some inputs in the brain with Class 1 properties provide a modulatory function. At present, the hypothesis that Class 1 and driver inputs are equivalent is a hypothesis requiring further testing.

Table 4.1
Class 1 and Class 2 general properties

Criteria	Class 1 (Driver)	Class 2 (Modulator)
Criterion 1	Activates only ionotropic receptors	Activates metabotropic receptors
Criterion 2	Synapses show paired-pulse depression (high p)*	Synapses show paired-pulse facilitation (low p)
Criterion 3	Large EPSPs	Small EPSPs
Criterion 4	Little or no convergence onto target	Much convergence onto target
Criterion 5	Thick axons	Thin axons
Criterion 6	Large terminals on proximal dendrites	Small terminals on distal dendrites
Criterion 7	Dense, well-localized terminal arbors	Delicate terminal arbors

*See section 4.1.4 for slight variation in type of thalamocortical input called "Class 1C."

4.1.2   Properties of Class 1 and Class 2 Glutamatergic Inputs

It is important to distinguish the glutamatergic Class 1 (driver) from the Class 2 (modulator) inputs. Many of the physiological properties of these different classes are shown in figure 4.1, which illustrates responses of cortical cells to different types of thalamocortical inputs in the mouse. These responses are fairly typical for these classes in both thalamic and cortical circuitry, although some variants are noted below. The robustness of this classification scheme is documented in figure 4.2, which shows both the clear clustering of selected parameters that distinguish these classes and also shows that the two classes are virtually identical in terms of the properties illustrated for various circuits in thalamus and cortex. A more complete summary is found in table 4.1 and shown schematically for an example of a geniculate relay cell in figure 4.3. Because of their importance, it is worth considering in detail each of the seven criteria listed in table 4.1, where they are listed in a rough order of perceived importance.

4.1.2.1   Criterion 1: Postsynaptic Receptors   Perhaps the clearest distinction between Class 1 and Class 2 inputs is that the former activate only ionotropic receptors, mostly AMPA and NMDA, whereas the latter activate metabotropic receptors in addition to ionotropic receptors (reviewed in Sherman and Guillery, 2006). Of particular importance here is the observation that, in general, activation of ionotropic receptors evokes a brief EPSP, with the underlying active conductance changes lasting 10 or so milliseconds,[2] whereas activation of metabotropic receptors evokes a prolonged EPSP, lasting hundreds of milliseconds to several seconds (see chapter 2). The fast, brief EPSPs activated by Class 1 inputs enable an individual action potential in the driver input to be encoded by one EPSP in the relay cell up to rates of presynaptic firing that begin to evoke temporal summation postsynaptically. Such temporal summation obliterates the one-to-one relationship between action potential and EPSP, but the presence of ionotropic receptors in the absence of metabotropic receptors allows for this not to occur up to much higher rates of presynaptic firing than would be the case with activation of metabotropic receptors. For Class 1 inputs, this can be up to 50–100 spikes/s or higher, which are rates commonly

---

2. The underlying synaptic conductance changes lead to current flow across the membrane. Because of the resistive–capacitative properties of the membrane, the voltage changes seen across the membrane last longer than the current flow, and this relates to the membrane time constant. Thus, measures of synaptic events recorded with the technique of voltage clamp, which measure postsynaptic currents (for example, EPSCs), produce briefer events than the same events measured during current clamp, which measure postsynaptic voltages (for example, EPSPs).

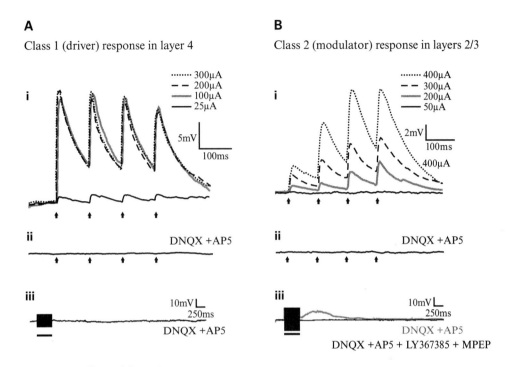

A

Class 1 (driver) response in layer 4

B

Class 2 (modulator) response in layers 2/3

Figure 4.1

Examples of Class 1 and Class 2 responses; recordings from mouse primary somatosensory cortex in response to stimulation of the ventral posterior medial nucleus; in vitro recording. (A) Class 1 response recorded from a cell in layer 4. (Ai) This cell responded with paired-pulse depression at 10-Hz stimulation rate. The EPSP amplitude did not change appreciably with increases in stimulation intensity. (Aii) Stimulation at 10 Hz (250 μA) in the presence of ionotropic glutamate receptor antagonists (DNQX and AP5) blocked the EPSPs. (Aiii) Stimulation at 120 Hz (250 μA) in the presence of DNQX and AP5 did not produce any membrane potential changes and thus no evidence of metabotropic glutamate receptor activation. (B) Class 2 response recorded from a cell in layers 2/3. (Bi) The cell responded with paired-pulse facilitation to stimulation of the ventral posterior nucleus at 10 Hz. Increasing the stimulation intensity produced increases in EPSP amplitude. (Bii) Stimulation at 10 Hz (250 μA) in the presence of DNQX and AP5 failed to produce any EPSPs. (Biii) Stimulation at 120 Hz (250 μA) in the presence of DNQX and AP5 produced a slow and prolonged membrane depolarization (gray trace) that could be blocked with a cocktail of type 1 (LY367385) and type 5 (MPEP) metabotropic glutamate receptor antagonists (black trace). Traces from 120-Hz stimulation are single sweeps, and all others represent the average of 10 sweeps. Scale bars in (Ai) and (Bi) apply to (Aii) and (Bii), respectively.

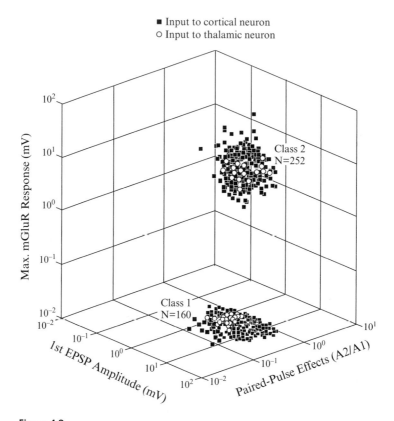

Figure 4.2
Three-dimensional scatterplot showing clustering of selected properties for different Class 1 and Class 2 inputs; data from in vitro slice experiments in mice (Reichova and Sherman, 2004; Lee and Sherman, 2009a, b, 2010; Petrof and Sherman, 2009; Covic and Sherman, 2011; Viaene et al., 2011b). The three parameters are (1) the amplitude of the first EPSP elicited in a train at a stimulus current level just above threshold; (2) a measure of paired-pulse effects (the amplitude of the second EPSP divided by the first) for stimulus trains of 10–20 Hz; and (3) a measure of the response to metabotropic glutamate receptor (mGluR) activation, which is the maximum voltage deflection (that is, depolarization or hyperpolarization) during the 300-ms period after high-frequency electrical stimulation and in the presence of AMPA and NMDA blockers to isolate any metabotropic glutamate receptor activation. Pathways tested here include various inputs to thalamus from cortex and subcortical sources, various thalamocortical pathways, and various intracortical pathways. Not only do the two classes of response separate into distinct clusters, documenting the classification here, but also different pathways overlap extensively within each class, indicating basic similarity on these parameters for these classes in thalamic and cortical circuitry.

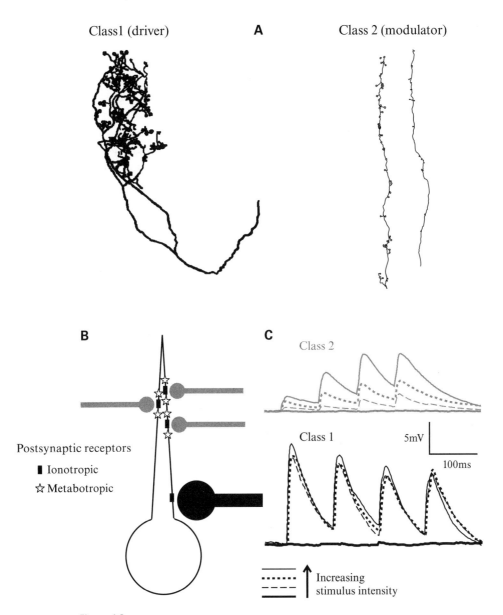

Figure 4.3
Schematic view of some differences between Class 1 (driver) and Class 2 (modulator) inputs to a thalamic relay cell. (A) Light microscopic tracings of a Class 1 afferent (a retinogeniculate axon from the cat) and a Class 2 afferent (a corticogeniculate axon from layer 6 of the cat). Redrawn from Sherman and Guillery (2006). (B) Class 2 inputs (gray) shown contacting more peripheral dendrites than do Class 1 inputs (black). Also, Class 1 afferents activate only ionotropic glutamate receptors, whereas those of Class 2 also activate metabotropic glutamate receptors. (C) Effects of repetitive stimulation on EPSP amplitude: For Class 2 inputs, this produces paired-pulse facilitation (increasing EPSP amplitudes during the stimulus train), whereas for Class 1 inputs, it produces paired-pulse depression (decreasing EPSP amplitudes during the stimulus train). Also, increasing stimulus intensity for Class 2 inputs (shown as different line styles) produces increasing EPSP amplitudes overall, whereas for drivers it does not or to a much less degree; this indicates more convergence of Class 2 than Class 1 inputs.

seen during active firing of thalamic and cortical neurons. Put another way, the sustained EPSPs seen with metabotropic receptors act like low-pass temporal filters that result in the loss of temporal information. Thus, information flow is maximized by having ionotropic receptors only.

The sustained EPSPs evoked by Class 2 inputs are consistent with their role as modulators. Such sustained EPSPs produce a prolonged shift in responsiveness of the relay cell, a clear modulatory role. Furthermore, as noted in chapter 2, thalamic relay cells (like neurons generally) possess a number of voltage- and time-dependent conductances for which the membrane voltage must be altered for sustained periods to control the state of the conductance (for example, at least ~100 ms); examples are I_T, I_A, and I_h (see chapter 2). The fast EPSPs associated with ionotropic receptors are ill-suited to control these conductances, whereas the sustained EPSPs associated with the metabotropic glutamate receptors are ideal for this control. Further details about how these metabotropic responses help control I_T are provided in chapter 2, and this control of I_T by metabotropic glutamate receptors serves as an excellent example of the significance of these postsynaptic receptors for modulatory functions. Finally, the lengthy metabotropic response typically outlasts activity in the Class 2 input, often by seconds (Govindaiah and Cox, 2004), which may be useful for modulation but distorts temporal information.

Application of agonists to metabotropic glutamate receptors has also been shown to affect the gain of synaptic transmission in thalamus (Landisman and Connors, 2005; Govindaiah et al., 2012) and cortex (Otani et al., 2002; Barbara et al., 2003; Trettel et al., 2004; Mateo and Porter, 2007; Hashimotodani et al., 2007; Marinelli et al., 2008; Tsanov and Manahan-Vaughan, 2009; Kiritoshi et al., 2013). However, these studies, by only applying agonists, failed to identify glutamatergic afferents that would be expected to activate these metabotropic receptors, and few such studies exist. One such study reports that Class 2 layer 6 inputs to cells in layer 4 that also receive Class 1 thalamic input activates presynaptic group II mGluRs on thalamocortical terminals to lower the amplitude of thalamocortical EPSPs (Lee and Sherman, 2012); another reports that local Class 2 inputs onto cells in layers 2/3 causes a reduction of EPSPs from Class 1 layer 4 input to these cells by activating postsynaptic group II mGluRs (DePasquale and Sherman, 2012).

Why do Class 2 inputs as a group (but see section 4.1.2.8) activate ionotropic in addition to metabotropic receptors? One explanation is that if these inputs to a target cell show considerable convergence (*criterion 4*) and as long as their firing is not synchronized, their summed EPSPs should produce sustained changes in membrane potential, which then serve to complement the activation of metabotropic receptors. The advantage of this is that the latency is much shorter for ionotropic receptors (see chapter 2), so the modulation can start earlier, and then the metabotropic response can kick in to sustain the change. Also, there are situations during which activation of ionotropic receptors on their own can produce modulatory effects (Chance et al., 2002).

4.1.2.2 Criterion 2: Paired-Pulse Effects and Probability of Transmitter Release

In chapter 2, the property of paired-pulse depression or facilitation and its relationship to probability of transmitter release were described: high for a depressing synapse and low for a facilitating one. With one curious exception noted in section 4.1.4 for a group of thalamocortical inputs that are referred to as a subclass called "Class 1C," the difference between classes 1 and 2 in this property is complete: Class 1 inputs show paired-pulse depression, and Class 2 inputs, paired-pulse facilitation. This is true for inputs to thalamic relay cells (Reichova and Sherman, 2004; Lee and Sherman, 2008, 2009b, 2010; Petrof and Sherman, 2009), thalamocortical inputs (Lee and Sherman, 2008, 2009b; Viaene et al., 2011b, c), and corticocortical inputs (Covic and Sherman, 2011; DePasquale and Sherman, 2011, 2012). Because these effects are so closely tied to the probability of release (p), it is assumed in the following that paired-pulse depression is associated with high p values and paired-pulse facilitation, with low p values. Furthermore, paired-pulse depression may play an important role in information processing by helping the system to adapt to ongoing levels of activity (Chung et al., 2002), thereby opposing response saturation at high input levels and extending the dynamic input–output range across the synapse (see section 2.2 of chapter 2). If so, this would be a useful property of main information inputs.

For Class 2 inputs, the property of paired-pulse facilitation leads to EPSPs that get larger and thus more effective during trains at higher rates of afferent firing. This may relate to the way in which metabotropic receptors are activated (see chapter 2) because activation of these receptors by Class 2 inputs becomes relatively more dominant at higher firing rates. That is, activation of these receptors can begin at low firing rates with as few as two afferent action potentials occurring at roughly 10 Hz or higher, but at higher and more prolonged epochs of afferent firing, the metabotropic response grows relatively more than does the ionotropic, even with facilitation of the latter (Viaene et al., 2013). The overall result is that Class 2 inputs are much less effective at lower rates of firing than at higher rates.

4.1.2.3 Criterion 3: EPSP Amplitude

Where Class 1 and Class 2 inputs to the same cell population, and often the same cell, have been successfully studied, the initial EPSPs evoked from the Class 1 inputs are significantly larger than those from the Class 2 inputs. This relationship holds for inputs to thalamic relay cells (Reichova and Sherman, 2004; Lee and Sherman, 2008, 2009b, 2010; Petrof and Sherman, 2009), thalamocortical inputs (Lee and Sherman, 2008, 2009b; Viaene et al., 2011b, c), and corticocortical inputs (Covic and Sherman, 2011; DePasquale and Sherman, 2011). However, because Class 1 inputs show paired-pulse depression, producing diminishing EPSP amplitudes in response to a train of high-frequency action potentials, and Class 2 inputs show the opposite property (*criterion 2*), EPSPs late in such trains can be larger for Class 2 inputs.

The best studied Class 1 and Class 2 inputs, retinal and layer 6 cortical inputs to geniculate relay cells, offer at least three ready explanations for this amplitude difference. First, retinal contact zones on relay cells tend to be much larger than those of the cortical inputs, and each retinal terminal produces roughly 10 such zones, whereas most cortical terminals tend to produce only one (Van Horn et al., 2000). Larger, more numerous contacts are consistent with more transmitter release and larger EPSPs. Second, retinal terminals are proximally located (see chapter 3, figure 3.8), where they are more likely to influence the soma and axon hillock, whereas the cortical inputs are distal; a similar relationship has been described for the ventral posterior nucleus, in which relay cells receive their lemniscal (Class 1) input much more proximally on their dendrites than they do their cortical (Class 2) input (Liu et al., 1995). Third, as noted in chapter 2 (also *criterion 2*), synapses can vary greatly with respect to the probability that an action potential invading the presynaptic terminal leads to transmitter release. Class 1 synapses have much higher probability of release than do Class 2 synapses (see chapter 2.2.2), and this is consistent with larger evoked EPSPs.

4.1.2.4 Criterion 4: Convergence onto Postsynaptic Target One way to estimate roughly the relative convergence of inputs to a recorded cell is to note how the postsynaptic response grows with the amplitude of electrical activation of the inputs. The idea is that if there are many convergent axons, more will be recruited with increasing activation strength, leading to a monotonic relationship of evoked amplitude with stimulation intensity over a fairly broad range from threshold to saturation. This is known as a "graded response" pattern. If, however, only a single axon can be activated, the response jumps from zero to maximum over a narrow range of stimulus intensities, with no further growth for higher intensities. This is an "all-or-none" response pattern. All Class 2 responses so far studied show a graded response pattern for inputs to thalamic relay cells (Reichova and Sherman, 2004; Lee and Sherman, 2008,

2009b, 2010; Petrof and Sherman, 2009), for thalamocortical inputs (Lee and Sherman, 2008, 2009b; Viaene et al., 2011b, c), and for corticocortical inputs (Covic and Sherman, 2011; DePasquale and Sherman, 2011). For these same target cells, Class 1 responses are more varied, ranging from all-or-none for thalamocortical relay cells and many thalamorecipient cells in cortex (Lee and Sherman, 2008, 2009b; Viaene et al., 2011b, c), to a graded response for some thalamocortical and corticocortical inputs (Covic and Sherman, 2011; DePasquale and Sherman, 2011). However, where the two input classes onto the same cell groups have been quantitatively compared in a study of connections between primary and secondary visual cortex in the mouse, Class 1 inputs show less of a graded pattern than do Class 2 inputs (DePasquale and Sherman, 2011). It thus appears more generally that Class 1 inputs show less convergence than do Class 2 inputs.

Independent evidence for this relationship has been demonstrated in the lateral geniculate nucleus. Lack of much convergence among retinogeniculate (Class 1) afferents has been documented from paired recordings of retinal axons and their geniculate cell targets, where typically one or two but occasionally up to six retinal axons converge onto a relay cell (Cleland et al., 1971; Cleland and Lee, 1985; Mastronarde, 1987b; Usrey et al., 1999). In contrast to this, for the corticogeniculate (Class 2) afferents, it has been estimated there are 10–100 layer 6 axons for each relay cell, clearly suggesting that there must be considerable convergence in this pathway (Sherman and Koch, 1986).

It is perhaps revealing that whereas Class 1 inputs appear to converge less than do Class 2 inputs, among the former there is evidence for more convergence in certain cortical circuits than in thalamus. This may be related to the differences between thalamic and cortical circuitry as regards receptive field elaboration, or if its equivalent is found in nonsensory areas it may represent a more general difference between cortical and thalamic processing. The role that the different input classes play in this function remains to be empirically defined, but as elaborated in the following sections, we suggest that Class 1 inputs provide basic receptive field information. In the thalamus, there appears to be little such elaboration, based on our knowledge of circuitry of the primary visual, somatosensory, and auditory thalamic relays and the relay in the anterior dorsal nucleus (see chapter 3). That is, in each of these nuclei, receptive fields of the relay cells closely match those of their Class 1 inputs, indicating little need for convergence. However, in cortex, elaboration of receptive field properties is a major function, and this implies that different Class 1 inputs must be combined in different ways to continue to elaborate ever more complex receptive fields as the signal passes up through the synaptic hierarchies. This would be consistent with the greater convergence among Class 1 inputs seen in many cortical compared to thalamic circuits.

One final note about numbers relates to layer 6 corticothalamic cells. Their axons are especially interesting because they branch to innervate thalamus and thalamorecipient cells in cortical layer 4 and in both places provide a Class 2 input (Reichova and Sherman, 2004; Lee and Sherman, 2009a, b). Thus, these layer 6 cells can affect thalamocortical transmission at both its source and target. In the case of the cat geniculocortical system, there is a remarkable correspondence of numbers regarding these inputs. Geniculate relay cells receive about 5% of their input synapses from Class 1 retinal axons and about 35% to 40% from layer 6 cortical axons (Erişir et al., 1997b; Van Horn et al., 2000). Thalamorecipient cells in layer 4 of cortex receive about 6% of their input from Class 1 geniculate axons and about 45% from layer 6 axons (Ahmed et al., 1994). These numbers are remarkably similar for thalamic and cortical cells, which could be a coincidence. Or perhaps this is telling us something important more generally about the quantitative relationships of Class 1 and Class 2 inputs to their target cells.

4.1.2.5   Criterion 5: Axon Diameter

Class 1 inputs to the thalamus include various subcortical sources (for example, retinal, lemniscal) as well as layer 5 corticothalamic axons, and these are clearly thicker and faster-conducting than are the Class 2 inputs from layer 6 of cortex. Unfortunately, an analysis of axon diameters for Class 1 and Class 2 inputs does not yet exist for cortical circuitry, so this criterion remains provisional. Axon diameter is clearly related to conduction velocity, and it thus follows that Class 1 inputs should conduct more rapidly than do Class 2 inputs. It may be argued that transmitting signals faster is more important for drivers (Class 1) than for modulators (Class 2). Axon diameter, however, may have other implications as well, such as the rate at which metabolic agents can be transferred from the cell body to the axon terminals or the number and size of the terminals that can be sustained by the axon, and this could conceivably also impact afferent function.

4.1.2.6   Criterion 6: Terminal Morphology

Wherever they have been measured in thalamus or cortex, synaptic terminals from Class 1 inputs are larger on average than those from Class 2 inputs, because, whereas both inputs produce some smaller terminals, the range of terminal sizes for Class 1 inputs has a tail of larger terminals (Covic and Sherman, 2011; Viaene et al., 2011c). The Class 2 inputs to thalamus from layer 6 are smaller than those from various Class 1 sources, such as retina, and we saw in chapter 3 that, in all thalamic nuclei, the driver inputs are characteristically large. In cortex, a similar size difference has been documented for Class 1 and Class 2 afferents, including corticocortical (Covic and Sherman, 2011) and thalamocortical (Viaene et al., 2011c) examples. This also applies to layer 4 cells in visual cortex, where inputs from the lateral geniculate nucleus are Class 1 (Lee and Sherman, 2008) and those from layer 6 are Class 2 (Lee and Sherman, 2009b), and the former have larger terminals (Ahmed et al., 1994). Furthermore, the thalamic terminals mostly contact spines, whereas those from layer 6 cells contact dendritic shafts (Ahmed et al., 1994), suggesting another interesting difference between Class 1 and Class 2 inputs, given many theories regarding spine function, including their ability to boost signal transfer (Sorra and Harris, 2000; Smith et al., 2003; Higley and Sabatini, 2008; Giessel and Sabatini, 2011).

In the thalamus, the Class 1 inputs with their larger terminals contact relay cells more proximally in the dendritic arbor than do the Class 2 inputs. The location of such large terminals on proximal dendrites has been documented for the lateral geniculate nucleus (Guillery, 1969a, b; Wilson et al., 1984), the ventral posterior nucleus (Peschanski et al., 1984; Ma et al., 1987; Liu et al., 1995), the pulvinar (Baldauf et al., 2005), the ventral lateral nucleus (Kultas-Ilinsky and Ilinsky, 1991), and the anterior ventral and anterior medial nuclei (Somogyi et al., 1978).

4.1.2.7 Criterion 7: Terminal Arbor Morphology Class 1 inputs produce denser terminal arbors than do Class 2 inputs. The clearest evidence for this comes from studies of thalamus. Class 1 inputs to the lateral geniculate nucleus, ventral posterior nucleus, and ventral lateral nucleus, as well as those from layer 5 of cortex to higher order relays, have dense arbors with many terminals well localized in a limited volume, whereas Class 2 inputs from cortical layer 6 have delicate arbors less densely populated with terminals (see figure 4.3A).

4.1.2.8 Technical Issues There is an important issue regarding the classification of glutamatergic afferents as described in the previous paragraphs. That is, when an input is activated to evoke EPSPs for study, the pathway so activated typically includes multiple axons. This has several implications. One is that variability among the afferents will largely go undetected, so, for instance, when such stimulation of Class 2 inputs activates ionotropic and metabotropic receptors, it is not clear if individual axons activate both receptor types or just one. Likewise, when multiple types of metabotropic receptor are activated (see section 4.1.4.2 below), whether this is true of individual axons is impossible to determine.

Another potential problem with this approach is the possibility that Class 1 and Class 2 inputs exist in an activated pathway, but it is difficult to detect both afferent types in the evoked response. One might think this particularly likely if the larger Class 1 responses obscured the smaller Class 2 EPSPs. However, in such cases, any significant Class 2 input would be revealed by the test for presence of metabotropic receptor activation done after applying pharmacological blockers of the ionotropic glutamate receptors, and such metabotropic receptor activation has not been detected in cases of inputs identified as Class 1. Furthermore, a mixture of Class 1 and Class 2 inputs has been only rarely seen, although such a mixture has been quite evident (Viaene et al., 2011b) when present. Indeed, it is remarkable that cases of mixture are quite rare, suggesting that Class 1 and Class 2 inputs to cells generally involve axons that rarely mingle along an input pathway.

We cannot absolutely rule out the possibility that mixture of input classes occurs more commonly than detected, but logic indicates that this would be so only if the contribution of one of the classes was exceedingly small. Even if so, this would not change the general conclusion that inputs can be classified as we have described.

4.1.3 Possible Functional Implications for Class 1 and Class 2 Inputs

Although in this chapter we have argued for a neutral terminology for these classes of glutamatergic input, mainly because the function for each in cortical circuitry is not established, it nonetheless seems reasonable to speculate regarding the functional correlates of Class 1 and Class 2 inputs for processing in thalamus and cortex and probably in other brain areas as well. In particular, we propose the hypothesis—which may be seen as an extension of our initial one regarding the driver/modulator classification of inputs to thalamus (Sherman and Guillery, 1998, 2006)—that Class 1 inputs represent the main information source for their target neurons, whereas Class 2 inputs act in a modulatory fashion.

4.1.3.1 Relationship of Class 1 and Class 2 Inputs to Receptive Fields One way of estimating information content for a sensory neuron is by determining its receptive field properties. For instance, a geniculate relay cell has a center/surround receptive field, and this describes the messages the neuron passes on to cortex as a function of different visual stimuli. By comparing the receptive field of a neuron with those of its inputs, one can estimate which subset of the inputs contributes directly to the receptive field properties, and are thus major sources of information, and which do not, suggesting another function, such as modulation.

It is clear that the basic receptive field properties of lateral geniculate relay cells are based on inputs from one or a small number of retinal (Class 1) afferents, and the same can be said for receptive fields of relay cells in the ventral posterior nucleus and ventral medial geniculate nucleus, where the Class 1 inputs arrive from the medial lemniscus and inferior colliculus, respectively. Class 2 inputs with known receptive field properties are rare, and one major example is that of the visual corticogeniculate input from layer 6; these layer 6 cells have elongated receptive fields with orientation and direction selectivity and are usually binocularly driven (Gilbert and Wiesel, 1979), properties the monocularly driven, center/surround receptive field structure of geniculate relay cells lack. The layer 6 feedback provides more subtle effects on geniculate relay cells that can be classified as modulatory, such as subtle adjustments to receptive field properties (Kalil and Chase, 1970; Baker and Malpeli, 1977; Schmielau and Singer, 1977; Geisert et al., 1981; McClurkin and Marrocco, 1984; McClurkin et al., 1994; Andolina et al., 2007) and control of the burst/tonic firing mode transition (Godwin et al., 1996b).[3]

Another example is the geniculocortical input to layer 4 cells, a Class 1 input (Lee and Sherman, 2008), and this input delivers to its layer 4 targets the basic information used to establish receptive fields in these cortical cells (Hubel and Wiesel, 1962; Ferster, 1987; Ferster et al., 1996; Alonso et al., 2001). One Class 2 pathway seen in visual cortex, from layer 6 to the layer 4 thalamorecipient cells (Lee and Sherman, 2009a, 2009b), is also consistent with the idea that such input is insufficient to transmit basic receptive field properties because these inputs seem to impart only the added subtle feature of "end-stopping" (Gilbert, 1993). This can be regarded as more of a modulatory function, as can the recent observation that layer 6 inputs to layer 4 cells can affect the size of thalamocortical EPSPs (Lee and Sherman, 2012).

Whereas analyses of receptive fields provide a powerful test to determine the information-bearing nature of an input, this has proved to be of very limited use. Even within sensory systems, and especially within cortex, there are few relevant data for tracing a receptive field from one cell to another. Although in principle such an approach could also be used in motor pathways, where

---

3. The receptive fields of cells of the visual sector of the thalamic reticular nucleus provide another example of inputs that are not candidates as a main source for the geniculate cell receptive field structure that is relayed to cortex, because these reticular fields are too large, diffusely organized with mixed on and off responses to light throughout, and are mostly binocular (Dubin and Cleland, 1977; Uhlrich et al., 1991).

one might substitute "movement fields" for receptive fields, data comparing such fields in inputs to a cell with those of that cell itself are lacking, largely because of the difficulty of obtaining such data; that is, whereas receptive fields can be obtained in anesthetized preparations, movement fields require active movements on the part of the subject, and this usually means more difficult behaving preparations.

With this proviso of very limited pertinent data regarding receptive field analyses, where they exist they support the notion that Class 1 inputs represent the main information source for a target neuron. We suggest that modulator inputs operate like other classic modulatory inputs, such as cholinergic or serotonergic inputs. The point is that while all of these inputs convey some information, a distinction should be made among glutamatergic pathways between those that are primarily information-bearing and those that are primarily modulatory. Furthermore, the Class 1 (driver) inputs carry information about specific, localized events in the body or the world, whereas Class 2 (modulator) inputs do not.

4.1.3.2 Implications of the Hypothesis In section 4.1.2, we explain how the criteria that distinguish Class 1 from Class 2 support the hypothesis that the former are driver inputs, carrying the main information for processing, whereas the latter are modulatory in function. The concept of Class 1 inputs as providing the main route for information has another important implication, at least for the thalamus, and here the example of the lateral geniculate nucleus helps to illustrate the point, although it applies to all thalamic nuclei. We normally consider a geniculate relay cell as driven by retinal afferents, but there are many other inputs to the cell: cortical, brainstem, and local GABAergic. Firing in any of the nonretinal inputs can conceivably evoke action potentials in the relay cell. Thus, even if a corticogeniculate or brainstem input produces a weak EPSP on its own, if the membrane potential of the relay cell is sufficiently close to firing threshold, this could evoke an action potential. Even an IPSP evoked from an interneuron or reticular cell input can lead ultimately to relay cell firing because strong inhibition can de-inactivate T channels, and the passive repolarization of the relay cell after this input can produce a low-threshold spike and burst of action potentials.

What does it mean to cortex if relay cell firing is caused not by retinal input but by such other input? We suggest that cortex must always treat relay cell firing as if it has been caused by retinal activity and retinal activity alone. This argument borrows from the concept of "labeled lines"[4] in sensory pathways.

4. Also known as Müller's law (Norrsell et al., 1999).

For example, any event that activates a photoreceptor is experienced as a visual stimulus, and so pressure applied to the eyeball that activates photoreceptors is always perceived as a visual stimulus. Available evidence suggests that virtually all geniculate relay cell firing is in response to retinal input and thus not to any of the nonretinal sources (Cleland et al., 1971; Cleland and Lee, 1985; Mastronarde, 1987a; Usrey et al., 1999), so mistaken signals are rarely a problem for geniculocortical transmission. Nonetheless, the point here is that, if our interpretation for Class 1 inputs to thalamus is correct, one implication of identifying such input to a relay cell is that all activity of that relay cell will be treated in cortex as if evoked by the Class 1 input alone.

This idea seems straightforward for thalamic relay cells that have Class 1 inputs with little convergence or combination of different information channels. This is consistent with the lack of significant receptive field elaboration so far seen in thalamic relay cells. For instance, the receptive fields of geniculate relay cells have the same center/surround properties as their ~1–3 retinal inputs, indicating very little receptive field elaboration. However, such elaboration is present in cortical circuitry, and a clear early example is among the converging Class 1 geniculocortical inputs, where ~30 geniculate afferents with offset receptive field locations, including both on- and off-center fields, converge to produce the elongated, orientation-selective receptive fields of their target cells in cortex (Hubel and Wiesel, 1962; Ferster, 1987; Ferster et al., 1996; Alonso et al., 2001). With such convergence of Class 1 inputs and the possibility that different combinations of these may be active during different times and conditions, the question of how the firing of such a cell in cortex relates to the activity in its inputs is more difficult to address. Consider, for instance, cells in layer 4, with Class 1 inputs both from thalamus and from other cortical areas (Lee and Sherman, 2008; Covic and Sherman, 2011; DePasquale and Sherman, 2011). Whereas we have argued that firing of a thalamic relay cell will be treated more centrally exactly like a response to its Class 1 inputs, firing of these layer 4 cells will be transmitted as a variable transformation of its multiple Class 1 inputs for the simple reason that different combinations of these inputs may be active at different times.

There is obviously much we still need to learn about the implications of Class 1 (driver) and Class 2 (modulator) inputs to regions of thalamus or cortex, especially in complex circuits as seen in cortex where integration of Class 1 inputs is common, and an extension of the analysis to other parts of the brain still remains to be explored. Nonetheless, perhaps the chief value of the hypothesis that the two input classes have distinct roles, with Class 1 involved in basic information transfer and Class 2, in modulation, is that it offers a new tool for understanding the functional role of circuits. That is, if the hypothesis proves valid, then relating the different classes of glutamatergic inputs to specific circuits will allow one to follow the route of information processing by following the links of Class 1 afferents and in addition to appreciate better the function of various inputs to a circuit by identifying each as Class 1 or Class 2 or other. Even if the specific hypothesis proves wrong, bringing attention to the fact that glutamatergic inputs are heterogeneous makes clear that we can no longer treat such circuits as some sort of anatomical democracy where the numerically largest inputs are the most important.

4.1.3.3 Topography of Glutamatergic Modulators As noted in section 4.2, thalamocortical circuits have at their disposal a number of conventional modulatory inputs using cholinergic transmitters, noradrenergic transmitters, serotonergic transmitters, and so forth. One general difference between the known glutamatergic Class 2 modulatory inputs and these classical modulators is topography. That is, whereas the classical modulatory inputs have relatively poor topography and provide a fairly diffuse innervation of thalamus and cortex, the Class 2 afferents so far described for thalamus and cortex have a high degree of topography. Thus, Class 2 inputs provide thalamic and cortical circuits modulatory functions having a degree of topographic control not possible with other types of modulatory input.

4.1.4   Variation within Classes

So far, the properties of Class 1 and Class 2 inputs in thalamus are consistent across examples that have been studied as regards the criteria in table 4.1, but in cortex, as shown in more detail later in this section, there is enough variability to suggest that different subclasses of each exist. It is important to note here that this classification scheme for glutamatergic inputs is far from complete, and the scheme proposed in the following sections is provisional.

4.1.4.1   Classes 1A, 1B, and 1C Differences among Class 1 inputs have been documented both for thalamocortical and corticocortical pathways, and these differences suggest three subtypes. Because these differences are subtle and all share most Class 1 properties, we propose that they are subclasses of Class 1 rather than three completely different types; we call these Class 1A, Class 1B, and Class 1C. What these have in common is a lack of any metabotropic glutamate receptor response, the presence of relatively large initial EPSPs, and the presence of responses that are predominantly depressing.

Basic properties of Class 1A inputs are shown in figure 4.4A (see also figure 4.1A). These exhibit the properties of Class 1 inputs to thalamic relay cells, a

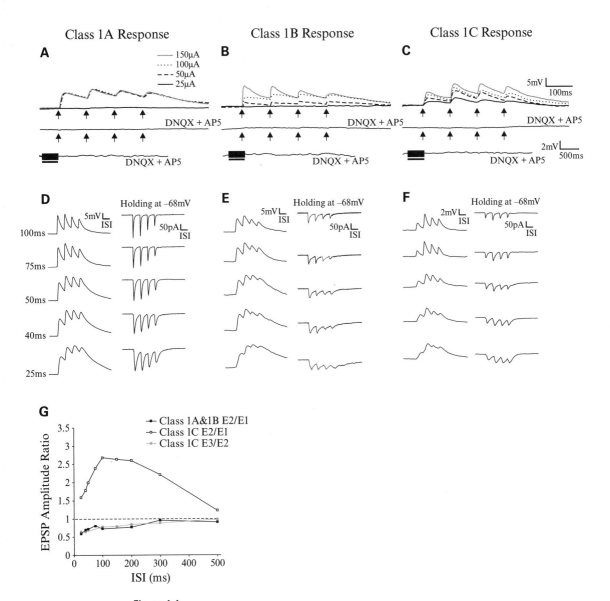

Figure 4.4

Examples of Class 1A, Class 1B, and Class 1C responses. (A) Class 1A response pattern. The top trace shows response to differing intensities of thalamic stimulation. The middle trace shows the absence of a response after thalamic stimulation at 200 μA in the presence of ionotropic glutamate receptor antagonists (DNQX and AP5). The bottom trace shows the absence of metabotropic glutamate receptor response after high-frequency stimulation of thalamus at 200 μA in the presence of DNQX and AP5. (B) Class 1B response pattern; conventions as in (A). (C) Class 1C response pattern; conventions as in (A). In (A)–(C), the arrows indicate the time of stimulation, and the black bars represent the duration of stimulation in high-frequency stimulation trials. (D) Class 1A responses to different stimulation frequencies at 150 μA in current clamp (left column) and voltage clamp (right column). The value of the interstimulus interval (ISI) is shown to the left, and these apply as well to (E) and (F). (E) Class 1B responses; conventions as in (D). Class 1C responses; conventions as in (D). Black bars represent the duration of stimulation in high-frequency stimulation trials. With the exception of high-frequency stimulation trials, all traces represent the average of 10 sweeps. Redrawn from Viaene et al. (2011c). (G) Changes in EPSP amplitude ratios of the second to first EPSP (E2/E1) or third to second (E3/E2) across stimulation frequencies for Classes 1A–1C.

prototypical example being retinal input to the lateral geniculate nucleus. Class 1A is distinguished from Classes 1B and 1C by showing paired-pulse depression (criterion 2 in table 4.1) and having an all-or-none activation profile (criterion 4 in table 4.1). In cortical circuitry, Class 1A inputs are found among thalamocortical inputs as seen in the mouse somatosensory and auditory cortices (Lee and Sherman, 2008; Viaene et al., 2011a, 2011c), and further details are given later in section 4.1.5.

Class 1B inputs differ only in terms of the activation profile (criterion 4 in table 4.1): these show a graded profile suggesting more convergence, but less than that seen for Class 2 inputs (DePasquale and Sherman, 2011). An example is illustrated in figure 4.4B. Class 1B inputs are the only Class 1 inputs so far seen in connections between the primary and secondary auditory areas and between primary and secondary somatosensory areas and are also seen among some thalamocortical inputs (see section 4.1.5. for details).

Class 1C inputs are rather unusual because they show paired-pulse facilitation for the first two evoked EPSPs (or EPSCs) in a train, and depression thereafter (figure 4.4C). Thus, the response is mostly depressing. The patterns of paired-pulse depression seen with classes 1A and 1B inputs are seen over a wide range of stimulus frequencies (figure 4.4D, E, and G), and this is also the case for the pattern seen in Class 1C inputs (figure 4.4F and G). This same pattern has been described among inputs to hippocampus and is thought to be associated with an intermediate initial probability of release, lower than that of purely depressing synapses and higher than that of purely facilitating synapses (Dittman et al., 2000; Sun et al., 2005). A possible function for this curious Class 1C response, shown in figure 4.4G, is that it acts as a temporal filter that favors intermediate afferent firing frequencies for which the facilitative component of the responses is enhanced, whereas at lower and higher stimulation frequencies, facilitation is diminished (Viaene et al., 2011c). Note also that Class 1C inputs also show a graded response pattern (figure 4.4C). This unusual Class 1 subtype has so far been seen among some thalamocortical inputs (see section 4.1.5 for details).

The scatterplot of figure 4.5 shows the differences among these Class 1 subtypes and provides the justification for this subclassification.

4.1.4.2 Class 2 The major variant among Class 2 inputs is the nature of the metabotropic glutamate receptors activated. These include group I (types 1 and 5) and group II (types 2 and 3 are usually lumped together; see chapter 2) metabotropic glutamate receptors. Of interest here is that activation of the group I receptors leads to depolarization, mainly due to closing of K^+ channels, whereas activation of the group II receptors leads to hyperpolarization, mainly

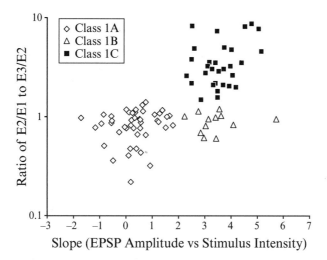

Figure 4.5
Two-dimensional scatterplot showing relationship between two parameters among subtypes of
Class 1 thalamocortical inputs based on recordings from single thalamic recipient zones in primary
somatosensory and auditory cortices. The abscissa plots the slope of the relationship between the
intensity of electrical stimulation and amplitude of the evoked EPSP; the ordinate plots the amplitude ratios of the second to first EPSP (E2/E1) divided by the third to second (E3/E2). Redrawn
from Viaene et al. (2011c).

due to opening of K^+ channels. The Class 2 corticothalamic inputs appear to
activate only type 1 metabotropic glutamate receptors on relay cells (Godwin
et al., 1996a; Vidnyanszky et al., 1996; Liu et al., 1998). This is also true for
the Class 2 inputs from the shell of the inferior colliculus to the dorsal division
of the medial geniculate nucleus (Lee and Sherman, 2010). No other Class 2
input to thalamic relay cells has yet been identified for such study. Also, all
Class 2 inputs so far seen in thalamocortical projections involve group I, and
therefore depolarizing, metabotropic glutamate receptors, although the distinction between types 1 and 5 was not made in these studies (Viaene et al., 2011a;
Viaene et al., 2011b; Viaene et al., 2011c).

However, within cortex both groups I and II metabotropic glutamate receptors are found among those activated by Class 2 inputs. Connections between
primary and secondary auditory areas and between primary and secondary
visual cortical areas activate both groups I and II metabotropic glutamate
receptors, often on the same cell (Covic and Sherman, 2011; DePasquale and
Sherman, 2011). Finally, the Class 2 input from layer 6 to 4 activates both
groups of metabotropic glutamate receptor (Lee and Sherman, 2009a).

From these Class 2 patterns, we emphasize three points. First, in cases
where an input activates more than one type of metabotropic glutamate

receptor, it is not clear if an individual axon can activate different types, because multiple afferent Class 2 axons were activated in these experiments (see section 4.1.2.8). Second, the fact that activation of group II metabotropic glutamate receptors is not uncommon in cortex means that there are glutamatergic inputs that can cause inhibition. Third, it is not clear if the type of the activated metabotropic glutamate receptor, group I or group II and their various subtypes, by itself warrants a subclassification of Class 2 inputs. More data are needed here.

The second point in the preceding paragraph is worth repeating and emphasizing. That is, some Class 2 inputs in cortex, which are *glutamatergic*, can produce inhibition via activation of group II metabotropic glutamate receptors. Inhibition, which plays an integral role in the formation of cortical receptive fields and may also be involved in such processes as gain control and synaptic plasticity, had been previously thought of as purely GABAergic (Berman et al., 1992; Hirsch and Martinez, 2006; Huang et al., 2007). We suggest that glutamatergic modulatory inputs can also contribute to these inhibitory functions.

4.1.5   Known Distributions of Class 1 and Class 2 Inputs in Thalamus and Cortex

A number of glutamatergic inputs have been identified for classification in thalamus and cortex. To date, all such inputs fall neatly into either Class 1 or Class 2 types, including the subtypes just noted. No others have yet been identified. We had expected that the more complex circuitry of cortex would yield other classes, but this has not yet occurred, possibly because investigation of glutamatergic circuits in cortex is still at a very early stage. In the following, we consider the distribution of Class 1 and Class 2 inputs for thalamus and cortex and consider some of the problems that remain to be resolved. An important proviso to this description is that much of the data are based on measuring evoked responses in single neurons to activation of inputs by various techniques, such as electrical stimulation of afferent axons and laser uncaging of glutamate to activate afferent neurons (Callaway and Katz, 1993); this means that multiple afferents rather than individual ones are tested.

4.1.5.1   Thalamus   All Class 1 inputs so far described for thalamus are Class 1A. For the first order nuclei, the Class 2 inputs other than those from layer 6 of cortex have not been adequately explored. The input from the superior colliculus, which innervates the lateral geniculate nucleus, is a glutamatergic input that has not been studied with regard to functional synaptic properties, and in cat and monkey, its termination is largely in the W cell layer [cat (Torrealba et al., 1981)] or koniocellular layer [primate (Lachica and Casagrande, 1993; Feig and Harting, 1994)]. However, ultrastructural studies of cats and primates suggest that the terminals of this pathway are small (Torrealba et al., 1981; Feig and Harting, 1994), suggesting the possibility that this is a Class 2 input.

For the higher order relays, there is only scattered information about the glutamatergic inputs beyond those from layer 6 of cortex. One question, still unresolved, is whether any given higher order relay cell receives Class 1 inputs from more than one source, which would suggest integration of distinct driver inputs (see section 3.2.1.3 of chapter 3) not known for any thalamic first order relay cell. A related question is whether some nuclei that we have classified as higher order have first order circuits running in parallel. For instance, many relay cells in the pulvinar and lateral posterior nucleus receive layer 5 input (Rockland, 1994; Bourassa and Deschênes, 1995; Guillery, 1995) that defines them as higher order relays, but these nuclei also receive input from the superior colliculus. Is this collicular input Class 1 or Class 2? There is electron microscopic evidence for the lateral posterior nucleus of the cat that these terminals are large and have the morphological characteristics and synaptic relationships consistent with a Class 1 input (Kelly et al., 2003), and this would suggest that some classically defined thalamic nuclei may contain first and higher order circuits.[5] This raises the possibility that some relay cells may integrate these two inputs, although the cat's lateral geniculate nucleus provides a nice example of Class 1 retinal X and Y inputs innervating the same layers but each targeting its own class of relay cell, without any integration there (Sherman, 1982; Usrey et al., 1999).

The inputs to the dorsal medial geniculate nucleus present an interesting pattern (figure 4.6). This dorsal, higher order, division receives a glutamatergic

---

5. There is a serious technical problem in identifying subcortical inputs to higher order thalamic relays. Layer 5 inputs to these thalamic targets involve branching axons (see chapters 1 and 6), with the extrathalamic branches targeting other subcortical sites, including often the superior colliculus. An example of the difficulty posed is presented by attempts to identify an input to the pulvinar or medial dorsal nucleus from the superior colliculus (for example, Sommer and Wurtz, 2004a; Sommer and Wurtz, 2004b; Berman and Wurtz, 2010; Berman and Wurtz, 2011). One method involves electrical stimulation of the superior colliculus while recording from thalamus, but this implies that the stimulated afferents arise from cells in the colliculus. They might, however, be branches of cortical layer 5 axons that innervate superior colliculus and send a branch to the pulvinar (or medial dorsal nucleus), which would mean that the stimulation actually activates the corticopulvinar input and not the collicular input. The same problem arises when injections into the superior colliculus appear to label colliculothalamic axons, because the thalamic label might be in the branches of layer 5 axons that have terminals in the colliculus and the thalamus, and this label could then travel retrogradely to the branch point and anterogradely from there to the thalamus.

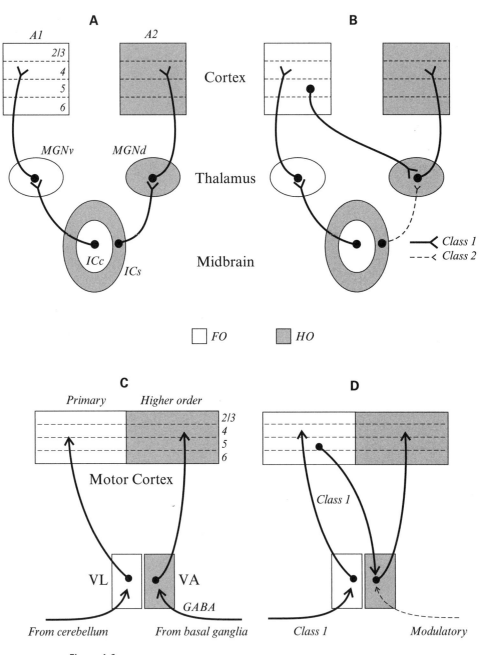

Figure 4.6
Application of the hypothesis that Class 1 inputs are drivers and Class 2 inputs are modulators, for auditory pathways (A, B) and for basal ganglia inputs to motor thalamus (C, D). FO, first order; HO, higher order; MGNd, dorsal medial geniculate nucleus; MGNv, ventral medial geniculate nucleus; VA, ventral anterior nucleus; VL, ventral lateral nucleus. See sections 4.1.5.1 and 4.2 for details.

input from the shell region of the inferior colliculus, and this has been previously thought of as part of a colliculo-thalamo-cortical auditory pathway running in parallel to the one coming from the central region of the inferior colliculus and relayed through the ventral medial geniculate nucleus (figure 4.6A). This implies that the collicular input is Class 1 (driving) to both divisions of the medial geniculate nucleus (Hu, 2003). The classification of inputs we propose suggests a very different scheme, because the dorsal medial geniculate nucleus receives layer 5 inputs (Ojima, 1994; Llano and Sherman, 2008), which provide an alternative Class 1 source, and recently it has been demonstrated that the input from the inferior colliculus to this nucleus is Class 2, or modulatory (Lee and Sherman, 2010), leading to the very different picture shown in figure 4.6B.

A similar issue arises for the higher order somatosensory relay in the posterior medial nucleus. The prevailing view is that this nucleus is part of a paralemniscal pathway that relays information to cortex from the spinal nucleus of the fifth cranial nerve in parallel with the relay of lemniscal information through the ventral posterior nucleus (Yu et al., 2006; Masri et al., 2008). However, the possibility remains to be tested that this input is largely or exclusively Class 2, which would imply a very different functional organization for these circuits.

The final higher order relay for which some information is available is the medial dorsal nucleus, which relays information to much of the prefrontal cortex. Because it receives characteristic large Class 1 terminals from layer 5 of prefrontal cortex (Schwartz et al., 1991), we have classified it as a higher order relay (Sherman and Guillery, 2006). However, there is evidence for inputs to one part of the medial dorsal nucleus from the amygdaloid nuclei (Price, 1986; Groenewegen et al., 1990), and the superior colliculus also projects to the medial dorsal nucleus (Sommer and Wurtz, 2004b). It remains to be determined which of these afferents are Class 1 and which are Class 2.

4.1.5.2 Cortex: Thalamocortical Afferents Whereas it seems reasonable to expect that the more complex circuitry of cortex would yield a larger number of classes of glutamatergic input, this has not yet been demonstrated. The same basic classes 1 and 2 of such inputs are all that are known at present, but there are two provisos. First, most connections in cortex are quite local, within a column, and most of the tested inputs so far are relatively distant, so that many more examples need to be tested. Second, as noted, nearly all of the relevant evidence comes from in vitro work on slices of mouse brain limited chiefly to somatosensory, auditory, and visual cortices; more examples from other areas and species are needed.

While thalamocortical projections are often described as if they terminate nearly exclusively in layer 4, actually cells in all layers from 1 to 6 in the mouse and rat receive thalamic input (White, 1978; Keller et al., 1985; Bureau et al., 2006; Meyer et al., 2010). However, a distinction needs to be made between the layers in which thalamic terminals are found and the layers in which recordings are made from the postsynaptic cells. For instance, a pyramidal cell with its cell body (and typical recording site) in layer 5 might be postsynaptic to thalamic input that actually innervates its apical dendrite in more dorsal layers. In the evaluation of laminar location of various inputs described here, it is usually the cell body location that has been determined.

With this in mind, the following pattern has been seen both for first order thalamic inputs to primary cortex (the ventral posterior medial nucleus to primary somatosensory cortex and the ventral medial geniculate nucleus to primary auditory cortex) and for higher order thalamic inputs to secondary cortex (the posterior medial nucleus to secondary somatosensory cortex and the dorsal medial geniculate nucleus to secondary auditory cortex) based on a series of in vitro studies (Lee and Sherman, 2008; Viaene et al., 2011b, 2011c). The pattern is summarized by figure 4.7. Layers 4–6 receive exclusively Class 1 inputs from the thalamus: layer 4 exclusively Class 1A, and layers 5 and 6 a combination of classes 1A, 1B, and 1C (Lee and Sherman, 2008; Viaene et al., 2011a, 2011c). However, layers 2/3[6] display a mixture, so that about three fourths of the cells receive Class 2 inputs and one fourth receive Class 1A inputs. Given our hypothesis for the function of Class 1 and Class 2 inputs, this suggests that these thalamic inputs to layers 4–6 and some to layers 2/3 are information-bearing drivers, but that most to layers 2/3 have a modulatory function.

This suggested modulatory function provides a new role for some thalamo-cortical inputs with two further implications. First, if most layer 2/3 cells receive modulatory input from the thalamus, what cell group, if any, provides a driving (Class 1) input? An obvious candidate is layer 4. When recording from a cell in layers 2/3, activation of layer 4 produces paired-pulse depression (Feldmeyer et al., 2002; Oswald and Reyes, 2008), indicative of a Class 1 input, which has now been confirmed (DePasquale and Sherman, 2012). This is also consistent with the older notion that thalamic information is mainly channeled through layer 4 to layers 2/3 (Gilbert and Wiesel, 1979). In this new view presented here, most of the thalamic input to layers 2/3 modulates the flow of information to those layers from layer 4. Exactly what form this

---

6. Layers 2 and 3 of cortex cannot generally be distinguished and are frequently lumped together as "layers 2/3." This is the terminology we use.

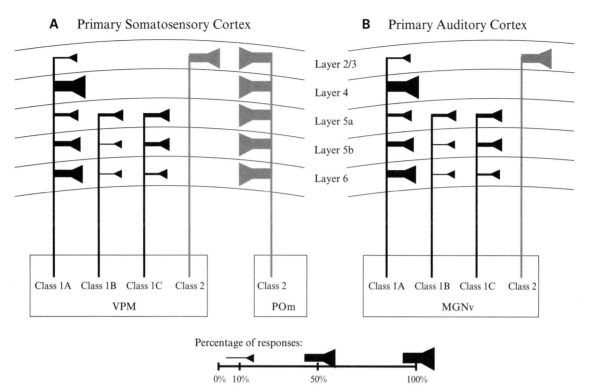

Figure 4.7
Schematic representation of projections from the ventral posterior medial nucleus (VPM, first order), the ventral medial geniculate nucleus (MGNv, first order), and the posterior medial nucleus (POm, higher order) to primary cortical receiving areas. (A) Thalamocortical projections to primary somatosensory cortex. (B) Thalamocortical projections to primary auditory cortex. Within each target layer in cortex, the line thickness and symbol size represent the percentage of cells receiving each response class in that layer. Redrawn from Viaene et al. (2011c).

modulation takes remains to be determined, as is the case generally for Class 2 inputs, although evidence exists as noted above for some examples: layer 6 cortical input has subtle effects on receptive field properties of geniculate cells (Kalil and Chase, 1970; Baker and Malpeli, 1977; Schmielau and Singer, 1977; Geisert et al., 1981; McClurkin and Marrocco, 1984; McClurkin et al., 1994; Andolina et al., 2007) and appears to help to control the burst/tonic firing mode transition (Godwin et al., 1996b), and the layer 6 input to layer 4 cells or nearby input to layers 2/3 cells reduces the amplitude of thalamocortical EPSPs (Lee and Sherman, 2012; DePasquale and Sherman, 2012). Second, although the above experiments on thalamic inputs to layers 2/3 were performed in mice, there is an interesting possible relationship to the cat and monkey visual systems. As noted in chapter 3, in both these animal groups the projection from the retina through the lateral geniculate nucleus to visual cortex involves at least three parallel pathways. The X and Y pathways in the cat and the M and P in the monkey terminate predominantly in layer 4, but the W pathway in the cat and the K pathway in the monkey terminate in layers 2/3 (Ferster and LeVay, 1978; Ding and Casagrande, 1997; Shostak et al., 2003; Anderson et al., 2009). This raises the possibility that whereas the X and Y and the M and P pathways are likely to be Class 1 and driver to cortex, the W and K pathways may be largely, but perhaps not exclusively, Class 2 and modulatory to cortex.

Another potential modulatory thalamocortical pathway has recently been described, and this is illustrated in figure 4.7A. Although the main thalamic input to the rodent primary somatosensory cortex derives from the first order ventral posterior medial nucleus, there is also one from the higher order posterior medial nucleus (Koralek et al., 1988; Lu and Lin, 1993), although until now we have concentrated on the projection from this higher order thalamic relay to secondary somatosensory cortex. Whereas, as just noted, this latter projection to the secondary area is mostly Class 1, that from the posterior medial nucleus to primary somatosensory cortex provides a Class 2 input to cells in all layers from 2 to 6 (Viaene et al., 2011a).[7] A similar analysis has not yet been performed for the projection from the dorsal medial geniculate nucleus to primary auditory cortex.

4.1.5.3 Direct Corticocortical Connections   Theories of macroscopic cortical functioning have emphasized the importance of projections between different cortical areas; examples are sensory processing (for example, Felleman and Van Essen, 1991; Van Essen et al., 1992; Kaas and Collins, 2001) and dynamic binding of different cortical areas for various cognitive functions, such as attention (for example, Moore and Armstrong, 2003; Fries et al., 2007; Fries, 2009; Gregoriou et al., 2009). Despite the importance of such connections, nearly all of the available studies are limited to morphological analyses of

---

7. The rat and mouse primary somatosensory cortex is also known as "barrel cortex" because layer 4 has prominent barrel-shaped regions that can be visualized with metabolic stains, and these "barrels" are separated from one another by smaller, poorly staining regions known as "septa." In their study of inputs to the rat's barrel cortex, Killackey and Sherman (2003) suggested that the septa actually represent secondary somatosensory cortex, arranged in a mosaic with the barrels, which were suggested to be primary cortex. Their reasoning was that the septa were innervated by the posterior medial nucleus, a higher order relay, which they suggested targets mainly higher order cortical regions. But this suggestion implicitly assumed that such input from the posterior medial nucleus to barrel cortex would be driver, or Class 1. This finding that all such input to barrel cortex, both to barrels and septa, from the posterior medial nucleus is Class 2 argues strongly against the hypothesis that the septa represent a higher order cortical area.

these pathways, and there is surprisingly little information about the synaptic properties of corticocortical connections that can be used to classify these inputs. The relevant data that do exist are limited to the connections in both directions between the primary and secondary cortices in the mouse visual and auditory systems (Covic and Sherman, 2011; DePasquale and Sherman, 2011). Several conclusions stand out from these studies (see also section 5.4.1.1 and figure 5.2 in chapter 5):

• Cells in all of layers 2–6 contribute to these projections, and cells in all of these layers receive such projections.

• Class 1B and Class 2 projections, and only these, have been found among these corticocortical inputs.

• There is a laminar pattern to these connections. On the postsynaptic side, cells in layers 2–4 receive both classes of input, but, with few exceptions, cells in layers 5a[8] and 6 receive only Class 2 inputs and cells in layer 5b receive only Class 1A inputs. This means that of the two major corticofugal projections, those from layer 6 that innervate thalamus with Class 2 inputs are themselves in receipt of mainly Class 2 inputs from the related cortical area, and those from layer 5b that form the corticofugal outputs and the afferent limbs of transthalamic circuits through Class 1 inputs to higher order thalamic relays receive Class 1A inputs from the related cortical area. On the presynaptic side, both classes come from all of the layers, although layer 5a produces predominantly Class 2 projections.

• The laminar patterns of connectivity by class of input are remarkably similar between the auditory and visual cortices, suggesting the possibility of a general pattern, although clearly many more examples are needed.

• The pattern in both directions, from primary to secondary cortex (feedforward) and the reverse (feedback) are also remarkably similar. This is surprising for the following reason. Ideas of hierarchical processing assume that information flows up the hierarchy (for example, from primary to secondary areas) and that the projections in the reverse, feedback direction serve to modulate information flow. This, in turn, would lead to the prediction that, at least in relative terms, Class 1 projections are more dominant in the feedforward than in the feedback projections. The available evidence does not support this idea.

---

8. Layer 5 can be readily divided into a dorsal layer 5a and ventral layer 5b. Most corticofugal cells of this layer are found in layer 5b and include corticothalamic cells. Recently, Reiner et al. (2003) showed in the rat that layers 3 and 5a contained cells that project to the basal ganglia, but these cells do not project to thalamus.

It is interesting to compare these results to the influential model of cortico-cortical connectivity proposed for monkey visual cortical areas by Felleman and Van Essen (1991). These authors divide cortical layers into supragranular (layers 2/3), layer 4, and infragranular (layers 5 and 6), and they propose different laminar patterns for each of three different types of projection. In their framework, feedforward projections (that is, that ascend the hierarchy) come from supragranular layers only or from supragranular and infragranular layers and target layer 4, feedback projections (that is, that descend the hierarchy) come from infragranular only or supragranular and infragranular layers and target supragranular and infragranular layers, and lateral connections (that is, that connect areas at the same hierarchical level) come from supragranular and infragranular layers and target all layers. The patterns described above for mouse somatosensory and auditory cortices match none of these patterns. There are many possible explanations for this discrepancy, including species differences; the fact that the mouse pattern is based on physiology data and the monkey on anatomical data; and so forth. This remains a puzzle, but it does underscore the vital need for more and better functional studies of corticocortical connections.

4.1.5.4 Cortex: Local Corticocortical Connections   Others have noted certain differences in synaptic properties in local corticocortical connections (for example, Reyes et al., 1998; Thomson and West, 2003), but these studies fall short of a systematic classification, and very limited data are available that we can relate to our Class 1 and Class 2 input types. One example is the layer 6 to layer 4 projection, which is Class 2 (Lee and Sherman, 2009b). As noted above, the projection from layer 4 to layers 2/3 is Class 1 (DePasquale and Sherman, 2012). Clearly, much more study of these local connections is needed, both because such analysis can provide important insights into the functioning of local cortical circuits and because it is here that one might search for classes in addition to classes 1 and 2.

4.1.5.5 Convergent Class 1 Inputs   Convergent Class 1 inputs, as noted, are common in cortex, although specific examples of convergence from different sources are rare, but several do exist. One is layer 4 of the primary and secondary somatosensory and auditory cortical areas, because cells there receive Class 1 inputs both from corticocortical connections (Covic and Sherman, 2011; DePasquale and Sherman, 2011) as well as from thalamus (Lee and Sherman, 2008). Another example is layers 2/3, which receive Class 1 inputs from both corticocortical connections (Covic and Sherman, 2011; DePasquale and Sherman, 2011) and locally from layer 4 (DePasquale and Sherman, 2012).

4.1.6 Possible Differences in Properties at Different Terminals of an Axon

Given that glutamatergic inputs can be classified into at least two main types, does it follow that all synapses from a single axon must be of the same type? This is an issue especially important for branching axons, and this topic of branching glutamatergic axons is taken up in more detail in all of the remaining chapters of this book. For instance, axons innervating thalamus with Class 1 properties typically branch to innervate extrathalamic targets, usually in the brainstem. Do the brainstem targets also receive Class 1 inputs from these branches?

An elegant study by Reyes et al. (1998) offers insight into this issue. These authors intracellularly recorded simultaneously from several cells in cortex with the arrangement that one of the recorded cells was found to be presynaptic to two or more of the other recorded cells, and all of the connections here were glutamatergic. It was found that the afferent cell could produce paired-pulse depression (a Class 1 feature) at one target cell and paired-pulse facilitation (a Class 2 feature) at another. This demonstrates that different branches of an afferent axon can quite plausibly evoke different synaptic properties at its different targets.

4.2 Nonglutamatergic Afferents

Both thalamic and cortical circuits involve many transmitter systems other than glutamate. These include GABAergic, cholinergic, noradrenergic, serotonergic, and dopaminergic inputs. Most of the GABAergic inputs to these circuits are from a local source, such as interneurons, although some inputs to thalamus come from GABAergic brainstem sources, such as the zona incerta (Barthó et al., 2002), pretectal nucleus, globus pallidus, and substantia nigra. The other transmitter systems stem from various sites in the brainstem, and, in addition, a prominent cholinergic input to cortex comes from the basal forebrain region. Details of these pathways are considered more fully in chapter 3 and can also be found elsewhere (thalamus: Sherman and Guillery, 2006; Jones, 2007; cortex: Descarries et al., 2004; Tseng and Atzori, 2007; Iversen et al., 2010).

The question for consideration here is the functional nature of these inputs. Cholinergic, noradrenergic, serotonergic, and dopaminergic inputs are generally thought to be modulatory in function. In this sense, they may be seen to have a function similar to that of Class 2 inputs, and like these inputs, they activate metabotropic receptors. There is one important exception to this functional similarity. That is, these other afferents, which arrive from various sites, appear to be relatively diffuse or widespread in their connections in thalamus and cortex, and they thus appear to provide modulation for general behavioral states such as drowsiness versus alertness. In contrast, the known Class 2 glutamatergic inputs, which include layer 6 corticothalamic inputs, inputs from the inferior colliculus to the dorsal medial geniculate nucleus, and various corticocortical and thalamocortical inputs, are all topographic in connectivity, suggesting that their modulatory effects are more specific. This may provide a distinct, more topographical modulatory role for Class 2 inputs compared to the less topographic modulation provided by the nonglutamatergic afferents.

What of the GABAergic inputs to thalamic and cortical neurons? We have argued elsewhere that GABAergic inputs are poor candidates for the transmission of information (Smith and Sherman, 2002), which by default leads to the suggestion that GABAergic inputs are modulatory in function. In both thalamus and cortex, $GABA_B$ receptors are common, suggesting that many and perhaps all GABAergic inputs in these structures activate such receptors, which are metabotropic, and as noted in chapter 2, activation of metabotropic receptors may be a key to several modulatory functions. In this sense, GABAergic inputs can act with other modulatory systems, which are usually excitatory, to provide modulatory function in a push–pull manner.

However, GABAergic inputs are often considered, at least implicitly, as operating in the same way as Class 1 inputs. One example is the input from the basal ganglia to the ventral anterior nucleus. In most textbook accounts (Kandel et al., 2000; Purves et al., 2008), this is seen as a key link in a loop between cortex and basal ganglia (see figure 4.6C), as if this input to thalamus were information-bearing, functionally comparable to the retinal input to the lateral geniculate nucleus or the cerebellar input to the ventral lateral nucleus. We offer an alternative model shown in figure 4.6D in which the basal ganglia input to thalamus is modulatory. In chapter 3, we have provided evidence that cerebellar inputs to the ventral lateral nucleus are drivers (Class 1), and thus the relay cells innervated by this pathway represent a first order relay. Evidence exists in the monkey (Sakai et al., 1996) and rat (Kuramoto et al., 2011) that the inputs from the cerebellum and from the basal ganglia occupy different territories, with the former mainly innervating the ventral lateral nucleus and the latter, the ventral anterior nucleus, with little overlap. There is a projection from cortical layer 5 to these thalamic nuclei, and axons from the motor cortex form characteristic large, localized terminals there (McFarland and Haber,

2002; Kultas-Ilinsky et al., 2003), which suggests a Class 1 input and thus defines a zone of higher order thalamic relay. Note that figure 4.6D shows the first order and higher order thalamic relays as separate, but this is only for illustration, because we do not yet know how the cerebellar and the layer 5 inputs relate to one another. However, a reasonable guess is that these two driver inputs (that is, cerebellar and cortical layer 5) mostly occupy substantially different thalamic territories, but with possibly some overlap, which in turn would mean that the regions rich in cortical layer 5 input largely correspond to the input from the basal ganglia, but there is not yet rigorous evidence for such a view. However, the available evidence suggests that the input from the basal ganglia serves to modulate or gate mainly the layer 5 input to the higher order portions of the ventral anterior and ventral lateral anterior nuclei. This represents a very different view than standard textbook accounts of the function of the basal ganglia and their influence on motor cortex.

4.3   Concluding Remarks

One of the first steps in trying to understand the brain is determining the nature of its component elements. For instance, an early step in unraveling the mysteries of structures like the retina or cortex was determining the various types or classes of the constituent neurons of each. For the retina, this meant not only appreciating the difference between types like bipolar cells and ganglion cells, but also determining the different classes within these types: the number of separate classes of ganglion cells so far identified exceeds 10 and is likely to continue to grow (Dacey et al., 2003; Farrow and Masland, 2011).

Although the value of a classification of neuronal types has been recognized for over a century (for example, Cajal, 1911), there is another classification that is important to recognize but that has not received much attention. That is, just as a classification of neuronal types is an early step in making sense of many brain structures, a classification of constituent elements in complex neuronal circuits should be seen as an important step in decoding functional connectivity among neurons. There is recognition that afferents in circuits can be classified by their neurotransmitters, and so we recognize glutamatergic, GABAergic, cholinergic, and so forth, inputs to a neuron or region and distinguish between them. But this is like noting that in retina, ganglion cells differ from bipolar cells. What about the equivalent of appreciating that many distinct classes of ganglion cell exist? That is, are there distinct classes within the population of glutamatergic (or GABAergic, and so forth) afferents?

Because glutamatergic inputs represent the dominant type for basic information transfer and processing in the brain, we have focused on the evidence that these inputs include different classes. There has been a tendency in both text-book accounts and modeling studies to treat all glutamatergic afferents as functionally equivalent, weighting the relative importance of each input on the basis of anatomical numbers, as if their postsynaptic effects summed simply and relatively linearly. We have previously pointed out the importance of recognizing different glutamatergic inputs for thalamic circuitry, which we called "drivers" and "modulators" (Sherman and Guillery, 1998, 2006), and there have been several studies noting differences among glutamatergic affer-ents, mostly based only on paired-pulse effects (Agmon and Connors, 1992; Reyes et al., 1998; Thomson and West, 2003; Tan et al., 2008; Hull et al., 2009). However, surprisingly little attention has been devoted to attempting a thorough functional classification of these afferents.

We have provided evidence that there are two major classes of glutamatergic input, which we have called Class 1 and Class 2, and, furthermore, that Class 1 inputs in cortex have at least three subclasses (see section 4.1.4). A major value of such a classification scheme lies in the insights it provides for under-standing the functional organization of complex circuits. For instance, we have argued that Class 1 inputs in thalamus are drivers, meaning that these represent the input route for information to be relayed to cortex, whereas the Class 2 thalamic inputs are modulators, acting like other modulatory inputs (for example, cholinergic and serotonergic) to affect many aspects of the thalamic relay of driver input. This knowledge has led to a novel interpretation of numerous thalamic circuits described above, including the inferior collicular input to the dorsal medial geniculate nucleus and pallidal input to the ventral anterior nucleus. It has highlighted issues involving tectal input to pulvinar and inputs from the trigeminal or spinothalamic pathways to the posterior medial nucleus. Furthermore, understanding the significance of the Class 1 driver inputs to much of thalamus has led to the rethinking of thalamocortical relationships as regards higher order thalamic relays, which is discussed more fully in chapter 5.

Finally, we offer the hypothesis that, as in thalamic circuitry, Class 1 inputs represent the main information routes in cortical circuitry and Class 2 inputs are modulatory in function. Given the more complex circuitry of cortex com-pared to thalamus, this hypothesis will be more difficult to test, but if verified, it does provide important insights into cortical functioning. This includes a better understanding of thalamocortical relationships, some being driver inputs and some modulatory; a verification that higher order thalamic relays are a transthalamic route for corticocortical communication running parallel to the direct corticocortical pathways (see chapter 5); and a critical means for analyz-ing cortical circuitry, both local and between cortical areas.

4.4 Outstanding Questions

1. Are Class 1 inputs always driver and Class 2 always modulator?

2. The Class 1 and Class 2 inputs represent the only glutamatergic afferents seen to date in thalamus and cortex, but will this continue to be the case as other circuits are studied in cortex and in many other parts of the central nervous system, including those in nonmammalian species?

3. Are there classes of driver input other than Class 1?

4. Do some thalamic relay cells receive distinct Class 1 inputs from two or more sources and integrate this information?

5. Can inputs using transmitters other than glutamate also be usefully classified?

6. Given that cortical circuitry suggests that different Class 1 (presumably driver) inputs converge onto single cells, may one role of Class 2 modulatory inputs be to determine which combinations of driver input are active at any one time?

7. What, if any, difference exists in the Class 1 versus Class 2 mix of afferents between cortical areas in the feedforward and feedback directions?

8. Do all cells in thalamus and cortex receive both Class 1 and Class 2 inputs or do some receive exclusively one of these inputs?

9. In thalamus, Class 2 inputs greatly outnumber Class 1, measured either by numbers of afferent axons or synaptic terminals, but what are the relative numbers for cortical circuitry?

10. In terms of Class 1 and Class 2 inputs, what are the different patterns seen in corticocortical connections between the feedforward and feedback directions?

5 First and Higher Order Thalamic Relays

As we have noted in the previous chapter, Class 1 inputs help to define a main function of a thalamic relay because they identify the driver inputs that carry the message that is sent to cortex, and thus, for example, the lateral geniculate nucleus is a relay of retinal messages, and the ventral posterior nucleus is a relay of messages from the medial lemniscus and spinothalamic tract. On the basis of the origin of the Class 1 input to any thalamic relay, the relay is first order if this input has an origin from a subcortical source, because this is typically the first relay to cortex of a specific type of information, and if this input is from cortical layer 5, the relay is higher order, because this represents information already in cortex that is relayed from one cortical area to another. This identification of higher order relays and their proposed role as participating in corticocortical processing has many crucial ramifications. Before considering these, we review the supporting evidence.

5.1 Evidence for Distinguishing First and Higher Order Thalamic Relays

In chapter 3.2.1.2, we described the difference between a higher order thalamic nucleus and a higher order thalamic relay. Briefly, the latter term is defined on the basis of individual relay cells that receive their driver input from layer 5 of cortex, and the former term describes a classically defined nucleus comprised mostly if not exclusively of higher order relays. This implies that some nuclei can have a mixture of first and higher order relays. As far as we know, all first order nuclei contain only first order relays.

5.1.1 Anatomical Evidence

The first suggestion of the existence of higher order thalamic relays came from anatomical studies (reviewed in Guillery, 1995; Sherman and Guillery, 2006; see also chapters 3 and 4). Historically, the first evidence was from electron microscopic studies. In general, these showed that projections from cortex to certain thalamic nuclei, which we now regard as first order (for example, the lateral geniculate and ventral posterior nuclei), ended in small terminals typically on peripheral dendrites, whereas such projections to other thalamic nuclei seen now as higher order (for example, the pulvinar and medial dorsal nucleus) ended in both small terminals on distal dendrites and in large ones on more proximal dendrites, often with more complex synaptic relationships in glomeruli as described in chapter 3. The small corticothalamic terminals had the morphological characteristics that were regarded as modulators, whereas the large ones had effectively the same morphological characteristics as Class 1 or driver inputs to first order relays, such as the retinal or lemniscal inputs to the thalamus. Later, when techniques allowed tracing of individual axons from source to target, it was found that layer 6 was the source of the small (Class 2/modulator) terminals to thalamus, whereas layer 5 was the source for the large (Class 1/driver) terminals and that first order nuclei received cortical afferents only from layer 6, whereas higher order nuclei received cortical inputs from both layers (Guillery, 1969b; Wilson et al., 1984; Bourassa and Deschênes, 1995; Bourassa et al., 1995; Rockland, 1996; Ojima ct al., 1996; Guillery et al., 2001; Kakei et al., 2001; Guillery and Sherman, 2011). That is, all thalamic nuclei receive a projection from layer 6 that is anatomically Class 2, modulator, but only a subset, the higher order nuclei, receive the additional, Class 1, driver input from layer 5.

The corticothalamic projections from both layers 5 and 6 demonstrate a high degree of topographic order (Guillery et al., 2001; Kakei et al., 2001), but there is an important difference in their organizational pattern. The projections from layer 6 are largely, but not exclusively, organized in a feedback manner, meaning that a region of a thalamic nucleus projecting to a cortical area receives a layer 6 projection from that same area with the same topographic order (Guillery, 1967; Van Horn and Sherman, 2004; Llano and Sherman, 2008). However, the layer 5 projection is not organized in this manner, because it goes to a thalamic region different from that providing input to the cortical area of the efferent layer 5 cells. Thus, the first visual area (area 17) receives inputs from the lateral geniculate nucleus, sends layer 6 outputs back to that nucleus but sends layer 5 outputs to the pulvinar or the lateral posterior nucleus, and the first somatosensory cortex comparably receives inputs from the ventral posterior nucleus, sends layer 6 outputs back to that nucleus and sends layer 5 outputs to the pulvinar (but see footnote 4 of chapter 1) or the posterior medial nucleus (Van Horn and Sherman, 2004; Llano and Sherman, 2008). That is, the layer 5 projection to the thalamus is feedforward, and further evidence in favor of this is provided below.

The observation that some thalamic nuclei receive a cortical layer 5 input and others do not is sufficient evidence to distinguish the two types of nucleus that we refer to as first and higher order. Given that these data include observations from rodents, carnivores, and primates (reviewed in Sherman and Guillery, 2006; Guillery and Sherman, 2011), they can be regarded as reflecting a general mammalian plan for thalamic relays, although the details for brains that have relatively few functionally distinct cortical areas have not been defined, and these could serve to provide useful clues to the evolutionary origins of these connections (see section 5.3.1).

5.1.2 Physiological Evidence

Complementary physiological support for this scheme has come primarily from studies of rats and mice. Li et al. (2003) noted that, in rats, synaptic inputs to the lateral posterior nucleus[1] showed different paired-pulse effects. One input showed paired-pulse depression (subsequently recognized as a Class 1 feature) and was presumed to derive from layer 5, and the other, facilitation (a Class 2 feature) and was presumed to derive from layer 6. This pattern was more directly confirmed for inputs to the somatosensory thalamus (ventral posterior medial and posterior medial nuclei) in thalamocortical slices from mice, permitting identification of the cortical source of the input as well as its synaptic properties (Reichova and Sherman, 2004). It was found that the inputs from layer 5 to the posterior medial nucleus had Class 1 features but those from layer 6 to both nuclei had Class 2 features (see also Turner and Salt, 2000; Landisman and Connors, 2007). Indeed, the Class 1 properties of the layer 5 input from S1 to the posterior medial nucleus were identical to those of the driver inputs from retina to the lateral geniculate nucleus (Reichova and Sherman, 2004), matching the functional characteristics to the structural features.

Functional evidence from monkeys, rats, and mice further supports the view that higher order relays are a link in corticocortical communication based on Class 1 projections. In the monkey, lesions of the striate cortex have subtle modulatory effects on geniculate neurons (Marrocco et al., 1996), but silence responses in pulvinar neurons (Bender, 1983; Chalupa, 1991), suggesting that this cortical area provides a driving input to pulvinar but not to the lateral geniculate nucleus. Lesions of S1 in the rat have little effect on receptive fields of the ventral posterior medial nucleus but silence neurons of the posterior medial nucleus, suggesting that S1 provides a driver input to the posterior medial nucleus but not to the ventral posterior medial nucleus (Diamond

1. See chapter 1, footnote 4.

et al., 1992). In the mouse, strong activation of S2 is seen after stimulation of layer 5 in S1 (the source of a Class 1 input to the posterior medial nucleus), and this S2 activation disappears reversibly during reversible inactivation of the posterior medial nucleus (Theyel et al., 2010b).

5.2   Some Differences between First and Higher Order Relays

One clear difference between first and higher order relays is the source of the Class 1 inputs. Otherwise, their cell and circuit properties are, so far as is currently known, quite similar with but a few exceptions that are considered in the following sections.

5.2.1   Percentage of Class 1 Synapses

In a quantitative electron microscopic analysis of first and higher order thalamic nuclei in the cat involving the respective relays for the visual, auditory, and somatosensory systems, a systematic difference was found in the relative number of Class 1 driver synapses (Wang et al., 2002; Van Horn and Sherman, 2007). These were less than half as numerous, in relative terms, in the higher order nuclei. Whether this is due to these nuclei having more synapses of other types (lumped together as modulatory) or fewer Class 1 synapses, or both, is unclear.[2] Nonetheless, it appears that higher order relays have relatively more circuitry devoted to modulation than do first order relays, and the same relationship is present in rats (Cavdar et al., 2011).

5.2.2   Sources of Afferent Input

There is scattered evidence that certain modulatory inputs target higher order thalamic nuclei selectively. Examples exist of different GABAergic inputs that appear to target higher order nuclei specifically. In the rat, inputs from the zona incerta fairly selectively innervate higher order nuclei (Power et al., 1999), and it appears that in the posterior medial nucleus this is largely a GABAergic input (Lavallée et al., 2005). In addition, there is a comparable GABAergic input to higher order nuclei from the anterior pretectal nucleus (Lavallée et al., 2005). Finally, the GABAergic input from the globus pallidus (Sakai et al., 1996; Kuramoto et al., 2011) and substantia nigra (Carpenter et al., 1976; Francois et al., 2002; Tanibuchi et al., 2009) appears selectively

---

2. The issue could readily be settled by counts of cells per unit volume and counts of synapses per unit volume obtained from material fixed and treated in a uniform manner with thin sections cut for the synapses and thicker sections cut for the cells.

to target higher order relays. In section 5.4.2.2, we discuss how this targeted GABAergic innervation may be important to the function of higher order relays. In addition, evidence from monkeys and humans shows that dopaminergic input from various sources also selectively targets higher order relays (Sanchez-Gonzalez et al., 2005; Garcia-Cabezas et al., 2007). Overall, these observations suggest that thalamic relays involved in corticocortical communication have an additional set of modulatory controls, and this fits with the results that show the driver inputs forming a lower proportion of synapses in higher order nuclei than in first order nuclei.

5.2.3 Thalamocortical Relationships

The importance of identifying Class 1 and Class 2 inputs in thalamocortical relationships has been described in chapter 4 (see figure 4.6), and this is briefly reiterated here from the point of view of the first and higher order relays and their cortical terminations. On the basis of data from the mouse somatosensory and auditory pathways, thalamocortical inputs from the first order ventral posterior medial nucleus to S1 and the first order ventral medial geniculate nucleus to A1 are arranged as follows: these inputs to cells in layers 4–6 are purely Class 1 and to cells in layers 2/3 are mostly (70% to 80%) Class 2, with the rest being Class 1 (Lee and Sherman, 2008; Viaene et al., 2011a, c). In the case of S1, this differs markedly from inputs from the higher order posterior medial nucleus, which are all Class 2 (Viaene et al., 2011a). However, inputs to layer 4 from this same higher order nucleus, the posterior medial nucleus, to S2 as well as from the dorsal medial geniculate nucleus to A2 are all Class 1 (Lee and Sherman, 2008). Input from these somatosensory and auditory higher order thalamic nuclei to other layers in S2 and A2 has not yet been studied. That is, so far as the thalamic input to layer 4 is concerned, the feedforward pathways involving higher order thalamic relays are Class 1, and limited data suggest that the feedback pathways are Class 2.

5.2.4 Cellular Responses to Modulatory Inputs

Cholinergic and serotonergic inputs from the brainstem represent two important modulatory inputs to the thalamus. In a study of rat thalamus, it was found that whereas all relay cells in first order relays respond to both inputs with depolarization, a substantial minority (15% to 20%) of relay cells in higher order nuclei instead respond to these inputs with hyperpolarization (Varela and Sherman, 2007, 2008). An interesting possible consequence of this relates to the relationship of these cholinergic and serotonergic inputs to behavioral state: these inputs are typically more active during vigilance and become increasingly less active as the animal becomes less alert and enters drowsiness. For all first order and most higher order relay cells, this means that the more alert the animal, the more depolarized the relay cells, but for a minority of higher order relay cells, increased alertness would result in hyperpolarization. As noted in the next section, this suggests the possibility that vigilance results in this minority of higher order relay cells being disposed to burst mode and may offer one explanation as to why higher order relays show more bursting, on average, than do first order relays (Ramcharan et al., 2005).

5.2.5   Burst versus Tonic Response Modes

As noted in chapter 2, an important property of relay cells is their ability to respond in two very different modes, burst or tonic, which affects how driver information is relayed to cortex. Two pieces of evidence point to more bursting among higher order relay cells. A study of relay cells recorded in behaving monkeys indicated that relay cells in higher order thalamic nuclei tended to burst more frequently than their first order counterparts (Ramcharan et al., 2005). This might relate to the observation noted above that in rats, scrotonergic and cholinergic inputs hyperpolarize many higher order relay cells and that these cells also receive extra GABAergic inputs not directed to first order relays (see section 5.2.2). As detailed in chapter 2, such hyperpolarization is a necessary prerequisite to bursting, because it is needed to de-inactivate the T-type Ca^{2+} channels that underlie the bursting. Other related evidence comes from a recent study of relay cells from the lateral geniculate nucleus and pulvinar of the tree shrew that reports more bursting among the pulvinar neurons, perhaps related to a higher density of the T-type Ca^{2+} channels (Wei et al., 2011). This is considered again in section 5.4.2.2 in the context of GABAergic inputs that selectively target higher order thalamic relays, thereby providing a hyperpolarizing input that would predispose these relay cells to burst mode.

5.3   Developmental and Evolutionary Differences

5.3.1   Phylogeny

We have pointed out in chapter 3 that the first order relays are well established early in mammalian evolution along with at least a single higher order copy of each sensory cortical area and that the full development of the main higher order cortical areas follows later. We know very little about the developmental mechanisms, phylogenetic or ontogenetic, that produce the enormous increase in higher order thalamic relays and cortical areas and that characterize many mammals, particularly primates. The very general tendency for many different groups of mammals to have developed a significant collection of distinct higher order cortical areas points to the functional significance of this evolutionary development and shows that there are some general patterns of thalamocortical interconnections common to all mammals (summarized in chapter 1.2). However, to understand how these several different cortical areas actually function, singly or as a group, we need to look at their relationship to the rest of the brain and to consider their origins from this point of view.

One important point about the simple mammalian brains is that we do not have the connectional information on the basis of which the relationships of the separate sensory areas for any one modality can be evaluated: we need to know whether, in the simplest brains, the feedback connection of the layer 6 cortical cells are distinguishable from the feedforward connections of the layer 5 cells. And if they are, then we need some details of these connections that would perhaps allow the identification of the earliest first and higher order pathways. It would prove of interest to study such pathways and to define the nature of the actions that the extrathalamic branches of the layer 5 cells are stimulating and thus the nature of the actions that the animal is able to anticipate before they occur.

The observation that there appear to be simple mammalian brains that have two or more areas of sensory cortex but lack a motor cortex suggests that sensory cortical areas must have had motor functions when they first appeared during evolution and, as we and others have argued, they continue to have such functions in rodent, carnivore, and primate brains. That is, because even in primate brains significant areas of somatosensory cortex contribute to the corticospinal tract (Brodal, 1981), and primary visual cortex links to the superior colliculus (Harting et al., 1992), we should not be surprised to find that primary sensory areas can make significant contributions to the control of movements (see Matyas et al., 2010). The phylogenetic history of the separation between granular cortical areas that have primarily sensory functions and agranular areas with primarily motor functions in relation to the connectivity patterns established by the cells of layers 5 and 6 of very simple mammalian brains could prove of considerable interest in showing how the functions of the cortical areas became separate and distinct.

5.3.2 Ontogeny

The early development of the first and higher order relays and their cortical areas was reviewed briefly in Sherman and Guillery (2006) and more fully in

Guillery (2005). The evidence shows that cortical areas receiving inputs from higher order relays mature later than those receiving first order inputs. It is clear that at birth the primary sensory areas and the motor cortex are significantly more mature than higher cortical areas in the rest of the cortex. Cortical maturation extends well into adolescence in the human and primate brain. Further, some of the mechanisms related to the formation of new connections in thalamus and cortex also survive for longer in rats during postnatal development in the higher order thalamocortical pathways than in the first order pathways (Feig, 2004; Feig, 2005). These observations relate closely to the functions that we discuss in more detail in chapter 6 concerning the role of the transthalamic connections that pass through higher order thalamic nuclei. These connections can serve to provide higher cortical areas information about the outputs that lower cortical areas are currently sending to subcortical motor centers. That is, the higher order cortical areas are connected so that they can monitor the motor outputs of lower cortical areas, and in terms of such a function, it would make sense for the monitors to develop their mature circuitry sometime after the circuits being monitored are relatively firmly established.

5.4   Role of Higher Order Relays in Corticocortical Communication

The connections and synaptic properties of higher order relays strongly support the hypothesis that they are a vital link in corticocortical communication. A great deal of anatomical and physiological evidence from various mammalian species, including rodents, carnivores, and primates, supports this hypothesis (see Guillery and Sherman, 2011; Sherman and Guillery, 2011). The differences in the functional roles of the direct and transthalamic corticocortical links are likely to be of importance and are considered in what follows. This involves not only comparing the possible roles of the two pathways but also, as a first step, determining the extent to which the function of either pathway is everywhere the same.

5.4.1   Direct versus Transthalamic Corticocortical Pathways

Until recently, ideas about corticocortical communication centered exclusively on direct connections, largely ignoring the role that indirect pathways might play (for example, Van Essen et al., 1992; Salin and Bullier, 1995; Wise et al., 1997; Lamme, 2003; Moore and Armstrong, 2003; Bond, 2004; Hilgetag and Kaiser, 2004; Womelsdorf et al., 2006). However, the functioning of these direct pathways is not at all clear, mainly because nearly all studies of these connections have been based entirely on neuroanatomical pathway tracing and have largely left unexplored the functional properties of the synapses involved. Given the hypothesis that these pathways, which are glutamatergic, can be classified as different functional types, that is, as Class 1 or Class 2 (or other), and that Class 1 and Class 2 can be regarded as indicative of driving or modulatory functions, respectively, which is supported by evidence from the mouse described in chapter 4 and below in this chapter, it will be important to define further the connections in terms of their functional types. Until we have such a functional analysis, we can say little more about the role in corticocortical communication played by the direct pathways.

5.4.1.1 Some Properties of the Direct Pathways It is important to be clear about the currently prevailing view of corticocortical processing. This asserts that most direct corticocortical connections can be divided into feedforward and feedback projections that reflect directions within a hierarchy of information processing, feedforward projections ascending the hierarchy and feedback, descending (reviewed by Felleman and Van Essen, 1991). Furthermore, the functional distinction suggested for these is that the feedforward direction represents the further processing of information, whereas the feedback direction is mostly modulatory.[3] This idea of feedback as modulation is very widespread in studies of cortical processing, involving actions such as processing of vision (Sandell and Schiller, 1982; Felleman and Van Essen, 1991; Friston et al., 1995; Wang et al., 2000; Wang et al., 2007a; Shushruth et al., 2009), somesthesis (Larkum et al., 2004; Shlosberg et al., 2006), hearing (Tang and Suga, 2008; van Atteveldt et al., 2009; Niwa et al., 2012), and attention (Lamme and Spekreijse, 2000; Martinez et al., 2001; Reynolds and Chelazzi, 2004; Simola et al., 2009). Even though this assumption appears to be so basic for studies of cortical processing, we have to question it on the basis of recent studies of corticocortical connections in the mouse.

Indeed, there is an enigmatic but striking difference in the laminar pattern of corticocortical connections described in monkeys on the basis of the anatomical connections and in mice on the basis of the physiological properties

---

3. Felleman and Van Essen (1991) provide a cogent description of how they view the functions of these modulatory, feedback corticocortical pathways. They state: "The massive descending pathways that are so prominent anatomically may subserve a different set of functions. One likely possibility is that descending connections contribute to a set of modulatory surround influences, in which stimuli well outside the classical receptive field can dramatically influence the responses to stimuli within the receptive field. Such modulatory effects have now been demonstrated in the analysis of motion (Allman et al., 1985; Saito et al., 1986), color (Zeki, 1983), form (Desimone and Schein, 1987), and texture (Van Essen et al., 1989). Another perspective is that descending pathways may contribute to the modulation of response properties by visual attention in area V4 (Moran and Desimone, 1985) and more generally, for dynamic control of the routing of information through each visual area (Anderson and Van Essen, 1987; Van Essen and Anderson, 1990)."

of the inputs (see chapter 4.1.5.3). If we assume that, in mice, projections from the primary visual or auditory cortical areas (V1 or A1] to the second areas, V2 or A2) are feedforward, and the reverse projections (V2 to V1 or A2 to A1) are feedback, then two issues are raised and summarized in figure 5.1 (data from Covic and Sherman, 2011; DePasquale and Sherman, 2011).

The first is that the laminar patterns for connectivity proposed for the monkey differ in the two directions: the laminar origin of the projections is similar, most arising from supragranular or infragranular layers, feedforward axons targeting mostly layer 4 and the feedback axons targeting mostly the upper or lower layers (figure 5.1, left column). In the mouse data, no pattern of feedforward or feedback connections matches this pattern, either in terms of the originating layers or of the target layers (figure 5.1), even when Class 1 and Class 2 projections are separately analyzed.

The second issue relates to the view described above that feedforward and feedback projections differ functionally, with feedforward involved mainly in information processing and feedback, in modulation. The mouse data do not show this difference, with Class 1 inputs suitable for information processing and Class 2 inputs unsuitable for information processing and probably acting as modulators amply represented in both the feedforward and feedback directions. Further, they show no significant difference in laminar patterns between directions, for either subset of projections, Class 1 or Class 2, although there are differences between the classes. For instance, in the mouse in either the feedforward or feedback direction, cells in layers 5a and 6 are almost exclusively the targets of Class 2 inputs (recall that layer 6 is the source of the Class 2 corticothalamic feedback), and cells in layer 5b are almost exclusively targeted by Class 1 inputs (and these layer 5b cells include the corticofugal cells that target motor structures in brainstem and spinal cord, some of which also provide Class 1 inputs to higher order thalamic nuclei via an axonal branch); cells in layers 2–4 have both kinds of input. Perhaps more importantly, the similarity of the projection patterns in the mouse for the feedforward and feedback directions provides no clear asymmetry that could provide for directionality in a hierarchy. Related to this, if we accept the hypothesis that Class 1 inputs represent the main information pathway and Class 2 are mostly modulatory, then the idea that feedforward should be more involved with projections mostly of information to be processed (that is, Class 1) and that feedback should be more involved with modulatory inputs (that is, Class 2), then the patterns of direct corticocortical projection in mouse cortex are not consistent with this widely accepted view of cortical organization. In fact, the patterns of corticocortical connections in the mouse suggest no special directionality in a putative hierarchy of information processing. However, as we

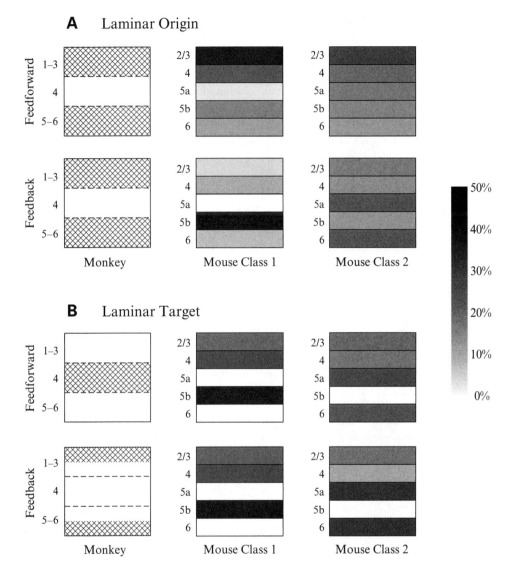

Figure 5.1
Comparison of laminar relationships between monkey and mouse in feedback versus feedforward corticocortical connections. The monkey pattern is redrawn from Felleman and Van Essen (1991), and the cross-hatching represents laminar location of origin or target of these connections without determination of quantitative extent; the mouse data are redrawn from Covic and Sherman (2011) and DePasquale and Sherman (2011). Connections in the mouse represent combined data, feedforward from V1 to V2 and A1 to A2, and feedback, from V2 to V1. The darkness of the layers represents the percentage of cells giving rise to a projection [(A) Laminar Origin] from that layer or the percentage of cells targeted by the input in that layer [(B) Laminar Target]; see scale to the right. The mouse data are separated into Class 1 and Class 2 projections.

point out below, this lack of directionality in the direct pathways may well depend on such directionality in the transthalamic pathways.

There are several explanations that might account for the apparent differences between the monkey and mouse data. First, as pointed out in chapter 4.1.5.2, whereas the anatomical accounts of laminar targets of inputs in the monkey data identify the layers in which terminals are found, the physiological approach used for the mouse data identifies the laminar locations of the post-synaptic cell bodies. That is, a pyramidal cell in layer 5 can receive inputs onto its apical dendrites in any of the layers above layer 5, and a good example of this difference is offered in a recent study (Mao et al., 2011). This could compromise the comparison of mouse and monkey data. The second possibility is that this discrepancy represents a species difference and that hierarchical organization of these early visual and auditory cortical areas in the mouse is simply not present. However, if this is the reason, then it places a significant obstacle in the path of using mouse data to understand cortical processing in other species, including human. A final possible explanation is technical: the monkey scheme is based on anatomy and the mouse, on physiology, and each may have different sampling artifacts that distort the final distributions.[4]

As noted, the pattern seen in the mouse suggests no clear hierarchical directionality in the direct corticocortical projections demonstrated so far for the transmission of information. However, if the thalamocortical projection patterns described in chapter 4.1.5.2 for the posterior medial nucleus hold for other higher order thalamic relays, the transthalamic projections do provide this directionality to cortical processing. That is, we can follow a Class 1, and thus presumably a driver or information route, from layer 5 of S1 to the posterior medial nucleus, and from there to S2. However, any sort of feedback route involving the posterior medial nucleus to S1 ends with the Class 2 (modulatory) thalamocortical projections to S1, and thus the routes between S1 and S2 have a clear directionality based on the Class 1 transthalamic pathway, where information travels from S1 to S2 and not in the reverse direction, whereas the direct connections between S1 and S2 or A1 and A2 lack this directionality.

Of course, we need to know the extent to which the patterns demonstrated for the transthalamic relays via the posterior medial nucleus generalize to other higher order relays, and recent data from monkeys suggests a parallel. This parallel is based on the view that the rodent posterior medial nucleus plays the

4. It is possible if the monkey anatomy were to include measures of terminal size for these corticocortical connections, given that Class 1 inputs produce larger terminals than do Class 2 (Covic and Sherman, 2011), the monkey and mouse laminar patterns might appear less different.

same role in processing somatosensory information, namely as a transthalamic corticocortical relay, as does the monkey pulvinar for visual processing. In the monkey, if the input from the lateral geniculate nucleus is eliminated, V1 cells become unresponsive to visual stimuli, as expected for a Class 1 input, but silencing the pulvinar only reduces responsiveness of V1 cells to visual stimuli, which would be expected for a Class 2 (modulatory) input (Purushothaman et al., 2012). This suggests the intriguing possibility that transthalamic pathways, in addition to other roles, are important for establishing the hierarchical order between cortical areas, a possibility that clearly needs further testing.

5.4.1.2 Some Differences between the Direct and Transthalamic Pathways   We have shown that at least two neuronal pathways exist to connect cortical areas: the direct ones and those that are relayed through higher order thalamus. This raises a series of questions, one of which is: What is different about these pathways besides the obvious routing difference? One possible answer is the nature of the messages each pathway carries, and a suggestion for this difference can be gleaned from another difference. That is, with few exceptions,[5] the axons involved in the direct connections do not have subcortical branches. In the mouse, there are virtually no cells in cortex that project both subcortically and to another cortical area (Petrof et al., 2012), so the messages carried by the direct corticocortical pathways represent information that stays within cortex. As discussed in chapter 6, the transthalamic route involves branching axons from layer 5 that also target subcortical sites, so the message in the transthalamic pathway is about messages being sent from cortex to lower levels that have significant capacities for contributing to the control of behavior. That is (see chapter 6), the transthalamic message is an efference copy of the command sent to the lower motor centers via the branching axon. This, in turn, implies that the transthalamic information is provided to inform higher cortical areas about motor commands sent out from lower areas. If so, then we suggest that a parsimonious explanation for the direct connections is that they may be involved primarily in basic information processing about the environment and that the transthalamic connections allow higher order cortical areas to monitor the outputs that lower areas are sending to motor centers, although the extent to which such a separation of action and perception does actually occur in the cortex remains an open question.

5. The Meynert cell is an interesting exception found in monkeys. These are large cells found in layer 6 of striate cortex that project both to area MT and subcortically, including to the superior colliculus, but it is not known if these cells also project to thalamus (Fries et al., 1985; Weisenhorn et al., 1995; Rockland and Knutson, 2001; Nhan and Callaway, 2012).

This idea can perhaps be best understood by an example from the somato-sensory pathways. The layer 5 projection from area S1 or barrel cortex in the rat or mouse involves one or more branches that innervate motor areas with a message to contribute to some form of movement, perhaps a whisker move-ment, and this same message is sent through another branch to the posterior medial nucleus where it is relayed to the higher cortical area, S2 (Veinante et al., 2000). This can be repeated up the cortical hierarchy (for example, S2 through the posterior medial nucleus to the tertiary somatosensory area). This provides each somatosensory cortical area with information about motor com-mands initiated by cortical areas lower in the hierarchy, and this is precisely what is needed to disambiguate self-induced motion (for example, an eye or whisker movement) from motion in the outside world. Meanwhile, the direct connections continue the analysis of the environment with the added informa-tion of motor commands initiated by lower areas in the hierarchy.

5.4.2 Significance of Routing the Indirect Pathway through Thalamus

One can ask: Why is the message in the transthalamic pathway routed through thalamus and not sent directly to the target cortical area? Figure 5.2 shows that a transthalamic pathway (figure 5.2A) is not an obvious requirement to project efference copies up the cortical hierarchy (figure 5.2B). The obvious conclusion is that there is some advantage to using the route through thalamus, where the form of the modulation or gating of this circuit considered in chapter 4 is available. This gating is not available in the direct pathway. Several such advantages described in the following sections can be proposed, among others, and these are not mutually exclusive.

5.4.2.1 Thalamic Gating As noted in chapter 3, the intrinsic circuitry of thalamus involves a combination of inputs to relay cells, including a variety of modulatory inputs as well as the drivers. The modulation of relay cells can take many forms, including affecting overall excitability and controlling volt-age-gated conductances, such as those that underlie the burst/tonic transition. Among the modulatory inputs are GABAergic afferents, mostly from local interneurons and cells of the thalamic reticular nucleus. These local GABAer-gic cells are themselves under the control of the same population of modula-tory afferents that innervate relay cells.

These GABAergic inputs are largely responsible for the gating properties of the thalamus. For instance, if the local GABAergic cells are very active, they will so strongly inhibit relay cells that the relay cells no longer fire action potentials in response to driver input. Under these conditions, the thalamic

A

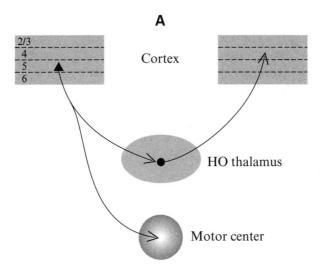

B

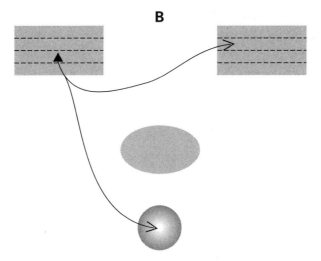

Figure 5.2
Schematic views of efference copies sent from one cortical area to another. (A) Example in which the efference copies are transmitted via a higher order thalamic relay. (B) Example in which the efference copies are transmitted directly.

gate is shut. Alternatively, if the local GABAergic cells are themselves inhibited, relay cells will respond vigorously to driver inputs, and the thalamic gate is open. Intermediate states can be imagined—the gate is variably ajar—when the level of activity among the GABAergic cells is intermediate.

A clear example of this process involves the cholinergic input from the parabrachial region to thalamus. These cholinergic inputs serve to inhibit the GABAergic interneurons and reticular cells via activation of M2 cholinergic receptors on these cells (McCormick, 1992). As a result, increasing activity of parabrachial neurons disinhibits relay cells, thereby opening the gate, and vice versa. These parabrachial cells become more active with increasing vigilance (Contreras and Steriade, 1995; Datta and Siwek, 2002), and more vigilance is associated with increased responsiveness of relay cells or more opening of the thalamic gate.

In addition, the layer 6 feedback to thalamus directly innervates relay cells and also affects them disynaptically through reticular cells (and also perhaps through interneurons), but the disynaptic inhibition is typically much stronger than the direct excitation (Lam and Sherman, 2010), and this could contribute to thalamic gating mechanisms. Furthermore, in some cases the detailed circuitry of the layer 6 feedback is arranged so that individual axons directly depolarize some relay cells and disynaptically hyperpolarize nearby ones, presumably via interneurons or reticular cells, allowing this pathway selectively to excite or inhibit different relay cells (Lam and Sherman, 2010). However, studies of these and other properties of local GABAergic circuits have effectively been limited to first order thalamic nuclei; we clearly need more information on these local circuits in higher order relays to begin to understand their possible role in gating transthalamic corticocortical circuits.

Finally, this gating of thalamic relay cells has another important consequence related to the burst/tonic firing modes of these cells (refer to chapter 2.1.3.2 for details). Sufficiently long hyperpolarization of these cells de-inactivates the T-type Ca^{2+} channels, placing the cells in burst mode. In this situation, whenever the gate opens, at least partially (for example, through activity of the layer 6 feedback or brainstem modulatory afferents), and a sufficiently strong new driver input depolarizes the relay cells, a burst will be evoked. As noted in chapter 2.1.3.2, the burst of action potentials represents a highly nonlinear response mode, but also one that will very strongly excite cortex, and so such a burst has been called a "wake-up call" for cortex (Sherman, 2001). Continued activity in the driver inputs will lead to a rapid switch in firing mode to tonic in the relay cells, and the continued message to cortex will more faithfully represent the message sent by the drivers because tonic firing lacks the nonlinear distortion of the stimulus representation seen in burst firing. Thus, opening a closed thalamic gate often provides an initially strong nonlinear signal to cortex followed by a linear message.

5.4.2.2   Gating of the Transthalamic Message by GABAergic Inputs   So far, we have concentrated on how local GABAergic inputs to relay cells serve a gating function. However, in addition, there are extrinsic GABAergic inputs to relay cells that can also contribute to this gating, and, as with the function of the local inputs, the thalamic gate can be shut if the external GABAergic inputs are sufficiently active, and vice versa. As noted in section 5.2.2, these external GABAergic inputs, from the zona incerta, pretectal region, and basal ganglia, fairly selectively target higher order thalamic nuclei and not first order.[6] This pattern of connectivity raises the likelihood that, under certain conditions, higher order thalamic relays are under specific modulatory control, and especially in the form of inhibition that would shut the higher order thalamic gate; the specificity of the connections means that this global control would not much affect first order relays.

It is instructive to take as an example the zona incerta projections and what we might infer in a frankly speculative manner from scattered, indirect evidence concerning the details of the innervation of thalamus and zona incerta from cortex. If these incertal cells are active, then their targets in the higher order thalamic relay nuclei will be profoundly hyperpolarized and inhibited (Trageser and Keller, 2004; Lavallée et al., 2005). This will shut down transthalamic circuits. However, there is no equivalent inhibitory input working to shut down first order relays, and so these remain generally open, although first order relay cells still receive other modulatory inputs (for example, local GABAergic and cholinergic, noradrenergic, etc., brainstem inputs) that can have more local, topographic gating effects. Thus, new information from the environment can still reach primary cortical areas, while transthalamic connections from these areas to areas higher up the cortical hierarchy are blocked. Under these conditions, there remains the possibility that direct corticocortical pathways enable higher cortical areas to monitor at least some aspects of activity at lower levels, although our profound ignorance regarding the nature of information relayed in the direct corticocortical pathways limit what can reasonably be proposed for this at present.

This state of affairs might exist when cortex is not actively controlling behavior, for example, for routine behaviors such as walking along a familiar

6. In addition to these GABAergic inputs, some higher order relay cells are hyperpolarized by cholinergic or serotonergic inputs (see section 5.2.4), but these represent only a minority (~15% to 20%) of higher order relay cells.

path, chewing gum, or calmly resting, leaving the control to lower motor centers in the brainstem and spinal cord (see also chapter 6.3), which also implies that layer 5 corticofugal cells in all the relevant cortical areas are not very active. Under such conditions, the zona incerta, whose cells can be silenced by activity in layer 5 corticofugal cells (Urbain and Deschênes, 2007), would be active and completely shut the gates of higher order thalamic relays with widespread and powerful inhibition (Barthó et al., 2002; Trageser and Keller, 2004; Bokor et al., 2005; Lavallée et al., 2005; Masri et al., 2006; Trageser et al., 2006). However, the first order thalamic relays, which are not significantly innervated by the zona incerta, would continue to relay information from the periphery to cortex. This allows cortex to monitor the environment continuously, and because direct corticocortical connections are not shut down by the zona incerta, passive monitoring of the environment can proceed up the hierarchy.[7] Under these conditions, an unexpected event in the environment can be recognized and, if important or threatening enough, can lead to cortical control of behavior. One way in which this could happen is as follows.

The unexpected or threatening event will lead to new firing of layer 5 corticofugal cells, starting with those in the primary cortical area and working its way up the hierarchy. Just how this occurs is not clear, but there are two relevant points that could play a role in this process. First, all cells in layer 5b, which include the corticofugal cells in question, receive direct corticocortical feedforward and feedback projections that are almost exclusively driver or Class 1 (Covic and Sherman, 2011; DePasquale and Sherman, 2011). Thus, if the new event leads to any new activity in the cortex, this will also include the layer 5 corticofugal cells in some or all cortical areas. Second, recall from chapter 2.1.3.2 that some first order thalamic relay cells are likely to be in burst mode. Evidence from receptive field studies of lateral geniculate cells in cats suggests that this happens when a dark (or bright) stimulus falls onto the receptive field of an on (or off) center geniculate cell, inhibits that cell, and the resultant hyperpolarization then de-inactivates the T-type Ca^{2+} channels needed for bursting; if then an excitatory stimulus suddenly replaces the inhibitory one (that is, a novel stimulus appears), a burst will be evoked (Lesica and Stanley, 2004; Alitto et al., 2005; Sherman and Guillery, 2006; Wang et al., 2007b). Such a burst will very strongly activate cortex, including the layer 5

---

7. It is interesting in this context that the cells that provide corticocortical projections and those that project subcortically from layers 5 and 6 are separate populations (Petrof et al., 2012), suggesting the possibility that direct corticocortical communication is not linked to activity in the layer 5 corticofugal cells.

output (Swadlow and Gusev, 2001; Swadlow et al., 2002), beginning the process described above.

In the scenario just presented, layer 5 cells from one or more cortical areas serve to inhibit the zona incerta and thus disinhibit higher order relays (Urbain and Deschênes, 2007), opening these thalamic gates, and they also provide a new input to these relay cells. Whether this reflects individual layer 5 cells that branch to innervate the zona incerta, brainstem motor centers, and higher order thalamic relay cells or whether different layer 5 cell populations are involved cannot be predicted on the basis of current knowledge about the branching patterns of the axons of these cells.

As noted in chapter 2.2, other GABAergic inputs from the globus pallidus, substantia nigra, and the anterior pretectal nucleus fairly selectively target higher order relays, and perhaps these inputs operate in a similar fashion to control gating of transthalamic information.

The main point of trying to put together an account of what may be happening is not to offer a clear explanation of events that we understand, but rather to stress that the pathways we know about have the capacity to produce a disconnection, allowing the first order cortical area to activate lower motor centers while blocking a copy of that motor instruction on its way through the thalamus to higher cortical areas. That is, the thalamic gate blocks the ability of higher cortical areas to monitor the motor outputs of lower areas, while still allowing some information flow between cortical areas. Here our ignorance about the nature of the information passed through the direct corticocortical channels is a crucial limit to what we can understand. However, it has to be recognized that there are so many currently unexplored relationships involved that for a detailed comprehension of what happens during particular behavioral situations, it will be necessary to explore several such situations, defining the conditions that change the properties of the relevant thalamic gates, exploring which of the direct corticocortical connections are active, and which of the active ones are Class 1 and which Class 2, and what the layer 5 cells are doing.

5.4.2.3 Other Modulatory Effects on the Burst/Tonic Firing Modes

As noted in chapter 2.1.2.3, relay cells throughout the thalamus, including higher order relays, can switch firing mode between burst and tonic, but bursting is more common in higher order relays (Ramcharan et al., 2005). There appear to be two possible reasons for this greater tendency to burst in higher order relays. One is the extra GABAergic inputs received by the relay cells just noted. Second, modulatory cholinergic and serotonergic inputs, which act to depolarize all first order relay cells, serve to hyperpolarize 15% to 20% of higher order relay cells by operating through appropriate metabotropic receptors, and this could contribute to more common bursting for these thalamic cells (Varela and Sherman, 2007, 2008).

5.4.2.4   Conjoint Activation of Parallel Direct and Transthalamic Corticocortical Pathways   In the few examples for which information is available, it appears that cortical areas connected directly to each other also have a transthalamic connection through a higher order thalamic relay organized in parallel. In the mouse and rat, this appears to be the case for connections between each of the primary visual, auditory, and somatosensory cortices with their secondary counterparts (Li et al., 2003; Lee and Sherman, 2008; Theyel et al., 2010b). In the cat and monkey, there is limited, indirect evidence for such a parallel organization in the visual pathways based mainly on the anatomical evidence that direct connections exist between visual cortical areas and between these areas and the lateral posterior-pulvinar complex and also that, in the monkey, lesions of the striate cortex silence pulvinar neurons (Bender, 1983; Chalupa, 1991). Whether this arrangement—that direct connections are paralleled by transthalamic ones—is typical or not requires evidence from more examples, but so far it has been demonstrated for at least some of these connections.

This pattern of parallel direct and transcortical links between cortical areas may have significance for the ability of different groups of cortical areas to form functional units as required by differing behavioral demands. For instance, several physiological studies in behaving monkeys have recently focused on situations where different cortical areas become functionally linked during particular cognitive tasks, and although the details of the relationships between cognitive needs and the corticocortical linking remain undefined, ideas about the underlying circuitry have focused on direct connections between cortical areas and have largely ignored thalamus as playing a role in this linking (Reynolds et al., 1999; Womelsdorf et al., 2006; Pesaran et al., 2008; Andersen and Cui, 2009; Fries, 2009; Gregoriou et al., 2009). If these areas have both direct and transthalamic connections, such a pattern may provide a basis for a form of AND gate or device for coincidence detection. That is, conjoint activation of both pathways, which would occur when the transthalamic gate is open, could lead to supralinear summation of inputs, producing especially strong activation of the target area, and such activation could lead to linking of the two cortical areas. In contrast, closing of the thalamic gate means that only the direct connection is available to activate the target area, which could produce a response that is too weak to support such linking. Thus, the thalamus may play an important role in this cortical process, including the selection of which among possible linkages are functionally connected in this manner, although details concerning controls of the thalamic gate and how conjoint activation of both the direct and transthalamic pathways leads to linking remain to be determined.

5.5   Concluding Remarks

The recognition of the distinction between first and higher order thalamic relays has great significance for the understanding of cortical functions. This can be seen from several different perspectives. One is the conclusion that higher order relays serve as an essential parallel link in corticocortical communication. Another is the point that these transthalamic pathways provide an opportunity for the gating and modulation that is present in the thalamus but unavailable to the direct pathways. Furthermore, the transthalamic pathway sends information to cortex about instructions for forthcoming movements, and the extent to which comparable messages may be present in the direct pathways is currently unknown. These connectivity patterns raise questions about the different information content of the two pathways and whether there may be reasons other than those outlined in this chapter for having one set of pathways connecting cortical areas directly to each other and another that involves the thalamic gate. Finally, it is probable that many connections between cortical areas involve parallel pathways, direct and transthalamic, which suggests the possibility that different functional purposes are served when only one or the other is active or when both are.

Generally, the cell and circuit properties of the higher order relays are quite similar to first order relays beyond the obvious difference in the source of driver input. But there are some subtle differences noted that suggest more modulation and gating functions taking place in the higher order relays: these have a higher ratio of modulatory versus driver synapses, and certain modulatory inputs, especially GABAergic, selectively target higher order relays.

In the final analysis, thalamus should no longer be seen as merely a means of relaying peripheral information to cortex. It now should be seen as incontrovertible that thalamus in the form of its higher order relays plays a critical role in ongoing cortical functioning in relation not only to the inputs that reach cortex but also in relation to the cortical outputs.

5.6   Outstanding Questions

1. Are there any pure higher order thalamic nuclei? Or are all higher order relays mingled with first order pathways?

2. Where the two types of relay, first and higher order, pass through a single nucleus, do they do so as parallel, independent pathways, comparable to the

X and Y pathways of the cat visual system, or is there integration within the thalamic relay?

3. For any one cortical area, can the nature of the message carried in the inputs from the direct and transthalamic pathways be compared to show how the transthalamic differs from the direct pathway?

4. How common is the pattern of two cortical areas connected in parallel by both direct and transthalamic pathways? Are some connected only by one or the other?

5. In simple mammalian brains that have few distinct sensory cortical areas and that can be regarded as representing an early stage of cortical evolution, are the feedback connections of the layer 6 cortical cells distinguishable from the feedforward connections of the layer 5 cells? That is, what do the connections reveal about the early, phylogenetic development of first and higher order pathways?

6. Given that some layer 5 corticofugal axons that innervate subcortical motor centers have branches that innervate higher order relays and others lack these branches, what are the functional differences between these two types of layer 5 cells? Do they send different messages to the lower motor centers?

6   The Dual Nature of the Thalamic Input to Cortex

6.1   A Brief View of the Phylogenetic Origins of Thalamocortical Inputs

The thalamocortical pathways, whether they are the highly complex and elaborate structures that characterize the primate brain or the simpler paired sensory cortical areas seen in representatives of early mammals and discussed in chapter 5.3.1, all depend for their inputs and outputs on the connections they originally established with the phylogenetically older circuits of the brainstem and spinal cord. That is, as the thalamocortical pathways evolved, these older circuits sent branches to the newly evolving thalamus, and the thalamus relayed this information to the cerebral cortex so that the cortex could monitor what the phylogenetically old circuits were doing in response to the inputs from the body and the world. In addition, the cortex could send its outputs back to the brainstem and spinal circuits and modify the activity of the older circuits. The cortex depends on these older structures for communication with the body and the environment, and all that the cortex can do depends on these links.

To understand the nature of these links with the phylogenetically older brainstem and spinal parts of the brain, we first look briefly at how the afferent and efferent mechanisms of the earliest mobile organisms related to each other and then at how these relate in early vertebrates to the mechanisms that feed into the first order thalamocortical relays. This will help us to find some clues that show how a shared functional pattern that characterizes all thalamocortical relays, both first order and higher order, may have developed during evolution. We can start very briefly with a very simple organism, such as a bacterium that first develops a receptor organ. The receptor will have very limited survival value if it is not linked to a motor that can move the organism in response to the input. The common occurrence of receptors on or closely related to mobile cilia reflects this relationship, which has survived for many of our receptors, including visual, auditory, vestibular, olfactory, and gustatory receptors, all related closely to cilia, although many of these cilia have lost their

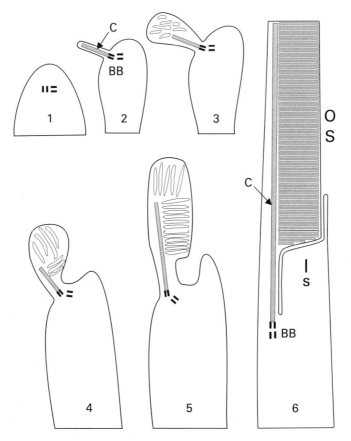

Figure 6.1
The development of a retinal rod to show the developmental relationship of the stacks of membranes that bear the receptor molecules (in the outer segment of the rod) to the early forming
cilium. BB, basal body from which the cilium develops; C, early forming cilium; IS, inner
segment; OS, outer segment of the rod. Numbers indicate the developmental sequence. Based on
a review and illustrations by Sung and Chuang (2010).

mobility. Figure 6.1 shows the development of a retinal rod in close relationship to a ciliary structure. The receptor molecules here are distributed on
membranous disks that become organized in close relation to the cilium, which
has lost its own mobility but plays a role in moving the membranous disks
from the base to the apex as they are replaced on a regular basis.

Cilia are useful for moving small organisms but are not suitable for moving
large creatures, and as larger organisms leading to the vertebrates evolved, so
neural connections, muscles, and bones replaced the link provided by a mobile
cilium. In vertebrates, the receptors are linked through the incoming axons to

the spinal and brainstem mechanisms, and these control movements through the outgoing axons of the spinal and brainstem motor neurons. With the exception of the olfactory and visual receptors, all of the information reaching the vertebrate brain from the environment and from the body is brought through the dorsal roots of the spinal cord or through their equivalents in the third to twelfth cranial nerves. This very sketchy account takes us to an essential relationship between the spinal segmental inputs and the motor machinery.

The spinal and brainstem mechanisms are not merely a collection of spinal reflexes that need to be integrated by higher centers before they can produce adequate movements and appropriate responses to complex inputs. Sherrington (1906) showed that a dog whose spinal cord has been separated from the brain retains a well-executed scratch reflex that is accurately aimed at a stimulus designed to mimic a flea crawling along the dog's flank. More tellingly, as mentioned in chapter 1, a number of detailed, more recent studies (Grillner, 2003; Grillner et al., 2007; Rossignol and Frigon, 2011) have shown that a cat whose spinal cord has been disconnected from all higher brain centers can walk on a mechanical walkway, changing its gait from a walk to a trot or a gallop as the speed of the walkway is changed, and can avoid objects placed in its path. The spinal mechanism that controls walking in this way has been called a central pattern generator. The rhythmic movements that it generates are responsive to the inputs that arise in the muscles and joints and also to the inputs from the walkway. These input messages travel in the dorsal roots of the spinal nerves and can be regarded not only as providing feedback information for the control of the movements but also as providing information about changes in the environment (for example, the speed of the walkway or an obstacle in the path of the animal's progress along the walkway) that can modify the actions of the central pattern generator.

These experiments have shown that a spinal central pattern generator, acting as an independent mechanism on its own, can generate the rhythmic discharges that produce the walking movements even when the spinal cord is isolated from higher centers and the dorsal root inputs have been cut, but that the detailed control of the circuits relating to the frequency of the rhythms and to the speed of the walkway depend on the input from afferent nerves in the dorsal roots and on descending inputs. These inputs provide an important part of the motor control that the central pattern generator exerts on the muscles of the limbs, with the dorsal root inputs representing the dynamic properties of the machinery that produces the movements (the properties of the muscles, joints, and bones that are moved) as well as the properties of the environment that may be changing as movements are executed; that is, they bring information about the body *and* the environment, and in that sense can be said to represent both, and this information plays an important role in the spinal control of movements.

The central pattern generator that controls walking movements in a cat represents a phylogenetically very old pattern of neural connections, remarkably similar to the pattern of spinal circuits that controls swimming movements in a primitive vertebrate, such as the lamprey (Grillner et al., 2007). Not only are the patterns of the connections similar, but also the transmitters and sensory receptors that control the activity of the central pattern generators in cats and lampreys are basically the same. Further, in both lampreys and cats, these pattern generators depend for their adequate functioning on the inputs from the dorsal roots and also depend on descending controls from higher centers. In lampreys, these controls come primarily from the brainstem, including the midbrain tectum and the vestibular nuclei, and in the cat, there are higher inputs coming from the basal ganglia, the superior colliculus, the cerebellum through the brainstem, and the cerebral cortex.

The central pattern generators in the spinal cord that control swimming, walking, or running represent examples of several different pattern generators present in the brainstem and spinal cord concerned with the control of basic actions like breathing, feeding, chewing, swallowing, scratching or defecating (figure 6.2). Each can produce a well-coordinated and efficient action but is also under higher control, and for each, the inputs from the peripheral nerves provide feedback information about the ongoing execution of the movements and new information about relevant changing aspects of the environment. The ascending messages from these pattern generators on the way to the thalamus inform higher centers about the ways in which the incoming sensory messages are producing movements. That is, the pattern generators and their inputs are, as we will see, sending messages to higher centers, including the thalamus and cortex, and are receiving descending motor instructions either directly or through the basal ganglia, superior colliculus, cerebellum, and brainstem.

6.2 Driver/Class 1 Afferents to the Thalamus Are Branching Axons

6.2.1 The Afferents to the First Order Relays

To understand how the inputs that pass from the periphery to the first order thalamic relays relate to the functions served by these relays and by their cortical targets, it is necessary to look closely at the connections that are established by the incoming axons on their way to the thalamus. In the following sections, we explore the prethalamic branching patterns of these axons, which we have already described in more detail earlier [Guillery and Sherman (2002, 2011); and see Lu and Willis Jr. (1999) for the spinal afferents]. We are concerned in

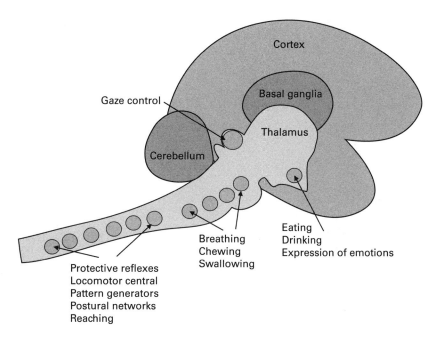

Figure 6.2
Schematic representation of some lower motor mechanisms in the hypothalamus, brainstem, and spinal cord. Based on details in Grillner (2003).

part with recording here enough of what is known about the branching patterns of the incoming axons to demonstrate that such branches play a significant role in linking action and perception, but in addition, and more importantly, we explore the functional implications of these branching patterns, including some earlier functional interpretations and misinterpretations of and disagreements about the significance of the known connections and branching patterns. These functional interpretations are crucial for understanding some of the issues that arise when one is looking at what must be regarded as the dual, sensorimotor functions of the axons that provide the driver[1] (Class 1) inputs to the thalamus.

6.2.1.1 The Somatosensory Pathways: Their Dual Function: Was Romberg Wrong?   Figure 6.3 shows the rich branching patterns of the incoming dorsal root axons as they enter the spinal cord. In this figure, Cajal (1911) shows that each dorsal root axon bifurcates to form an ascending and a descending branch, and these branches send many inputs to the local spinal circuitry, contributing

1. We define what is meant by a driver in the second paragraph of chapter 1 and in section 1.3 of that chapter discuss drivers and Class 1 inputs; we discuss this more fully in chapter 4.

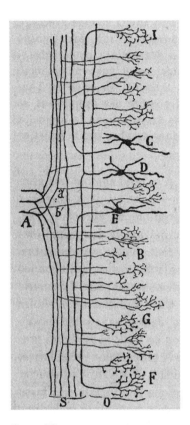

Figure 6.3
Drawing of dorsal root axons (labeled A) entering the spinal cord on the left and bifurcating to form ascending and descending branches in the spinal cord each of which gives off several local branches to supply spinal mechanisms. Many of the ascending branches continue through the spinal cord to the caudal medulla where they terminate in the gracile or cuneate nuclei, which in turn send axons to the ventral posterolateral nucleus of the thalamus. a,b, branches of dorsal root fibers sending terminal arbors like those labeled B,G, F and I into the spinal gray; S, descending axons in the white matter of the spinal cord; O, gray matter of the spinal cord. From Cajal (1911).

to the central pattern generators at spinal levels. Further, for many of the dorsal root axons, the ascending branch establishes a first link in the ascending pathway to higher control centers, such as the cerebellum and the tectum, and, in mammals, to the thalamus and cortex. In mammals, these ascending branches run in the posterior columns of the spinal cord, relay in one of the two brainstem nuclei at the head of the posterior columns, the gracile and cuneate nuclei, and then, after crossing the midline at the caudal levels of the brainstem, they ascend to the thalamus, where they enter the ventral posterior nucleus. Cajal (1911) has pointed out that very few of the incoming dorsal root axons lack a local branch, and has added the comment that he has only found two such axons that lack the bifurcation in the spinal cord of chickens. He also cites evidence that this branching pattern is found in vertebrates generally, including fish, amphibians, reptiles, and mammals.

Figure 6.4, another figure from Cajal (1911), shows incoming axons that carry a different type of information (pain and temperature) in the trigeminal nerve from the skin, mucous membranes, and the muscles of the head. These axons are shown in the figure descending to a group of cells that lie in the lower brainstem, in the midportions of the spinal nucleus of the trigeminal nerve. This is a part of the first relay for the incoming afferent fibers. They synapse with local cells as shown in the figure and relay through a crossed pathway that passes to the ventral posterior medial thalamic nucleus of the other side; only the first, brainstem part of this pathway to the thalamus is shown in the figure. Cajal (1911) shows these axons, before they pass to the thalamus, giving off branches that go to three important motor nuclei of the brainstem, those of the fifth, seventh, and twelfth cranial nerves. These nerves play a significant role (among other functions) in chewing and swallowing and also in the control of the rhythmic whisking movements by means of which many rodent species explore their environment. That is, these axons of the trigeminothalamic pathways also contribute branches to important lower motor centers before they continue to their thalamic termination for relay to the sensory cortex, and it is the relationship between the functions of the branches to thalamus and those to the pattern generators that we need to explore.[2]

2. It may be worth noting that the connections between the intermediate portion of the spinal nucleus of the trigeminal nerve, illustrated by Cajal (1911) in figure 6.4, link, through the branched axons shown in the figure, to the seventh (facial) cranial nerve, which also plays a major role in controlling the whisking movements by means of which rats and mice explore their environments, and that this part of the spinal trigeminal nerve also links to the region of the thalamus that receives inputs from the whiskers. The neural circuitry that controls whisking is not understood in the detail in which the spinal mechanisms concerned with walking, running, or swimming have been defined, but it may be instructive to see this trigeminal afferent as linked to a mechanism that produces the exploratory movements of the whiskers, thus forming another pattern generator.

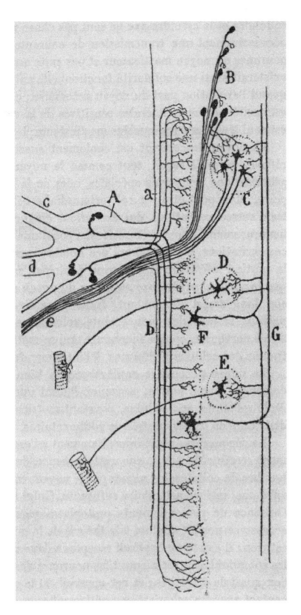

Figure 6.4
Drawing of trigeminal axons entering the brainstem on the left from the trigeminal ganglion (labeled A) and bifurcating to form ascending branches to the sensory ganglion of the trigeminal nerve and descending branches to the spinal or descending root of the trigeminal nerve. Two nerve cells in the descending root (the upper one labeled F) send their axons to the right (contralaterally) and form an ascending and a descending branch. The ascending branches, which will continue to the ventral posteromedial thalamic nucleus, each sends branches into three motor nuclei of the brainstem: C, the trigeminal motor nucleus for the muscles of mastication; D, the facial nucleus for the muscles of facial expression and, in rodents, the muscles that move the whiskers; E, the glossopharyngeal nucleus for the muscles of the pharynx concerned with eating, swallowing, and speaking. a, ascending branch of the sensory trigeminal nerve; B, mesencephalic root of the trigeminal nerve; b, descending fibers of the sensory trigeminal nerve;, c,d,e, ophthalmic, maxillary and mandibular divisions of the trigeminal nerve; C, motor nucleus of the trigeminal nerve; D, Facial nerve nucleus; E, glossopharyngeal nucleus; F, nerve cell in the spinal nucleus of the tri-geminal nerve; G, ascending trigeminothalamic fibers. From Cajal (1911).

For both of these pathways to the thalamus, the posterior column pathway and the trigeminothalamic one, the terminations of the ascending axons in the thalamus have been demonstrated to have the characteristic features that we described in chapter 2 for the driver axons. That is, these are large terminals, many of which lie in glomerular structures that are free of astrocytic processes and form many synaptic junctions with interneuronal and relay cell dendrites, often in the form of triads (Jones and Powell, 1969; Ralston III, 1969; Peschanski et al., 1984); also they are glutamatergic (Salt, 1987; Salt and Eaton, 1989), and the synapses they form on relay cells activate only ionotropic and not metabotropic postsynaptic receptors (as explained for the Class 1 inputs described in chapter 4). That is, they have Class 1 properties and are known to relay information from the periphery to the cortex.

Cajal (1911) recognized that the pattern of axonal branching that he illustrated in these and in many other figures was telling something important about how these afferent fibers might function to distribute the same information to many different sites. He suggested that, "At any point along an axon and its collaterals, the amount of energy associated with a particular input is proportional to the diameter of the process in question." That is, writing many years before the electrical conduction of the action potential had been demonstrated, he thought of axonal conduction on the basis of a hydraulic model. He considered that most of the axonal action would be through the many small local, spinal branches and that only the single long axonal branch ascending to the thalamus would be relaying "that part of the activity giving rise to conscious sensations." Today, we have to reinterpret his insight on the basis of what is known about the conduction of action potentials along branching axons.

It has now been clearly established that an action potential, when it reaches a branch point, generally passes down both daughter branches (Cox et al., 2000; Raastad and Shepherd, 2003), and the temporal pattern of the impulses is essentially the same in both branches. There may be branch point failures, but such failures are not common, and if they occur, at a constant temperature they will occur consistently. That is, because for each incoming dorsal root axon the pattern of axonal impulses that reaches the thalamus bears a constant relationship to the patterns of impulse going to the spinal mechanisms, the thalamic relay cell is receiving information not only about the nature of the stimulus acting on the peripheral receptor that innervates that dorsal root axon but also about the nature of the information concurrently being sent to the spinal mechanisms. If we now think of the spinal circuitry producing a walking rhythm or a scratch reflex, then it will be evident that the thalamic relay cell is sending information to cortex about instructions that relate to the production of movements *but that these instructions have not yet been carried out* and may not be carried out if unexpected circumstances arise.

This dual role of the dorsal root input in the control of walking and of postural adjustments, on the one hand, and in perceptual processing, on the other, is well illustrated by an example that appears in almost any textbook of neurology or neuroscience and is instructive for exploring how the relationships between sensory and motor functions have been treated in the past: tabes dorsalis is a clinical condition that involves severe degenerative changes of the dorsal root ganglion cells, and these lead to a major loss of axons in the posterior columns that, in turn, produces a loss of the sensory inputs from the limbs. This includes a loss of the sense of position (kinesthesis) and touch so that there is no sense of where the limb is if it cannot be seen. Romberg, a German neurologist who wrote what was essentially the first textbook of neurology in the 1840s, described a test, Romberg's test, that can be used to check for damage of the dorsal root input and is also commonly used by some police forces to test for inebriation (Lanska and Goetz, 2000; Khasnis and Gokula, 2003). It has the patient (or suspected drunk) stand with legs together and then close both eyes. The test is positive if the patient or suspect starts to show a tendency to fall, moving from side to side and eventually losing balance if not prevented by assistance.

Spillane (1981), in his history of neurology, notes that the loss of nerve fibers in the posterior columns had been recorded before Romberg's account and adds that a modern reader (knowing this) will have the feeling "that it was not, even then, necessary to push the classification he (Romberg) adopted to such an absurd degree, one which eventually led Romberg to consider tabes dorsalis as a motor disability, despite his description of its essential element—a sensory ataxia." The ataxia, that is, the loss of coordinated movements, has long been associated not only with the sensory losses reported by the patient, but also with the loss of the axons that ascend to the thalamus through the posterior columns of the spinal cord, a loss that can be clearly seen in postmortem studies. The affected axons are relatively thick and closely grouped together in the posterior columns, and as Brodal (1981) has pointed out, their loss is clearly evident in postmortem sections through the spinal cord. Brodal (1981) remarks that this is in contrast to the loss of the more scattered branches to the spinal cells, a loss that is far less evident in the postmortem sections, because these axons lie among many other axons that innervate the same regions and do not degenerate. The idea that tabes dorsalis demonstrates a "sensory ataxia" thus relates to both the sensory loss reported by the patient and to the loss of the fibers in the posterior columns that carry messages to the thalamus and cortex. It is not uncommon for the primary loss accounting for both this condition and the positive Romberg sign to be regarded as a posterior column loss. However, as Brodal (1981) points out, a posterior column lesion by itself, sparing the dorsal root ganglia, does not produce a positive Romberg sign. Lassek (1954) had shown that the motor deficits produced by a posterior column lesion are much less severe than those produced by loss of the dorsal root ganglia, and Brodal (1981) cites more recent clinical and experimental evidence supporting this observation. The interruption of the ascending pathway to the thalamus by itself is either not relevant for the production of the characteristic motor losses or else is not sufficient. It is reasonable to question Spillane's view of the condition as a *sensory* ataxia and to regard Romberg's views of the motor losses more positively.

The important point about this condition and the disagreements that its study has produced is that there are two separate functions, one served by each of the two branches of the inputs of the dorsal root axons to the spinal cord. One branch, ascending in the posterior columns, carries messages to the thalamus and cortex, providing the information that is needed for perceptions about tactile inputs and about the position and movements of the limbs. The loss of sensation in the limbs in patients with tabes dorsalis is clearly linked to the loss of axons in the posterior columns. The other branch, the one that innervates the spinal pattern generators is not, strictly speaking, "sensory" at all from the point of view of the patient or the clinician. It provides the same afferent information to the central pattern generators of the spinal cord, serving both as a feedback in the control of the movements and also providing information about changes in the environment. Its inputs to the cord act to modify the movements that are produced by the central pattern generator. It is a motor axon in terms of its spinal actions, and it does not provide information that can lead to sensory perceptions. These rely on the posterior columns as indicated by the posterior column lesions described by Lassek (1954) and Brodal (1981). However, because the two messages are essentially the same, each terminal zone receives the same message and in a sense "knows" the instructions that are being sent to the other center. The messages that go to the thalamus are copies of the instructions that are on their way to the spinal pattern generators. That is, they are "efference copies" in a sense that has played an important role in theoretical and experimental accounts of how accurate movements can be produced by brains or robots (Sperry, 1950; von Holst and Mittelstaedt, 1950; Wolpert and Miall, 1996; Sommer and Wurtz, 2008). By receiving a copy of an instruction that is currently on the way to motor centers, the thalamus and through it the cortex gets a preview of what is likely to happen next in the motor apparatus. We discuss this feature in more detail in section 6.3 and stress here that the patterns of branching described in the following sections indicate that essentially all thalamic nuclei are relaying such efference copies to the cortex. An important distinction is between efference copies that come to the thalamus from cortex and efference copies that reach the thalamus from a subcortical origin. In later discussion, we will distinguish them as cortical or subcortical efference copies. Our focus on the thalamic relay of efference copies should not be taken to imply that efference copies are limited to the thalamocortical pathways.

Although the sensory and motor functions of the two parts of the dorsal roots are clearly evident to the patient and the neurologist, we have seen that the agreement on the distinction is not easy. Indeed, the physiology of somato-sensory perception is often presented to students in quite separate lectures from those that deal with the spinal control of movements, and the two subjects are commonly treated in different chapters in a textbook, although they deal with messages that enter the nervous system through the same dorsal root axons. The important issue here is that the single message entering through the dorsal roots produces two distinct but related outcomes in the central nervous system.

The extent to which it has proved possible to read the messages that pass along any neural pathway and to interpret the meaning of any one message has made significant advances during the past century. It has been particularly powerful in the sensory pathways, demonstrating how particular patterns of axonal action potentials relate to particular environmental actions on sensory receptors and thus to particular sensory experiences. That is, the experimental neuroscientist recording from one or a few cells has been able to start to get a view of the perceptual experiences that these neural events are generating. The point about the branched axons is that either the perceptual experiences are more complex than neuroscientists have believed them to be (and in chapter 9 we will consider views that, indeed, our perceptions are not the pure sensory events that are commonly represented) or that there are other correlates of the recorded neural events that yet remain to be explored—probably both.

In the 1920s, Adrian (1928) developed techniques that allowed him to record the action potentials passing into the spinal cord from the dorsal root axons, and he was able to show that the spatial and temporal distribution of the action potentials passing to the spinal cord served as a code for the nature of the sensory experience produced by natural stimuli applied to the relevant receptors. Not only did the individual nerve fibers serve as a "labeled line" for a particular type of receptor, but the frequency of the impulses and their temporal distribution in relation to the onset and offset of a stimulus was characteristic for different types of sensory receptors and events and, thus, for different sensory stimuli. Adrian summarized many of his findings in a small and wonderfully lucid book in 1928 (Adrian, 1928). He wrote: "As far as the brain is concerned the function of the sense organs, or receptors, is to construct in it a map of certain physical events occurring at the surface of the body so as to show what is taking place in the world outside us." This is a neat expression of a view of the sensory nerves serving a function that can be described as "pure sensation." It has been an extremely powerful way of looking at sensory pathways for vision, hearing, taste, and olfaction as well as for the sense of touch and of limb position (kinesthesis) that is lost in the tabes patients. On this view, the brain is seen as a passive recipient of messages from an external world of solid objects and events and from the body itself. However, in the same book, Adrian also presented a deeper and more thoughtful view, based on the writings of his younger colleague, Craik (1945). Adrian wrote: "Thought models or parallels reality, its essential feature being not 'the mind,' 'the self'[3] but symbolism of the kind which is familiar to us in mechanical devices which aid thought and calculation, symbolism like that involved in the use of words and numbers set into patterns so as to correspond to a particular event. In this way the organism carries in its head not only a map of external events but a small scale model of external reality[4] *and of its own possible actions*: it is able to try out various alternatives, react to future situations before they arise, and utilize the knowledge of past events" [italics added]. We return to this view later in this chapter (section 6.3), but here need to recognize that the branching axons present us with a terminological problem that does not arise if we think in terms of pure sensations. It does provide a challenge if we consider the possibility of more complex relationships that include instructions for upcoming actions as a part of the perceptual process.

Although both branches of the entering dorsal root axons are carrying the same information, and although each branch can serve to produce movements, one rather directly at spinal levels and the other through thalamus and cortex and further connections, so far as the dorsal root ganglion losses are concerned, not only the patient but also the clinician report a clear distinction, which, as we have seen, they describe as "sensory" for the ascending pathway and "motor" for the spinal pathway. One reason for considering the disagreement

---

3. There is a puzzling use of words in this passage when it seems to question whether an essential feature of thought can be "the self." We argue later that when an organism is carrying in its head "a small scale model of external reality and of its own possible actions," insofar as this model includes some significant information about forthcoming actions, there exists the possibility, indeed the necessity, that this knowledge about forthcoming actions must be ascribed to an agent who will be performing the actions. Naturally, that agent must be the self; it would be insane to ascribe the information, no matter how remotely one is aware of it, to anyone other than the self.

4. Compare this with Clark (1998), who wrote more recently about "detailed world models" that provide a necessary knowledge base for robots.

raised by the structural and functional losses produced by these lesions has
been to focus on issues that are not easily resolved for anyone who tries to
think clearly about a brain in which many of the classical "sensory" or "motor"
pathways turn out to carry messages that represent features of the environment
(or the body) as well as messages that can serve as instructions for movement.
These are the sensorimotor contingencies (see O'Regan and Noë, 2001) that
govern our cognition and behavior, and we return to them later in this chapter
(section 6.3) and in chapter 9.

6.2.1.2 The Retinal Afferents   The driver axons that enter the lateral geniculate nucleus from the retina are, like those to the ventral posterior nucleus,
branches of axons that go to a center concerned with the production or control
of movements. Figure 6.5 shows the axons of the optic tract that are heading
toward the midbrain giving off branches to the lateral geniculate nucleus as

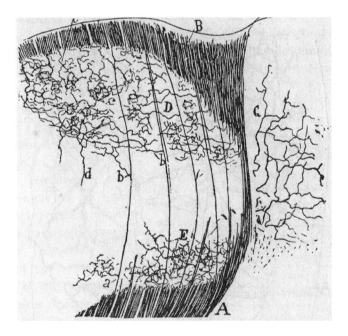

Figure 6.5
Drawing of optic nerve fibers of a mouse in a sagittal section of the thalamus. As the densely
packed fibers of the optic tract are passing toward the midbrain, they curve around the lateral
geniculate nucleus (they are shown cut at the top and the bottom of the figure) or pass through
it, giving off branches that innervate the lateral geniculate nucleus itself. From Cajal (1911).
   A, ventral part of optic tract; B, dorsal part of optic tract; b, branch leaving an optic tract axon
and passing to the (dorsal) lateral geniculate nucleus; C, medial geniculate nucleus; D. (dorsal)
lateral geniculate nucleus; d, axon entering dorsal lateral geniculate nucleus from an unidentified
source; E (ventral) lateral geniculate nucleus.

they pass around it and through it. In the midbrain, they innervate the superior colliculus and the pretectal nuclei. Evidence summarized earlier (Guillery and Sherman, 2002, 2011; Guillery, 2003), based on injections of retrograde tracer molecules into the visual pathways, on electrophysiological recordings, or most convincingly on orthograde tracing of individually labeled retinofugal axons, has demonstrated that for non-primate mammals, essentially all of the axons that innervate the lateral geniculate nucleus are branches of axons that travel to the superior colliculus or to the pretectal nuclei of the midbrain, the former concerned with control of head and eye movements, representing a control of gaze, and the latter concerned with pupillary control and accommodation (lens control).

For primates, the conclusion is likely to be the same but, whereas for two major, functionally distinct groups of retinogeniculate axons, terminating in the magnocellular and the koniocellular geniculate layers, the experimental evidence is the same as for non-primates; for the group of retinal cells that are generated earliest (the parvocellular component), the evidence is not adequate for an entirely definite conclusion. There is evidence that *some* axons of the parvocellular component branch, and, given the difficulty of the techniques that were used, the clear possibility for false negative results and the reasonable expectation that all parvocellular axons are likely to show the same axonal pathways, it is not unreasonable to conclude that primates may be like the other mammals that have been studied. However, better evidence would be useful, because if a "pure" thalamocortical pathway with no extrathalamic branches were revealed, this could have interesting functional implications. The important conclusion for us, here, is that the central visual pathways resemble the somatosensory pathways in having many branched axons each of which sends one branch, the classical "sensory" branch of the textbooks, through the thalamus to the cortex and another, a branch that provides an input to a motor center. However, there is a difference, because the dorsal roots feed into a spinal pattern generator, whereas the visual pathways feed into a center that already in early vertebrates played a role as a higher center connecting to the pattern generators (see Grillner, 2003; Grillner et al., 2007). Specifically, in mammals the branches go to centers in the midbrain concerned with the control of movements that are relevant for vision: eye movements, head movements, accommodation, and pupillary control.

There has been a view expressed to us on many occasions that the midbrain branches of the retinofugal pathway should be regarded as "sensory" inputs on their way through the colliculus along colliculothalamic axons for relay to cortex. This argument has not been applied to the pretectal component, but because there has been an earlier suggestion that there are two retinal pathways to the cerebral cortex, one going through the lateral geniculate nucleus and the other through the superior colliculus and then through the lateral posterior nucleus or the pulvinar to higher visual cortical areas, the role of the superior colliculus needs to be considered in more detail. The view of the mammalian superior colliculus as a second "sensory" relay to the cortex has a long history (Schneider, 1969; Sprague, 1972; Diamond, 1973) and has more recently been revived by Berman and Wurtz (2011). This is based on the fact that the superior colliculus receives inputs from the retinofugal axons in its superficial layers and sends outputs to the lateral posterior thalamic nucleus and the pulvinar and also to the medial dorsal nucleus (see chapters 4.1.5.1 and 9). These thalamic cell groups send many of their thalamocortical axons to higher visual cortical areas (or to frontal cortex concerned with the control of eye movements) so that the superior colliculus became a candidate for a second visual relay between the retina and the cerebral cortex.

To evaluate this proposal, we are here not concerned with denying that a colliculo-thalamo-cortical pathway exists, but rather we are primarily concerned to consider the extent to which the branches to the superior colliculus are comparable to the branches of the dorsal roots to the spinal motor apparatus. For this, we need to consider the motor functions of the phylogenetically ancient and well-established tectal structures that serve as a higher motor center in many simple, ancient vertebrates. We limit ourselves to an introduction of some of the relevant evidence for treating the superior colliculus as a structure with a significant motor role, justifying the view that the extrageniculate branches of retinal axons are sending messages to a motor center.

In the first place, it is important to recognize that the geniculate limb of the visual input to the visual cortex (area 17) is regarded as the classical "sensory" limb because patients who have suffered an extensive lesion that includes all of the primary visual cortex in one hemisphere report a complete loss of vision in the contralateral visual field, and that is how the attending physicians report it. There are also some abnormalities of ocular movements that can be regarded as secondary to the reported visual losses but that may also reflect the loss of the motor component to the superior colliculus, either because there is a loss of retinal ganglion cells after cortical lesions (Van Buren, 1963; Cowey et al., 1999) or because of the loss of the layer 5 input to the superior colliculus from many areas of the visual cortex that will have lost a significant part of their normal visual functions. However, extensive studies of the responses that can be elicited from patients with such lesions of the visual cortex indicate that even though the patient reports a loss of vision, messages from the retina still reach other visual centers and can produce accurate motor responses. Whether these responses, described as "blindsight" in the literature on the subject, depend on the direct innervation of extrastriate visual areas by the lateral geniculate nucleus (Yukie and Iwai, 1981; Sincich et al., 2004), on connections that involve the superior colliculus acting directly through its descending outputs, or on the transthalamic link of the superior colliculus to higher cortical areas is still the subject of debate (see Tamietto et al., 2010). Damage to the superior colliculus, which receives the midbrain branch of retinal input, does not produce a significant loss of visual capacities (Pasik et al., 1966; Sprague, 1972; Diamond, 1973; Mishkin et al., 1983). There are abnormalities of eye movements, but no significant loss of vision comparable to the losses reported after lesions of visual cortex (Dumont et al., 1974; Pierrot-Deseilligny et al., 1991).

One major conclusion from these accounts is that, for a patient, damage to the visual cortex produces significantly greater losses of visual perception than do lesions to the superior colliculus and that lesions of the superior colliculus produce more marked ocular movement deficits. The view that the retinofugal axons to the superior colliculus should not be regarded as a motor connection because the superior colliculus can be regarded as a sensory relay thus does not easily accord with the losses that are produced by lesions to cortex on the one hand and to the colliculus on the other. This second, midbrain pathway is afferent to the cortex through either the pulvinar or the lateral posterior nucleus (Schneider, 1969; Sprague, 1972; Diamond, 1973) or, according to more recent studies, the medial dorsal nucleus (Sommer and Wurtz, 2004a; Sommer and Wurtz, 2004b), and this more recent evidence suggests that it is carrying to cortex copies of motor instructions (subcortical efference copies, see sections 6.2.1.1 and 6.3 and chapter 9.9) generated by the superior colliculus, not messages about visual inputs from the retina. We conclude that the retinogeniculate axons are branches of midbrain axons that innervate an important motor center and are, therefore, most likely to have significant motor functions.

6.2.1.3 The Cerebellar and Vestibular Pathways   The cerebellar inputs to the thalamus come from the deep cerebellar nuclei and go to the ventral lateral thalamic nucleus. Cajal (1911) showed these axons branching on their way to the thalamus (figure 6.6), sending one branch to cells in the midbrain and the other branch to the thalamus. These connections have been described more recently by methods that do not show the branching patterns (Stanton, 1980, 2001) and also by study of individually labeled cerebello-thalamic axons that were shown to give off branches to the red nucleus on their way to the thalamus (Shinoda et al., 1988). McCrea et al. (1978) had earlier traced individually labeled axons from the nucleus interpositus in the cerebellum to the thalamic ventral lateral nucleus and showed that these axons give off branches to the

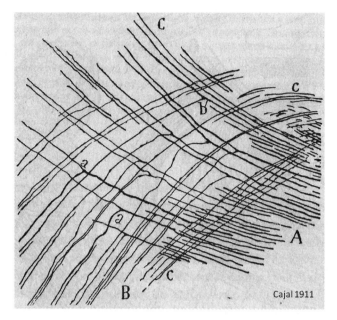

Figure 6.6
Axons of the superior cerebellar peduncle that arise in the deep nuclei of the cerebellum are shown
at A passing rostrally toward the thalamus from the right of the figure and giving off branches at
a, a and b to the midbrain reticular nuclei and the red nucleus. B and c, fibers heading to the
brainstem. From Cajal (1911).

red nucleus. The same authors also showed electrophysiological evidence that
there are branching axons that go to either the thalamus and the inferior olive
or the thalamus and the pontine reticular formation.

The vestibular pathways arise from cell groups innervated by the vestibular
nerve, whose axons on entering the brainstem divide like the dorsal root axons
into an ascending and a descending branch (figure 6.7) (Isu et al., 1991;
Bacskai et al., 2002). The ascending branches provide the input to the thalamus, and the descending branches project to the spinal cord. Activity in the
thalamus is relayed to the cerebral cortex through several scattered thalamic
cell groups, some of which lie near the borders of the ventral posterior and
ventral lateral nuclei (Meng et al., 2007; Marlinski and McCrea, 2009). That
is, these thalamic pathways from the vestibular inputs to thalamus, like the
thalamic inputs from the cerebellar, somatosensory, and visual pathways, come
from branching axons that innervate lower motor centers with one branch and
send information to thalamus for relay to the cortex with the other branch.

In addition to these vestibular inputs to the thalamus, there is another group
of vestibular afferents that is relayed through the dorsal tegmental nucleus to

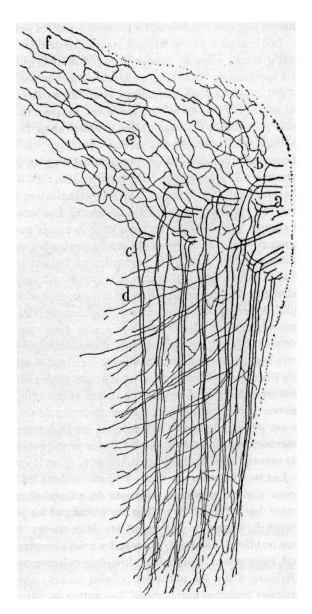

Figure 6.7
The axons of the vestibular nerve are shown entering the brainstem from the left and bifurcating to form an ascending and a descending branch. From Cajal (1911). A, central processes of the vestibular ganglion; c,d, descending branches; e,f, ascending branches,

the lateral mamillary nucleus, which in turn sends its axons in the mamillo-thalamic tract to the anterior dorsal thalamic nucleus (figure 6.8A). Extensive studies by Taube and colleagues (summarized by Taube, 2007) have demonstrated that in rats this pathway is concerned with passing information about the direction of the animal's head in relation to the environment. These "head direction cells" have been recorded in the lateral mamillary nucleus, in the anterior dorsal thalamic nucleus, and also in the retrosplenial cortex, which receives afferents from the anterior dorsal thalamic nucleus. The retrosplenial cells, in turn, appear to contribute to the place cells that are found in the hippocampus (Calton et al., 2003).

At first sight, this looks like a pure sensory system, but as has been noted above in this section (see also Taube, 2007), the vestibular nuclei receive from axons that divide to form ascending and descending branches, and whether these ascending branches provide the inputs that reach the dorsal tegmental nucleus is not known. More direct evidence for branched inputs to the thalamic part of the relevant pathways has long been known from Golgi preparations illustrated in the 19th century (Kölliker, 1896; Cajal, 1911) (figure 6.9). The axons in the mamillothalamic tract are branches of mamillotegmental axons that run caudally to the pons. They distribute terminals to the dorsal and deep tegmental nuclei (Kölliker, 1896; Cajal, 1911) and also, according to more recent evidence, to the medial pontine reticular nucleus (Guillery, 1957; Cruce, 1977; Torigoe et al., 1986; Hess et al., 1989). The medial pontine reticular nucleus is concerned with the control of gaze (Hess et al., 1989) so that the branches that ascend to the anterior dorsal thalamic nucleus can be seen as branches that carry copies of motor instructions related to the movements of the head and eyes.

The responses of the head direction cells in the lateral mamillary nucleus anticipate the actual head position by about 40 ms, but neither the functional significance nor the source of this anticipation is clear, because, according to Bassett et al. (2005), this anticipatory time interval is still seen when the rats are moved passively. This leaves the functional role of the mamillotegmental inputs to the medial pontine reticular formation unexplained, but in terms of the anatomical evidence, there can be little doubt that a copy of the message that is going to this important center for gaze control is being sent by the lateral mamillary nucleus to the anterior dorsal thalamic nucleus for further relay to the retrosplenial cortex and hippocampus. Thus, there is an important question that remains unanswered: what is the contribution that activity in the mamillotegmental pathway is making to the pontine reticular circuits concerned with gaze control? And a related question would be about the relationship of this activity to the role that the anterior dorsal thalamic nucleus plays.

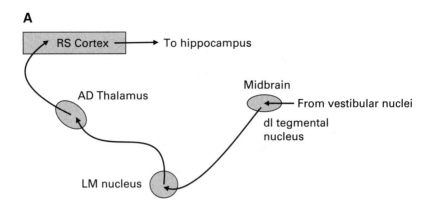

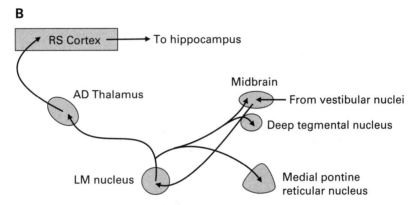

Figure 6.8
(A) Schematic view of the pathways that provide information about head position to the cells of the lateral mamillary nucleus (LM nucleus) and the anterior dorsal thalamic nucleus (AD thalamic nucleus). (B) Schematic view of the pathways that form as branches from the axons of the mamillothalamic tract (see also figure 6.11) and distribute to the dorsal and deep tegmental nuclei and to the medial pontine reticular nucleus, a cell group concerned with the control of gaze. Details in the text.

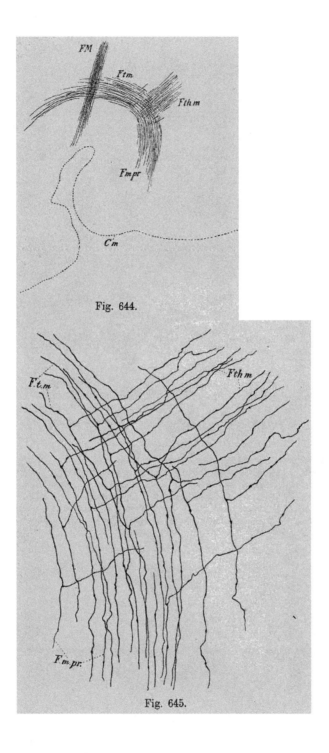

Fig. 644.

Fig. 645.

Figure 6.9
Drawing of the principal mamillary tract (Fmpr), which arises from the mamillary bodies (Cm) and passes toward the midbrain on the left of the figure forming the mamillotegmental tract (Ftm), which crosses the habenulopeduncular tract (FM) without any interchange of axons. As the axons are passing from the principal mamillary tract into the midbrain, they give off the branches of the mamillothalamic tract (Fthm), which end in the anterior thalamic nuclei. From Kölliker (1896). The lower part of the figure shows the thalamic branches leaving the fibers that pass from the principal mamillary tract into the tegmentum.

6.2.1.4 The Auditory Pathways Although a great deal is known about the prethalamic auditory pathways, the branching patterns of the individual axons that carry the information from the auditory nerve to the cochlear nuclei and through several other brainstem nuclei to the inferior colliculus and medial geniculate nucleus have not, to our knowledge, been documented. There is a group of cells that lies within the cochlear nerve before it enters the brainstem, the cochlear root neurons, and these receive collateral branches from axons passing onwards to the cochlear nucleus (Harrison and Warr, 1962). They lie on the pathway of the startle response (Lee et al., 1996), a response to a sudden loud noise commonly used to check hearing in infants. The descending pathway of the startle response goes through the reticular nucleus of the caudal pons and rostral medulla to the motor neurons of the spinal cord. Whether these pathways play any role in other motor reactions to auditory inputs is undefined at present. The fact that the cochlear nerve axons continue to the cochlear nucleus after the branches to the cochlear root neurons have been given off provides the only evidence we have been able to find that at least some of the branches passing into the ascending connections of the auditory brainstem centers and heading to the thalamus also have descending branches of the type we have seen in other afferent pathways to the thalamus.

6.2.1.5 The Gustatory Pathways The incoming gustatory axons in the 7th, 9th, and 10th cranial nerves appear not to bifurcate after they enter the medulla (Cajal, 1911), apparently going straight to the rostral parts of the nucleus of the solitary tract where they form the first central synapse. The postsynaptic cells send their axons to the parabrachial nucleus, which in turn projects to a small medial, small-celled part of the ventral posterior medial thalamic nucleus (the thalamic taste area) and also to the amygdaloid nuclei and the lateral hypothalamus (Tokita et al., 2010). These authors, and others earlier (Voshart and Van der Kooy, 1981), described a relatively small number of parabrachial cells that show evidence of double labeling when two distinct retrogradely transported labels are injected into the thalamic taste area and either the amygdala or the lateral hypothalamus. Thus, these branched axons appear to be sending one branch to the amygdala or hypothalamus for visceral responses to taste and sending the other branch to the thalamus as an efference copy of this motor instruction.

6.2.1.6 Summary of Points for Section 6.2.1   We have presented some of the parts of this section in considerable detail and others more briefly, in part because the evidence that is available varies greatly between the different pathways that terminate in the thalamus, but also because in the early sections, a consideration of some of the functional details that relate to the branching patterns of these thalamic inputs served to illustrate a key point of our presentation: that the branched axons on the route to the thalamus serve two distinct functions. One is the classical "sensory" function for perceptual processing, and the other is a contribution to a lower motor center or to a neural center that passes messages to a lower motor center. The relationship between these two functions of the axons that provide the thalamic inputs presents a challenge that has been almost entirely ignored in previous considerations of thalamic functions. Where a branching input is identifiable on a thalamic input, and we have seen that this is a very common feature of many inputs to first order thalamic relays, new questions arise about the precise content of the message that the thalamus is passing to the cortex.

It has to be stressed that obtaining evidence for a branched input is often difficult. Lu and Willis (1999), who reviewed the evidence for branching axons on the ascending pathways from the spinal cord, stressed how difficult it is to obtain clear evidence that such branching axons exist. Failure to find a branch point does not prove its absence. The Golgi methods are notoriously fickle, and where they clearly show a branching axon, the positive evidence is clear. The physiological methods for identifying a branching axon in vivo or in vitro in the central nervous system are difficult because both branches have to be found and stimulated. Again, the positive evidence is much more telling than the negative evidence. More modern methods for demonstrating branching axons have either used intracellular filling of the cell with a marker that allows the branches to be directly visualized or used two distinguishable retrogradely transported marker molecules injected into two separate regions each of which is known to receive afferents from a single, third cell group. Again, the positive evidence from such experiments does provide powerful evidence of branching axons, whereas not finding evidence for branching axons cannot be regarded as convincing evidence for the absence of the branches.

6.2.2 The Afferents to the Higher Order Relays

Evidence that the driver corticothalamic inputs to the higher order thalamic relays are branches of axons that pass to brainstem or spinal centers is more recent than most of the evidence considered in the previous sections. It is, so far as we know, entirely based on the careful tracing of individual axons through serial sections after injections of tracers such as biocytin or biotinyl-ated dextran amine into cells of cortical layer 5. Some of these studies were based on the identity of the individual layer 5 cells at the injection site, usually of single cells labeled by "juxtacellular" injections. The relatively thick axons were traced to their terminal sites, where the appearance of the thalamic ter-minals demonstrated the typical appearance of the driver axons, which, as we showed in chapter 2, differ strikingly from the modulator afferents that come from thinner axons of layer 6 cells. Further, because there is little evidence that the layer 6 cells project to regions of the brainstem beyond the thalamus,[5] the presence of long descending branches into the brainstem provided a further identification of these axons as coming from layer 5 rather than layer 6. Other studies, based on larger injections, identified the corticofugal drivers by the thickness of the axons, the appearance of the thalamic terminals, and the post-thalamic course of the axons, including the fact that they were not sending reciprocal connections back to the thalamus.

Figure 6.10A shows eight axons traced from individual cells in layer 5 of the visual cortex of a rat (Bourassa and Deschênes, 1995). All eight axons go to the superior colliculus. Two of the axons (1 and 2) send no branches to the thalamus, distributing to the superior colliculus, pretectum, or ventral lateral geniculate nucleus. The other six axons send branches to the thalamus and continue to the pretectum and to the colliculus. This figure illustrates an important point that was also seen for corticofugal axons arising in the visual cortex of the cat (Guillery et al., 2001): not all of the long descending corti-cofugal axons send branches to the thalamus, and in the rat and the cat, the ones that lack such a branch take a distinct course that avoids the thalamus, suggesting that the lack of the thalamic branch is not an artifact of the method. However, it appears that all of the axons that have terminals in the thalamus also have a long descending axon that goes to the midbrain. Guillery et al. (2001) labeled several cells in nine cats, either in area 17 or area 18, and traced the thicker layer 5 axons through the thalamus (figure 6.11) demonstrating that, for more than 50 thick axons that entered the lateral posterior nucleus at the lateral border of the thalamus, almost every one sent one or more branches to a single well-localized terminal focus within that nucleus and could then be traced caudally into the midbrain. Similarly, Veinante et al. (2000) traced axons from individually labeled layer 5 cells in the barrel cortex of the rat and

5. One curious exception is a projection described as originating in layer 5 with a sparser com-ponent from the deepest parts of layer 6 of auditory cortex and terminating in the inferior colliculus (Schofield, 2009). It is not known if these layer 6 cells have branching axons that target others structures, such as thalamus.

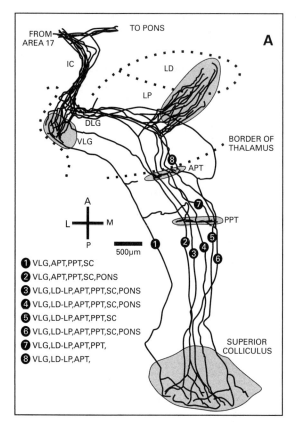

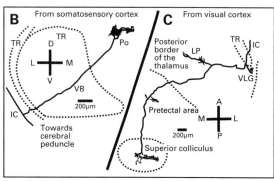

Figure 6.10
(A) Drawing of axons coming from cells in layer 5 of area 17 of a rat and going past the thalamus to the midbrain. Axons 1 and 2 give no branches to the dorsal thalamus, whereas axons 3–8 send branches to the lateral posterior and lateral dorsal thalamic nuclei. All the axons send branches to the ventral lateral geniculate nucleus and to the midbrain to one or more pretectal nuclei and for axons 1–6 also to the superior colliculus. From Bourassa and Deschênes (1995). (B) A single axon from the somatosensory cortex of a rat descending through the internal capsule on its way to the brainstem and giving off a branch that has large terminals in the posterior nucleus. From Deschênes et al. (1994). (C) An axon from area 17 of a rat passing to the superior colliculus and on its way giving off branches to the ventral lateral geniculate nucleus, the lateral posterior nucleus, and the pretectal area. From Deschênes et al. (1994).

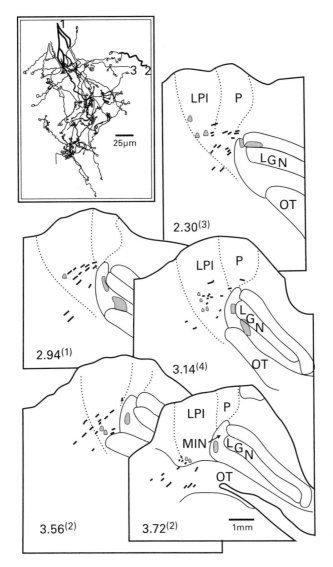

Figure 6.11

Outlines of five coronal sections through the thalamus of a cat to show the distribution of corti-cofugal fibers that had been filled by a small injection of axonally transported marker into area 18 of a cat. Dorsal is up and medial to the left. The large number in each section outline indicates the distance in millimeters from the rostral tip of the lateral geniculate nucleus. The coarse fibers that enter through the pulvinar (labeled P) rostrally (level 2.30) and exit to enter the midbrain caudally (level 3.72) are shown as short black lines and were traced through the serial sections to complex terminal arbors similar to that shown in the top left inset. Each of these terminal arbors is shown as a small dark outline in the lateral posterior nucleus (LPl), and the small numbers in parentheses represent the numbers of these terminal arbors that were present. Note that they lie along a narrow column that passes obliquely through the medial parts of the lateral posterior nucleus from dorsorostral to caudoventral. The finer axons that send terminals to the lateral geniculate nucleus (LGN) and medial interlaminar nucleus (MIN) are not shown, but their terminal distributions are indicated by the paler larger patches in these two nuclei. OT, optic tract. Scale bar: 1 mm. The top left inset shows one of the terminal arbors formed by the coarse axons. The numbers 1, 2, and 3 indicate what appear to be separate axons contributing to this terminal arbor, but each axon was traced back to a single parent axon each taking an entirely different course to the arbor. Scale bar: 25 μm. Modified and redrawn from Guillery et al. (2001).

reported that the "vast majority (approximately 95%) of layer 5 axons that innervate the somatosensory thalamus are collaterals of corticofugal fibers that project to the brainstem."

Figure 6.10B shows another corticothalamic axon that came from a layer 5 cell in the somatosensory cortex of a rat (Deschênes et al., 1994), and on its way to the brainstem through the internal capsule (IC) sends a branch to the thalamus, passing through the first order somatosensory relay nucleus of the thalamus (VB in the figure) without giving off any terminals and then passing to the higher order posterior medial nucleus where it has a group of typical Class 1 driver terminals.

Another corticothalamic axon arising from the visual cortex of a rat and taken from a study by Bourassa et al. (1995) is shown in figure 6.10C. It enters the thalamus from the internal capsule (IC), gives a branch to the ventral lateral geniculate nucleus (which does not project to cortex), and, without giving any branch to the thalamic reticular nucleus or to the first order visual relay in the lateral geniculate nucleus, it continues to give large terminals to the lateral posterior thalamic nucleus and then to the pretectal nuclei and the superior colliculus. Other comparable corticothalamic axons have been demonstrated in rats by Bourassa et al. (1995), who traced axons that formed one or two well-localized clusters of terminals in the posterior medial nucleus and the central lateral nucleus in the thalamus and then continued caudally into the brainstem. One other important distinction between the layer 5 corticothalamic axons with driver terminals in the thalamus and layer 6 corticothalamic axons with modulator terminals there—a distinction repeatedly reported in the studies cited above in this section—concerns branches that terminate in the thalamic reticular nucleus: whereas these are commonly, possibly always, present for the layer 6 axons, they are consistently absent for the layer 5 axons.

We have seen that the corticothalamic driver afferents commonly arise from long descending axons that enter the brainstem. This has been demonstrated for several different cortical areas: in rats for primary visual, somatosensory, and motor cortical areas, and in cats for visual areas 17, 18, and 19. In addition to this, Rockland (1998) has described corticothalamic axons with large (Rockland's R type), well-localized thalamic terminals, resembling the driver terminals described in chapter 2, and has noted that these axons also send branches to the superior colliculus and the pretectum.[6]

---

6. Some years ago, one of the authors (R.G.) during a visit to K. Rockland's laboratory, then in Tokyo at the Riken Institute, had an opportunity to examine sections from several monkeys in each of which a few cells in different areas of temporal and parietal cortex had been injected and in which essentially all of the axons that had characteristic Class 1 thalamic terminals (Rockland's R type) terminating in different parts of the thalamus could be traced through the thalamus, giving off the thalamic branch and then continuing caudally toward the midbrain. Professor Rockland has indicated that the material is available for scholarly use.

6.3   Efference Copies and Forward Models

The idea that copies of motor instructions play an important role in the control of movements is significant for appreciating the significance of the branching axons. The idea has a long history fully reviewed by Grüsser (1995) and relates to studies of animals as well as robots. Two studies have played a significant role in more recent discussions relating to the control of movements and their relation to perception. These were based on studies of fish and flies by von Holst and Mittelstaedt (1950) and on studies of fish by Sperry (1950) in all of which an eye had been rotated through 180°, and the animal developed a circling behavior. The basic idea is illustrated in figure 6.12, which shows that

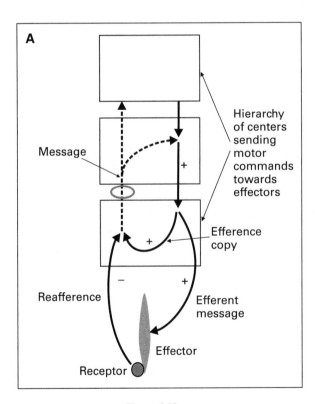
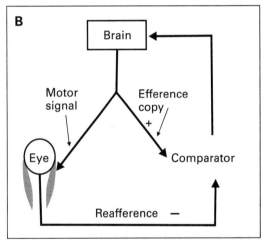

Figure 6.12
Schematic representation based (with modifications) on von Holst and Mittelstaedt (1950) (A) and on Perrone and Krauzlis (2008) (B) to show how the efference copy is thought to relate to the reafference signal. Both figures have been modified from the original to a limited extent. Whereas (A) shows the message going back to a motor center, the original of (B) showed the message going to a separate center labeled "perception of movement."

the efference copy, often called the corollary discharge,[7] is a copy of a motor instruction that can be compared with the reafference signal, which is the change in the sensory input produced by the movement. This comparison allows the command centers to distinguish sensory changes due to a movement of the sensory receptor produced by the organism from sensory changes due to movements in the environment. That is, when we move our eyes, the world does not appear to be moving because the efference copy signal matches the reafference signal. When we produce a movement of the eye by pressing on the eye with our fingers, there is a mismatch between the reafference signal and the efference copy and the world does appear to move. An important part of the pattern of connections that includes an efference copy is the fact that the copy of the instructions on the way to the muscles will reach centers in the brain before the muscle itself acts.[8] This is illustrated in figure 6.13, which shows the axon from the dorsal root ganglion branching to provide the thalamic and cortical inputs and also innervating the spinal mechanisms for control of the muscle. The ascending axon is shown carrying the afferent message and the efference copy at time t1 and the reafference signal at time t2. If the afferent message and the reafference signal match, then the cortex receives information that is expected; if they do not match, then there is an unexpected input that needs to activate a correction.

If we consider a message to the spinal motor mechanisms, no matter whether it comes from the dorsal root or from higher centers in the brain, that particular input will produce not just a particular muscle contraction, but will engage the whole of the spinal or brainstem pattern generator and lead to movements that can be anticipated to a significant extent by a center in receipt of a copy of the instructions. A striking example of such an anticipation has been recorded in areas of the visual cortex by the demonstration of a "forward receptive field," also described as a "re-mapping" of the receptive field (Colby et al., 1996; Sommer and Wurtz, 2008); this is a shift of the receptive field of a cell in a higher visual cortical area that precedes a saccade but is in accord with the shift that will be produced by the saccade.[9]

7. We use "efference copy" because it specifically refers to a copy of an output (motor), whereas a corollary discharge could refer to a copy of an input (sensory) although usually not used in that sense.

8. We stress that any instruction on the way to a muscle, when copied and sent to a central relay, can be regarded as an efference copy and can provide early information about forthcoming movements. The nonmotor ascending branches of the dorsal root inputs can serve this function and invariably provide information ahead of the information produced when the muscle contraction moves the receptor. Similarly, the thalamic branches of layer 5 corticofugal axons can serve the same functions for movement instructions coming from cortex.

9. Another example may well be the 40-ms anticipation recorded for head direction cells (see section 6.2.1.3).

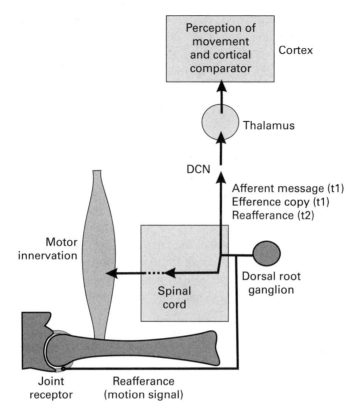

Figure 6.13
Schematic view of a dorsal root axon branching to form the ascending branch that provides inputs to thalamus and cortex and the spinal branch that provides inputs for the spinal motor mechanisms. The sequence in which the messages reach cortex is indicated (t1) for the afferent signal and the efference copy and (t2) for the reafference signal. DCN, dorsal column nuclei. Note that in contrast to the pathways proposed in figure 6.12, these signals are all passing along the same axon. This stresses the temporal relationships that apply to both figures.

The efference copies produce messages that allow nervous systems to create what have been called forward models of how the body relates to the environment (Wolpert and Miall, 1996; Wolpert and Ghahramani, 2000). When all of the many active efferent pathways and the efference copies that they send centrally, only some of which are described in this chapter, are taken into consideration, the brain can be seen to receive a full view of how the organism relates to the world *and what the organism and the world are likely to do next.* We generally know what we are likely to do next; our actions only rarely surprise us, and we generally have a quite accurate idea of how the world will react. The forward models can include more than the information that comes from the neural centers and the sensory receptors. Where an action relates to features of the environment that themselves have regularities that can be learned, those too can form a part of a forward model.

The importance of the forward model is not simply that we commonly know our upcoming actions, but also that we can distinguish our own actions from those of others; as has been pointed out by Blakemore et al. (1998), we cannot tickle ourselves. The forward models, based on the efference copies in all parts of the thalamocortical connections we have described, provide a model of ourselves in the world that Adrian was writing about when he was quoting Craik (section 6.2.1.1). Whereas there has been a clear recognition that efference copies play an important role in this way, the identification of the relevant neural connections has been missing or has proved elusive. The apparently ubiquitous presence of efference copies in the thalamocortical pathways may seem surprising, but we would claim that they provide a firm neural basis for an important and well-recognized functional property of mammalian nervous systems: the capacity to anticipate the actions of the body and also, where they are reasonably consistent, the reactions of the environment.

When we consider the ways in which efference copies can be produced, it looks as though a branching axon may be by far the simplest and most secure. It may be possible that one cell linked to another could produce a faithful copy over a long term, but given the very common occurrence of axonal branches and the important role they play in exploring new territories during development, it may prove that most or all efference copies are produced by branching axons. The distribution of axonal branches in pathways that may be playing a role as an efference copy merits close investigation in the future.

In the past, we have often encountered argument when we describe the ascending axons on their way to the thalamus as efference copies,[10] because as figure 6.12 shows, efference copies are often regarded as copies of motor instructions coming from a higher center, not uncommonly as copies of instructions for voluntary movements. We stress that, in principle, the organism needs information about all instructions on their way to motor centers. It would be odd if the forward models were based on voluntary or higher-level instructions only; that would leave out of any account the actual performance of the system, focusing instead on the higher instructions no matter how well the instructions were being carried out by the lower centers. We indicated at the beginning of this chapter that it is reasonable to regard the earliest ascending pathways in the spinal cords of early vertebrates as sending efference

10. As one referee put it, our ideas were thought likely to produce some raised eyebrows and a snicker or two.

copies of the sensorimotor relationships established at the segmental levels to higher centers and stress that this is a function that can be seen to have continued throughout the evolution of the vertebrates. The basic studies of efference copies by Sperry (1950) and von Holst and Mittelstaedt (1950), involved subcortical or, for the fly, noncortical circuits. The suggestion made in section 6.2.1.1 to distinguish cortical from subcortical efference copies may clarify the situation to some extent. Efference copies that involve, for example, the superior colliculus or the cerebellum or lower motor centers on the basis of connections that do not include thalamus and cortex are subcortical, play but a minor role in our review of thalamus and cortex, but cannot be ignored in the study of movement control.

Exactly how one describes or classifies efference copies may yet need significant discussion or may eventually prove unnecessary. The important point to recognize is that efference copies are very common in the brain. It looks likely that all areas of cortex receive efference copies from the thalamus and that most, possibly all also send efference copies back to the thalamus for relay from one cortical area to another.

6.4 Overall Conclusions about the Branching Driver Axons

In summary, wherever appropriate methods have been used to demonstrate the branched origin of a driver input to the thalamus, it has been found. Table 6.1 summarizes the major ones. However, such branching patterns have not yet been studied for many cortical areas, and further evidence about the branching patterns of these axons for species including and extending beyond mouse, rat, cat, and monkey, and for many driver afferents to higher order thalamic relays, is needed to show that possibly all driver afferents to thalamus, to first order as well as higher order relays, are branches of axons that innervate circuits concerned with the control of movements. At present, we will treat this possibility as a hypothesis that merits further study. It will be important to define whether all driver inputs to first order and to higher order relays show this relationship or whether there are some that lack an extrathalamic branch, and if there are, then to define how those with a branched origin differ functionally from those without such an origin.

Up to this point, we have written about single axons and their branches. For any particular situation, a cat or a person walking, a dog scratching a flea, a child running to catch a ball, or someone writing a manuscript, there will be very many branching axons sending many concurrent messages along the spinal or brainstem branches to the relevant motor centers on the one hand and, possibly, if the thalamic gate is open, to the cerebral cortex through the

Table 6.1
Summary of evidence that major thalamic nuclei receive inputs for relay to cortex from branched axons whose extrathalamic branches relate to motor actions

First Order Thalamic Nucleus	Cortical Area	Modality/ Function	Prethalamic Branching Axons
Ventral posterior nucleus	Primary sensory cortex	Somatosensory	Entering dorsal roots and lemniscal and trigeminothalamic pathways with branches to brainstem and spinal motor circuits
Ventral lateral nucleus	Motor cortex	Motor	Cerebellothalamic axons from deep cerebellar nuclei with branches to brainstem motor centers
Lateral geniculate nucleus	Primary visual cortex (area 17)	Visual	Retinofugal axons with branches to superior colliculus and pretectum
Ventral medial geniculate nucleus	Primary auditory cortex	Auditory	Cochlear nerve with branches to cochlear root neurons
Anterodorsal nucleus	Retrosplenial cortex	Head direction	Principal mamillary tract arising in mamillary bodies with branches to midbrain and pontine nuclei
Vestibular recipient cells near or in ventral posterior and ventral lateral nuclei	Parieto-insular cortex	Vestibular	Vestibular axons entering brainstem with descending branches to brainstem motor centers
Medial small-celled part of ventral posteromedial nucleus	Insular cortex	Taste	Parabrachial axons to thalamus with branches to hypothalamus and amygdala

Note: The major higher order nuclei that receive branched inputs from the cerebral cortex include: lateral posterior (cat and rat) from visual cortex, pulvinar (monkey and cat) from visual cortex, posterior medial (rat) from somatosensory cortex, dorsal medial geniculate nucleus from auditory cortex, ventral medial, central median, and parafascicularis (rat) from motor cortex, and medial dorsal nucleus from frontal cortex. Details regarding species and references in the text.

thalamic branches on the other. They can all contribute to the forward model. If the messages in these thalamocortical branches reach the cortex, this can then lead to further cortical outputs to the motor centers, and copies of those outputs may, in turn, if the thalamic gate is open (see chapter 5.4.2.1), be passed to higher cortical areas for further processing; this may, again, produce another set of motor instructions, probably also copied to another cortical area. That is, there will be a great many messages passed back and forth between the receptors, lower motor centers, and the thalamocortical pathways, so that the overall structure can be seen as a hierarchy of cortical areas and lower motor centers, with each level monitoring the instructions being sent out by lower levels. An important point to bear in mind is that any one of the axons on these pathways will represent one very small part of the relevant motor instruction. Each motor neuron in the lower motor center will be receiving a great many different inputs at any one time, and each of these inputs must be integrated to make a (probably small) contribution to the output of the motor neuron and an even smaller contribution to the final total movement pattern that will depend on many ventral horn cells.

There are several reasons for drawing attention to the many efference copies that will be active at any one time and the corresponding complexity of the messages that reach the cortex. One is that this in itself provides a serious challenge for any search for such messages on their way to the cortex or within the cortex if each message makes but a small contribution. Another reason is that when a well-practiced activity is progressing smoothly, it may depend primarily on the efficient performance of the pattern generators and lower motor centers. The complex of corticopetal and corticofugal efference copies may be present only fleetingly when there is activity suggestive of errors or surprises that call for the thalamic gate, discussed in chapters 4 and 5, to be opened. That is, most of the time when we are engaged in any activity that we are not undertaking for the first or second time, much of what we are doing is done automatically.[11] As stressed by O'Regan and Noë (2001) in their account of driving a car, many even quite complex activities have a large automatic component to them that can be taken care of largely by the lower motor centers with minimal supervision from the cerebral cortex. When one of the authors (R.G.) was very young, someone tried to teach him to knit. It is a difficult skill for a 6-year-old child to learn. However, when many years later he watched his wife or daughter knitting while reading or talking or watching television, it looked very much as though cortical involvement must be at a minimum, and that there cannot have

11. And this is almost certainly true of many well-learned activities performed by experimental animals.

been much conscious awareness of the movements . . . until something went wrong. The brain and the body can interact efficiently with the environment with what often appears to be minimal involvement of cortex.

It is probable that there are many situations where the cortex, particularly the higher areas of cortex, are not involved in ongoing motor activities, and the individual may be unaware or only marginally aware of the motor performance. As discussed in chapter 5.4.2.1, the higher order relays receive several GABAergic inhibitory inputs that do not act on the first order relays, and if this shuts these thalamic gates, then it may take an input about an error or an unexpected situation from the first order cortical area to send a wake-up call through the thalamic relay in the burst mode to the higher order cortical areas, which can then switch to the tonic mode to record what is happening in the lower centers and deal with it. That is, the thalamocortical pathways will not only be playing a role capable of reporting what is happening at any one time, but will also be contributing to an ongoing monitoring and supervision of the instructions that are on their way to control the changing activity patterns in the lower motor centers.

We stress that the pattern of connections we have described lead us to conclude that, on present evidence, all of the messages that the cortex receives from the thalamus (and the cortex receives very few messages from other sources) can be regarded as efference copies. This leads us to view the cortex as having two basic major functions that appear to underlie all the other things that the cortex undoubtedly is capable of doing. One is to receive efference copies from all parts of the brain that have a thalamic relay, that is, most of it, and from this generate a forward model of the body in relation to the environment. Cortex monitors what the rest of the brain (including other parts of cortex) is doing in order to anticipate what the brain and the body is likely to do next. The other is to send out instructions that serve to control the activities about which information is being received. We have written these last few paragraphs as a summary of all of cortex, which may be far too general to be useful. The statement can become more interesting if it is considered in terms of individual cortical areas, because then it is about the messages that any one area is receiving from the thalamus and the relation of those messages to the outputs that that cortical area is sending to lower centers including other (lower) cortical areas.

Understanding the power of forward models must be based not only on what they tell us about the capacities of the brain to anticipate events and to distinguish self-generated actions from the actions of others, but must also be firmly based on the recognition of their role in essentially all of our actions, probably even those involving the highest cortical functions, but many of the relevant pathways need yet to be defined.

We have presented the thalamus and cortex as a piece of neuroscience and not as a study of the mind. Our aim is to move toward an understanding of the mechanisms of the brain in terms of what the nerve cells and their interconnections can produce to guide our interactions with the world in our daily lives.

The organism, no matter how simple or complex, must interact with the world. This is far more important than that it should see the world, hear the world, feel or smell it. The first and essential step is in the interaction that can be represented, in our brains or in the brains of any vertebrate animal, by the first simple reflex response produced by the stimulation of a receptor and its link to an effector. It is this first, apparently simple interaction that is needed for survival right from the start of the organism's evolutionary history, and it is a *record* of this first interaction in some higher part of the nervous system that represents the first sign of the nervous system creating a view of the past and looking ahead to the future. We can see an example of this in the ascending spinal axons that send messages to the brain. These are not about the world or about the body, but about the interactions of the body with the world. These are the basic messages from which the nervous system creates its view of what the world is like, storing some of the information and creating the capacities to recall the past and anticipate the future in terms of perceptual and cognitive functions (seeing, hearing, feeling, smelling, and more) that represent our view of what the world is like.

6.5   Outstanding Questions

1. For any of the pathways that terminate as drivers in the thalamus, what are the actions produced by the structures innervated by the extrathalamic branches? The answers are known for only a few and generally have made little contribution to our understanding of the role of the pathway in perception.

2. Where a thalamic relay can be matched to a structure innervated by the extrathalamic branch (as in the previous question), how does the message relayed to the cortex relate to the motor action of the extrathalamic branch?

3. Are there any thalamic nuclei that are not first order and that receive no cortical driver inputs?

4. Are there driver axons to thalamus that lack extrathalamic motor branches, and if there are, how does their role in perceptual processing differ from those that have motor branches?

5. Are efference copies always, usually, or only sometimes produced by branching axons?

7 Linking the Body and the World to the Thalamus

7.1 Introduction

This chapter and the next will be an exploration of the pathways that link the thalamus and cortex to each other and to the rest of the brain, considering each in terms of the major points about the neural connections raised in earlier chapters. These connections provide the essential links between cortex, lower centers, action, and perception. Chapter 4 considered the pathways as either drivers (Class 1) that transmit a message or modulators (Class 2) that change the way in which a message is or is not transmitted. The messages transmitted by the drivers need to be understood in terms of the informational content that they contribute to the computational functions of the neural stations they serve, and this will relate to events in the world or in other parts of the nervous system. The modulators (Class 2) need to be understood on the basis of how they affect the transmission of messages. Understanding the messages and knowing the functional nature of the relays is basic to understanding how the organism relates to the world.

In chapter 6, we showed that many of the axons carrying messages to thalamus for relay to cortex have branches that innervate motor centers so that the cortex receives a great many messages that have a dual significance, representing events in the body or the world on the one hand and instructions for upcoming movements on the other. It is important to recognize that, although we raise the issue in a few instances only, this possible duality is relevant for all of the pathways that carry messages to the thalamus for relay to cortex. It is relevant for first order thalamic relays, which relay messages from the sensory periphery or from lower (mamillary or cerebellar) centers, and also for higher order relays, which relay messages from one area of cortex to another.

Corticofugal axons from a great many and probably all cortical areas, including classical sensory areas and areas far beyond the areas that are labeled *motor* in textbooks, carry motor outputs from layer 5 to lower centers with functions that are largely unknown and unexplored, and many of these outputs have branches to the thalamus. For a full view of the messages that thalamus sends to cortex, we need to recognize that there is this rich corticofugal output from all areas of cortex and that many of these outputs also send branches to the thalamus for relay to other areas of cortex (chapter 6). In this way, higher cortical areas can be informed about the outputs of lower areas. Each thalamic relay, whether first order or higher order, sends to the cortex messages that can have two meanings, one about the environment or about activity in other neural centers and the other about instructions currently on the way to motor centers.

We start by exploring the inputs to the thalamus that provide the cortex with its view of the world and of the body in relation to the world, then we look at the nature of the thalamocortical pathways that distribute this information to the cortex, and finally we consider the corticofugal pathways that provide the executive link between the cortex, the phylogenetically older lower motor centers and pattern generators (see chapter 6), the body, and the world. This is important not only for showing the extent to which any one cortical area can act directly on the lower centers but also for stressing that much of our behavior can depend on subcortical centers acting with minimal cortical controls. The question for understanding the role of cortex in movement control is not so much the classical question as to whether cortex controls muscles or movements, but one about the particular lower motor centers through which any cortical area can act.

The aim of chapters 7 and 8, as for most of this book, is to identify areas of ignorance as much as to demonstrate new facts or relationships and, although it may be no surprise that the former dominate the latter, we consider that an overall view of these connectivity patterns must form a basis for understanding cortical functions. The long-term aim is to understand how thalamus and cortex, or how any one thalamic nucleus and the cortical areas it innervates, relate to cognition and behavior, not only in terms of their neural connections but also in terms of the nature of the messages that are passing along these connections. This can then be related to information that is currently more readily available about the conditions under which a particular cortical area becomes active or about losses produced by silencing or removing one area. Such information is, at present, most commonly reported on the basis of functional imaging or lesion effects for particular cortical areas and produces a view of functional localization in particular cortical areas. This, however, has limited explanatory value. It provides information that can be regarded as "evidence-based phrenology"; it reveals nothing about the mechanisms concerned or the nature of the messages involved. We need to learn how higher cortical functions relate to the particular messages that are arriving over defined pathways from other centers or from the peripheral sensory receptors and then to assess the nature of the cortical actions and their outputs. The following is an exploration of some thalamic relays that can serve as examples of the type of knowledge that eventually one should hope to obtain about all thalamic relays.

7.2   The Inputs to Thalamus

We start this review at the thalamic levels for which we have the most information, that is, with the classical primary sensory relays, and then use this information to move to other thalamic relays where more information is still needed.

7.2.1   The Inputs to the First Order Thalamic Relays

7.2.1.1   The Inputs to the Major Classical Sensory Relays   We discussed the first order classical sensory relays in some detail in previous chapters (chapters 3 and 6). They provide examples of the information that needs to be obtained for any thalamic relay if the functions of the cortical area it supplies are to be understood. For the pathways to the primary sensory nuclei in the thalamus, we know the driver inputs and we can be reasonably clear about the messages that they carry about activities of the peripheral sensory receptors. This involves knowing first the nature of the sensory event and then knowing how this is converted to specific patterns of action potentials in the peripheral nerves or the ascending central pathways. For the somatosensory pathways, which we will use as a relatively simple example here, the early work of Adrian (1928) showed how particular sensory events are encoded in the peripheral nerves, and later studies, such as those of Mountcastle and Henneman (1952) provided evidence for the thalamus, where the activity was interpreted as representing the way in which the world is acting on the sensory receptors. However, as stressed in chapter 6, the activity must also be interpreted in terms of the actions that those same patterns of activity are currently about to produce through branches of the ascending axons innervating lower motor levels.

For inputs such as those passing toward the thalamic ventral posterior nucleus from the dorsal roots, the nature of the incoming sensory information from the receptors is clear: it may be muscle stretch or a nociceptive stimulus, and for these the neural code has been defined (see Adrian, 1928; Kandel et al., 2000). The instructions for future actions that these same incoming axons are passing to the spinal circuits are also known to some extent and were discussed at the end of chapter 6. They provide information for the forward models that allow higher centers to anticipate forthcoming actions. That is, the information that is passed from the ventral posterior thalamic nucleus to the cortex must be expected to relate not only to the message produced by the changes in the sensory receptor but also to some aspects of upcoming movement patterns.

In addition to the branches that the incoming dorsal root axons send to the spinal cord, there are other branches given off at higher levels of the pathways to brain structures that include the superior colliculus, the periaqueductal gray, the midbrain reticular formation, the inferior olive, and the hypothalamus (summarized by Guillery and Sherman, 2002; Guillery, 2003; Sherman and Guillery, 2006). The details of the actions produced by such branches can all be considered as relevant for a full appreciation of the messages that the ventral posterior nucleus can pass to the somatosensory cortex. For example, the details of the inputs to the periaqueductal gray, which sends important descending pathways to spinal centers concerned with pain, will be of particular interest for a fuller appreciation of the thalamocortical involvement in sensory responses to nociceptive stimuli, whereas those to the hypothalamus, whose outputs innervate spinal visceral centers, will relate particularly to forthcoming autonomic responses. The functional role of these extrathalamic inputs will thus also be present in the information available to cortex. The cortical circuitry that receives these ascending messages will have information about the messages that are on their way to periaqueductal gray and hypothalamus. That is, they will have information about upcoming events in the nervous system, contributing to a view of the future that can play a crucial role in generating a model of the future needed in the brain of any organism before it can interact successfully with its environment.

Defining these relationships in the thalamus or cortex may prove elusive and will necessarily depend on studies of an unanesthetized preparation, but knowing that information about these spinal and brainstem events is necessarily available in thalamus and cortex should at the very least change the expectations of investigators and help to modify views of pure sensory functions in thalamus and cortex.

For the major sensory thalamic nuclei other than those receiving from the ascending spinal pathways, comparable relationships produced by branched incoming driver afferents arise, and we refer to chapter 6 for details of the connections on which this statement is based. We have stressed these conclusions from the previous chapter here, because they are essential not only for understanding the nature of the message that each primary sensory nucleus is sending to the relevant cortical area(s) but also for understanding the functional organization of any thalamocortical pathway that is supplied by a branched axon; and it appears that most or all of them do have such branches (see chapter 6). The important point is that the functions of the extrathalamic branches are everywhere an essential part of understanding the nature of the messages that the thalamic nucleus is relaying to its cortical area(s). For example, knowing that the retinal inputs to the lateral geniculate nucleus have branches that innervate midbrain structures concerned with the control of gaze, of pupillary size, and of accommodation can serve as a clue for raising new questions about how visual perception can relate to action (Churchland et al., 1994; O'Regan and Noë, 2001) and can raise issues relevant for understanding how such links between actions and perception can be learned (Bompas and O'Regan, 2006a, b).

7.2.1.2   The Inputs to the Anterior Thalamic Nuclei   We next consider the first order thalamic nuclei receiving inputs for which evidence of sensory functions are less obvious or lacking. These are the two large anterior thalamic nuclei receiving afferents from the mamillothalamic tract and the ventral lateral nucleus receiving afferents from the deep cerebellar nuclei. For the anterior thalamic nuclei, we showed in the previous chapter that the smallest, the anterior dorsal nucleus, is a relay on the pathway of vestibular messages about head position to the retrosplenial cortex and hippocampus (Taube, 2007). This part of the pathway through the anterior thalamus is comparable to the first sensory level. It raises comparable outstanding questions about the extent to which the messages about head position also relate to messages to motor centers and about their origin either from branches at the entrance of the vestibular nerve to the medulla or from the descending branches of the mamillotegmental tract to the medial pontine reticular formation; these were raised in chapter 6. However, we have no comparable information about the pathways to the limbic cortex through the two larger anterior ventral and anterior medial thalamic nuclei (figure 7.1), and current views about the functions of these pathways often depend on what is known about the functions of the cortical recipient areas, the anterior and posterior limbic areas, rather than information about the nature of the messages that the thalamic cells receive and send on to cortex. We consider the known evidence about the inputs to these nuclei in some detail because this illustrates the importance not only of understanding the nature of the messages that any one thalamic nucleus is relaying to cortex but also of knowing whether the input is a driver or a modulator. It also demonstrates the distinction between localizing particular functions to a cortical area (the "evidence-based phrenology" mentioned in section 7.1) as opposed

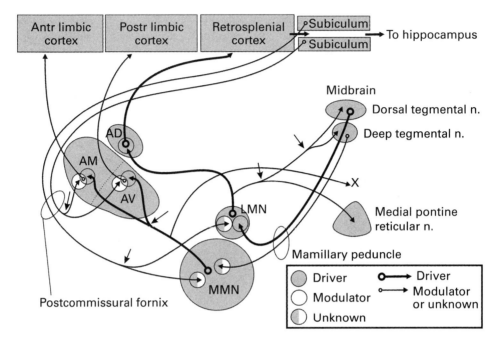

Figure 7.1
Schematic representation of the connections of the anterior thalamic nuclei. AD, AM, and AV are
the anterior dorsal, anterior medial, and anterior ventral thalamic nuclei, respectively; LMN is the
lateral mamillary nucleus; MMN the medial mamillary nucleus. The small arrows indicate regions
where pathways that may not be made up of branching axons are shown as branched in order to
reduce the number of lines in the figure; they may or may not be branched. X shows mamillotegmental branches that have not been traced to their midbrain terminal zones. The known driver
pathways discussed in the text through the mamillary nuclei and the anterior thalamic nuclei are
shown as thicker lines, whereas all of the other pathways shown as thin lines are not defined as
drivers or modulators nor do we know about the messages that they are carrying except for the
fornix inputs to the anterior thalamic nuclei, which are most reasonably regarded as modulators
on the basis of the appearance of their terminals (see Somogyi et al., 1978). n., nucleus.

to knowing the functions of the connections and the nature of the messages
that are transmitted: the difference between functional localization and comprehension of the neural machinery.

The ascending inputs to the two larger anterior nuclei come from the medial
mamillary nucleus and travel in the mamillothalamic tract (figure 7.1), which
is the bundle that also carries inputs for relay to the anterior dorsal nucleus
from the lateral mamillary nucleus. The three anterior nuclei are often treated
as a group, which may not be justified. Not only are the cells in the anterior
dorsal nucleus noticeably larger than those in the other two nuclei, but also
they relate to the large-celled lateral mamillary input with thicker axons, and
there may thus be significant functional differences. Further, the large cells in

the small lateral mamillary nucleus receive their driver inputs from a midbrain source that appears not to be shared by the smaller cells in the medial mamillary nucleus.

The mamillary bodies receive inputs from two sources (figure 7.1). One is the mamillary peduncle, which brings afferents from midbrain nuclei, and the other is the postcommissural fornix, which brings afferents from the hippocampal formation (the subiculum). The midbrain afferents to the medial mamillary nucleus come from the deep tegmental nucleus, whose functions are currently undefined (see Saunders et al., 2012); those that go to the lateral mamillary nucleus come from the dorsal tegmental nucleus, bringing vestibular information to the head position pathway (Taube, 2007).[1] Although significant numbers of GABAergic neurons[2] have been reported in these two nuclei (see Allen and Hopkins, 1989; Wirtshafter and Stratford, 1993; Gonzalo-Ruiz et al., 1999), there can be little doubt that the dorsal tegmental nucleus has a driver input to the lateral mamillary nucleus, because the transfer of information about head direction is known to be passed from the vestibular nerves to the anterior dorsal thalamic nucleus, first through the dorsal tegmental nucleus, then through the lateral mamillary nucleus, and then from the anterior dorsal thalamic nucleus to the retrosplenial cortex (Taube, 2007), and the inputs to the anterior dorsal nucleus have also been defined as Class 1, or driver (Petrof and Sherman, 2009). It would be reasonable to expect that the axons from the deep tegmental nucleus to the medial mamillary nucleus would also be acting as drivers on the route to cortex, like those from the dorsal tegmental nucleus to the lateral mamillary nucleus, but there is at present no experimental support for this view, and it is more common to consider the hippocampal input that reaches the mamillary bodies in the postcommissural fornix as the driver input to the mamillary bodies (Aggleton et al., 2010). These authors provide a detailed review and analysis of the pathways that link the hippocampus, the mamillary bodies, the anterior thalamus, and the midbrain nuclei, and they describe the memory losses that can be produced by lesions in these pathways. However, they do not raise any questions about the nature of the messages that are being passed along the pathways. On the basis of current knowledge, it is not possible to follow the message or identify the drivers to the medial mamillary nucleus in a way comparable to Taube's studies for the lateral mamillary nucleus. We lack the relevant information.

1. It is important to record that at least some of the axons in the mamillary peduncle, not included in figure 7.1, branch to supply the lateral mamillary nucleus with one branch and the medial mamillary nucleus with the other (Cajal, 1911).

2. Which, as noted in chapter 4.2, are not suited to act as drivers.

The morphological evidence suggests that the fornix (but not the mamil-lothalamic tract) provides a modulatory input to the anterior thalamus (Somogyi et al., 1978). It could be argued that if the fornix provides modulators for the anterior thalamus, it might also provide modulators for the mamillary bodies, but that is a weak argument. In the first place, the mamillary bodies are not a part of the thalamus: they have a separate developmental history, and the rules that apply to thalamic nuclei may not apply to the mamillary bodies. Further, Wright et al. (2010) have shown that the thalamic and the mamillary inputs come from two distinct cell groups in the subiculum (see figure 7.1). In addi-tion, the axon terminals from the fornix that have been described in the mamil-lary bodies (Allen and Hopkins, 1989) are somewhat larger and contain rather more mitochondria per terminal than the terminals from cortical layer 6 that are generally seen in the thalamus, and these larger terminals are features associated with Class 1 (driver) inputs at other sites (Covic and Sherman, 2011; Viaene et al., 2011b). These terminals also appear to make multiple synaptic contacts, another feature that is not expected for the layer 6 inputs to thalamus and is more like the appearance expected from a Class 1 input (see chapter 5).

The conclusion is that the drivers providing the inputs to the cells of the medial mamillary nucleus need to be defined. We need to learn about the details of their actions in terms that allow a distinction between Class 1 (driver) and Class 2 (modulator) inputs (see chapter 4), and, as for the lateral mamillary nucleus, we need to learn about the nature of the messages that the medial mamillary nuclei pass to the anterior thalamus and define where those mes-sages come from. They may come from the fornix or from the mamillary peduncle, or from both. The mamillary peduncle would be likely to bring a link with lower centers and the world, but the nature of that input needs to be defined on the basis of the properties of cells in the deep tegmental nucleus. If the fornix provides the drivers, then this would probably be a link with established memory circuits. If both turn out to be drivers, then there would be an interesting question about exactly how the two relate to each other in the medial mamillary nucleus.

There is one further related unknown in these pathways, and that concerns the action of the fornix input on their target cells in the lateral mamillary nucleus. A reasonable guess would treat these as modulators of the head direc-tion relay in the lateral mamillary nucleus, comparable to the input from the fornix to the anterior thalamic nuclei, and acting rather like layer 6 inputs to thalamic nuclei. However, at present this is an unknown. They could provide a second driver input.

We have explored the anterior thalamic nuclei in some detail, because they serve to illustrate two important questions. One is about the relation of the thalamic relay to the lower centers and the world, a question that is answered for the anterior dorsal nucleus by the demonstration of head direction cells that depend on vestibular inputs, shown as thick lines in figure 7.1. That is, we know where the messages that are relayed to cortex are coming from and we know about the nature of these messages. The second question concerns the identification of drivers and modulators in the pathways. The transfer of information about head direction through the anterior dorsal nucleus also tells us about the drivers in this part of the pathways (thick lines), except that they leave open the possibility of a second driver input to the lateral mamillary nucleus from the postcommissural fornix (thin lines). In contrast to this, our current knowledge of the pathways through the medial mamillary system and the two large anterior thalamic nuclei does not provide clear information about which inputs are drivers and which are modulators, demonstrates no clear functional link with the world, and at present provides no information about the messages that are being passed along the pathways to the cortex; they are all shown as thin lines in figure 7.1.

Whereas in the previous section (7.2.1.1), we stressed the need for information about the function of the extrathalamic branches of the afferents, in this section we have focused on the need to distinguish the drivers from the modulators and then to identify the message that the drivers carry. From this point of view, the anterior thalamic nuclei provide an example that can be usefully compared with the evidence discussed for the dorsal medial geniculate nucleus and ventral anterior thalamic nucleus in chapter 4.2 and 4.3. However, whereas for those nuclei the functional properties of the relevant pathways have been identified, for the mamillothalamic pathways that information still remains largely undefined for the two largest nuclei of the group. It should be noted that, in the mamillary pathways, the branching patterns are also important and that their functions are still unexplored (see chapter 6).

One alternative approach for elucidating the nature of the messages that are relayed by a thalamic nucleus to the cortex is to look at the functions of the cortical area(s) that receive the thalamic inputs and then use this information to gain insights into the messages carried by their inputs. For the anterior medial thalamic nucleus, studies of the anterior limbic cortex can suggest a range of possible functions. On the basis of human and animal experiments and of imaging and postmortem studies, the functions or functional losses that have been reported for the anterior limbic cortex relate to autism (Simms et al., 2009), geriatric depression (Gunning et al., 2009), borderline personality disorder (Whittle et al., 2009), psychosis (Fornito et al., 2008), and to mechanisms for predicting aversive events and terminating fear (Hayes and Northoff, 2011; Steenland et al., 2012) or evaluating the cost of foraging (Kolling et al., 2012). This cortical area has also been described as serving functions relating to spatial working memory in rats (Mendez-Lopez et al., 2009), has been proposed as an area that relates to emotional processing in general (Etkin et al., 2011), as a mechanism supporting the selection and maintenance of learned options (Holroyd and Yeung, 2012); it has also been proposed as an area that relates to the production of a natural smile, as opposed to a smile posed for a photographer (Damasio, 2006). This is a list from a larger set of claims for the anterior limbic cortex, and a comparable list might be prepared for the posterior limbic cortex.

None of the items in this list is readily related to anything known about the mamillothalamic pathways except that degeneration of the mamillothalamic tract, fornix, or mamillary bodies is associated with memory losses (Aggleton et al., 2010) and that a transient Korsakoff's syndrome has been reported after a bilateral removal of the anterior limbic area (Brodal, 1981). That is, the listed functions may all be related in different ways to memory functions, and that would suggest that the fornix afferents coming from the hippocampus are the drivers in the mamillary bodies on the basis of the lost memory functions and probably modulators in the anterior thalamus on the basis of the fine structural evidence (Somogyi et al., 1978) (see figure 7.1). But at present this is a guess. We have no good evidence to come to any clear conclusions about the nature of the messages that the two larger anterior thalamic nuclei transmit to cortex or receive from the midbrain. We do not know whether the medial mamillary part of the system has any links with the world through the deep tegmental nucleus, and in terms of the activity within the pathways, we know nothing about the messages that are being transmitted. The contrast between the experimental and clinical reports for the anterior medial nucleus and the anterior limbic cortex on the one hand and the experimental reports for the anterior dorsal nucleus and the retrosplenial cortex on the other represents an important conceptual difference in approaches to study of the cortex. The former is an attempt to localize functions to cortical areas without concern for the thalamic inputs, or for any inputs, which may be bringing messages relevant for those functions to the cortex, whereas the latter involves tracing the driver inputs to thalamus and understanding the messages carried in those inputs from the body and the world through several defined relay stations.

7.2.1.3 The Inputs to the Ventral Lateral Nucleus   The ventral lateral thalamic nucleus is included here, because it is another first order thalamic relay nucleus whose inputs do not come directly from sensory pathways but from the cerebellum, whose links to the world are complex. They need to be defined if we wish to understand the nature of the message that the cortex is receiving from the thalamus. The projection of the ventral lateral nucleus to the motor cortex has often played a more important role in interpreting the nature of the thalamocortical message than has the cerebellar input. That is, the cerebellar input can be interpreted as the route for cerebellar influences on the motor cortex, often without specific questions being raised about the nature of the information sent to the cortex. However, it may well be that the message(s) can be understood on the basis of what the cerebellum sends to the thalamus rather than on the basis of the functions of the motor cortex itself: as already indicated, looking back from cortex to thalamus for a view of what the thalamus is transmitting to cortex may be more difficult than tracing the message the other way.

The inputs to the ventral lateral nucleus come from the deep cerebellar nuclei, mainly the dentate nucleus and also to some extent from the interpositus nucleus (Hoover and Strick, 1999; Teune et al., 2000). However, these inputs to the thalamus represent only a small part of the outputs from these cerebellar cell groups, which also send a rich innervation to many parts of the brainstem, including the red nucleus and the pontine reticular formation as well as more caudal parts of the brainstem concerned with motor control. The extent to which the thalamic inputs come from branches of these brainstem afferents is largely unknown, although we saw in chapter 6 that the Golgi method demonstrates a rich pattern of cerebellofugal axons giving rise to thalamic inputs with one branch and to brainstem centers with the other. One issue that will be important for understanding the message(s) that the ventral lateral nucleus passes to the motor cortex is to learn, in the first place, which of the cerebello-thalamic axons has branches to brainstem and, in the second place, to define these termination sites and the actions at those brainstem centers.

The inputs to the relevant deep cerebellar nuclei come from the lateral cerebellar hemisphere, and it has been shown that one of the functions of this input to the ventral lateral nucleus is a role in guiding limb movements to distant, visual targets (Stein and Glickstein, 1992). The cerebellar discharges related to visually guided movements precede activity in motor cortex by about 20 ms (Thach, 1975; Liu et al., 2003). On the basis of this delay, it has been suggested that the visual trigger for the movements passes from the parietal cortex to the cerebellum and is then passed through the thalamic relay to the motor cortex.[3] Miall and King (2008) have proposed that this part of the cerebellum produces a "forward model" that predicts the actions that will be produced by currently active motor commands. They consider that "The

---

3. It may be important to recognize that a transthalamic connection from parietal cortex to motor cortex may prove to be more direct (see chapter 5).

cerebellum receives ascending proprioceptive inputs and efferent copies of descending motor commands, and it outputs to cortical and brainstem motor nuclei." The forward model is thus seen here as created in the cerebellum by messages coming in corticopontocerebellar pathways, and from ascending axons from the spinal cord that are likely to be efference copies as well (chapter 6.2).[4] That is, efference copies are established in the cerebellum before the branched cerebellothalamic axons (see figure 6.6) send one branch to brainstem motor centers and another to the thalamus. Two points may be relevant for this schema. One is that that there are many possible ways of feeding copies of motor instructions into the pathways that eventually reach the thalamus and themselves represent copies of motor instructions as can be seen for the cerebellothalamic pathways that give off motor branches before entering the thalamus (figure 6.6). Another is that there is a rich transthalamic corticocortical pathway through the ventral anterior thalamic nucleus from and to motor and premotor areas in the frontal cortex and that the corticothalamic driver branches from layer 5 in this link are all likely to be branches of long descending corticofugal axons, including corticopontine axons. The cortico-pontocerebellar pathway may thus represent but one small part of the circuitry that is providing a forward model to the frontal cortex; the transthalamic cor-ticocortical pathway through the ventral anterior nucleus may provide another and more direct link for visually guided movements. However, the extent to which the final smooth movement is based on components added by the cer-ebellum is largely unexplored in terms of the several other pathways that are likely to be relevant. These include, on the input side for vision, the retinal afferents to the tectum and pretectum as well as those to the cortex via the lateral geniculate nucleus, and for the arm, the spinocerebellar and the lemnis-cal pathways through the ventral posterior nucleus to the somatosensory cortex. In addition, there are the links established between vision and the arm movement not only in the parietal cortex but also in the superior colliculus and in brainstem motor centers, including the red nucleus. The importance of the tectum is easily ignored once we forget that frogs can catch flies; the projections from the cortex to the superior colliculus (Harting et al., 1992) and the tectopontine projections (see Schwarz et al., 2005) will also be relevant. We introduce these several pathways here, because, as indicated in chapter 6, several of the corticotectal and corticopontine axons also send branches to

4. Not only do the axons that supply Clarke's column (which gives origin to the spinocerebellar tract) form as branches that are likely to be innervating other spinal centers, but also the spino-cerebellar axons themselves give off branches in the caudal medulla before they enter the inferior cerebellar peduncle (Cajal, 1911).

higher order thalamic relays, and these are considered in the next section. The interim conclusion is that there are a great many pathways that are able to provide information about instructions for upcoming movements and contribute to forward models. The thalamocortical pathways are everywhere carrying efference copies, and contributions to forward models are available in many different pathways. Once a functional role is accepted for the branching axons, efference copies that can contribute to forward models will be seen as extremely common in all parts of the central nervous system.

7.2.2   The Inputs to the Higher Order Thalamic Relays

Our treatment of the higher order nuclei can be brief, primarily summarizing the limited knowledge about what amounts to the largest and still most mysterious parts of the primate thalamus. Higher order relays are defined as those that receive their driving inputs from layer 5 of cortex (see chapters 3 and 5). In several higher order thalamic nuclei, there is a possibility that first and higher order relays are present (see chapter 5), and to allow for this possibility, when we speak of first order and higher order nuclei, instead of relays, we regard first order nuclei as lacking corticothalamic drivers, whereas higher order nuclei have them. So far as the nature, driver or modulator, Class 1 or Class 2, of the corticothalamic axons is concerned, as indicated in chapters 3 and 5, in the thalamus we regard as drivers all of the axons that form large, excitatory terminals making many synaptic junctions often not close to astrocytic processes and often forming serial synaptic junctions (triads) and that have the physiological properties of Class 1 inputs as defined in chapter 4. Where an origin from cortical layer 5 rather than layer 6 is known, this strengthens the interpretation. However, for most higher order relays, this interpretation is still tentative, awaiting detailed information either about the nature of the message and its transmittal to cortex or about the properties of the transmission at the thalamic relay that would identify the input as Class 1 or Class 2 (or other). We need to ask questions (generally expecting few answers yet) about the nature of the messages that these layer 5 axons are carrying, and we expect that the message that reaches the thalamus will be relayed to cortex unless blocked at the thalamic gate (see chapter 5). We showed in chapter 3 that there are higher order relays that transmit visual messages through the lateral posterior nucleus or the pulvinar in rodents, cats, or monkeys, that higher order somatosensory messages are relayed through the posterior medial nucleus of the mouse and higher order auditory messages through the dorsal medial geniculate nucleus of the mouse, but for most of the other higher order nuclei, such as the lateral dorsal, medial dorsal, and central medial nuclei, we have little or no information about the nature of the message that is relayed, nor can we be certain that a message is relayed, although the possibility that some thalamic nuclei transmit no message to cortex seems extremely unlikely. (We exclude the thalamic reticular nucleus, because it is part of ventral rather than dorsal thalamus; see chapter 3.2.)

Where a higher order pathway has been identified, information about the nature of the message that is passed to the thalamus may be gleaned from evidence about the activity in layer 5 of the cortical area from which these axons arise or, where a descending branch of the corticothalamic axon has been identified, from studies of the functional relationships established by these axons at their terminal sites. However, these are questions for future experiments; they have been explored to only a limited extent. We can take the layer 5 cells in area 17 as an example. They send axons to the superior colliculus with branches to the lateral posterior nucleus or pulvinar. They have complex receptive fields and are mainly binocular. Their axons terminate in the most superficial layers of the superior colliculus, far from the major motor outputs, which arise in the deeper layers of the colliculus (Harting et al., 1992). This suggests that these axons are likely to play a minor role in defining the output of the colliculus. However, in the superficial layers they overlap with the terminals of corticotectal axons from many higher visual and other cortical areas (including the frontal eye fields) that also have deeper terminals and that are likely to play a more significant role in the collicular control of gaze. These many visual inputs into a phylogenetically old center for the control of movements (Grillner et al., 2007) suggest that the inputs from layer 5 of area 17 are likely to make a minor contribution to the control of gaze, and that a message about that contribution will be passed from the higher order thalamic relay (pulvinar or lateral posterior nucleus) to higher visual areas, contributing to the relevant forward model. The observation that minimal localized stimulation of the deep layers of area 17 produces ocular movements (Tehovnik et al., 2003) lends some support to such a view.[5]

It should be stressed that we currently have very limited information about the extent to which thalamocortical pathways are drivers (Class 1) or modulators (Class 2); see chapters 4 and 5. As indicated in those chapters, the evidence that some thalamocortical relays are class 2 and thus not drivers transmitting messages raises major questions about many of the transthalamic cortical con-

---

5. The extent to which the strength of the motor outputs increases for higher levels of the corticortical hierarchies and may perhaps match the increase in the frequency with which forward receptive fields can be observed in the visual hierarchy merits investigation (Melcher and Colby, 2008).

nections considered here and in the following sections with the exception of those where there is clear evidence that a defined message is transmitted from the thalamus to the cortex or where the thalamic inputs have been identified on the basis of their morphological criteria as drivers, or functionally as either Class 1 or Class 2 (see chapter 4). At present, these are all pathways that are transmitting messages about the classical sensory inputs to the brain. There is an urgent need for information about which of the many remaining transthalamic pathways are relaying messages to the cortex and also for identification of the nature of the message. This is an important point to bear in mind in the rest of the book, but one that will not be repeatedly raised.

7.3   Outstanding Questions

1. For the ascending and cortical afferents to thalamus, the thalamocortical pathways, or the corticofugal axons to motor centers, which are drivers (Class 1) and which are modulators (Class 2) for each group of axons?

2. What exactly can we say about the messages carried in any of the pathway groups in question 1 apart from those that involve first order inputs to sensory nuclei? For example, what are messages that the ventral lateral nucleus is sending to the motor cortex or the anterior medial nucleus is sending to anterior limbic cortex?

3. Where we can identify the structures innervated by the nonthalamic branch of a driver input to thalamus, can we use information about the nature of the structures innervated by that branch, and their actions in response to stimulation of those branches, to provide clues about the nature of the message that the thalamic branches bring to the thalamus?

4. Where a cortical area has some descending layer 5 axons that send a branch to the thalamus and some that do not, what other functional characteristics distinguish these two types of axon arising from the same area of cortex?

5. Where the functions of a cortical area have been defined on the basis of losses produced by lesions or of conditions that produce activation of that area, is there a strategy that will allow us to trace the thalamic input to those functions back to a thalamic relay and its afferents?

8 The Inputs to the Cortex from the Thalamus and the Cortical Descending Outputs

8.1 Early Studies of Thalamocortical Relationships

Early studies, well summarized by Walker (1938) and by Polyak (1957), were based on the Golgi method and on Nissl preparations of normal brains (for example, Kölliker, 1896; Cajal, 1911) and were extended by experimental methods that used the Marchi method for tracing degenerating axons or recorded retrograde degeneration in thalamic cells after localized cortical lesions. Walker's own experimental studies have significantly influenced our current views about thalamocortical relationships; contemporary textbooks often still owe much to his account, as does our introductory material in chapter 1. Walker used the method of retrograde degeneration, which is based on localized cell losses in the thalamus produced when thalamocortical axons are cut by removal of limited cortical areas. The thalamic changes can be very severe, as illustrated in figure 8.1, showing clearly localized, heavy cell losses in a well-defined sector of the lateral geniculate nucleus after a small lesion in the visual cortex. However, the degeneration is not always as severe and clearly outlined as this, and Walker's drawings (figure 8.2) provide some rough indication of the differences. He linked each major architectonically defined thalamic nucleus to a particular cortical area defined on the basis of cyto-architectonic studies and in his summary figure (figure 8.3) showed a high density of thalamic input to the classical sensory and motor areas and sparser or even absent input in other areas, a view more recently presented and reviewed critically by Macchi and Bentivoglio (1999). Specifically, Walker showed the temporal lobe, as a white area, receiving no thalamic afferents.

The severity of thalamic degeneration illustrated in the literature by photographs[1] rather than drawings, which inevitably carry a large subjective component, vary greatly, but it is clear that the cell loss is often less severe and

---

1. Many of these figures reproduce poorly from the Web sites and are not shown here.

Figure 8.1
Photograph of retrograde degeneration in the lateral geniculate nucleus of a mandrill (*Mandrillus sphinx*) showing a sector of retrograde degeneration produced by a lesion in the striate cortex (area 17). From Kaas et al. (1972) with permission from S. Karger AG Basel.

lacks the clear boundaries shown in figure 8.1; see for example the degenerative loss illustrated by Scollo-Lavizzari and Akert (1963) in the monkey's medial dorsal nucleus after a frontal lesion. This raises an important point for interpreting much of the evidence on thalamocortical projection patterns. The severity of the degeneration may relate to the density of the projection as suggested by Walker's results (figures 8.2 and 8.3), but the severity may also represent the susceptibility of the thalamic cells to damage of their axons, quite unrelated to the density of the projection. We return to this problem in section 8.2.2.

Ten years after the publication of Walker's book, Rose and Woolsey (1948a, b) used the method of retrograde degeneration to study thalamocortical projections to the limbic cortex and the orbitofrontal cortex. Their studies were an attempt to define a single thalamic nucleus in terms of its connections to an architectonically defined area of cortex, generalizing a relationship that had earlier been most clearly established for the lateral geniculate nucleus and the primary visual cortex (see Minkowski, 1914) and was most striking in

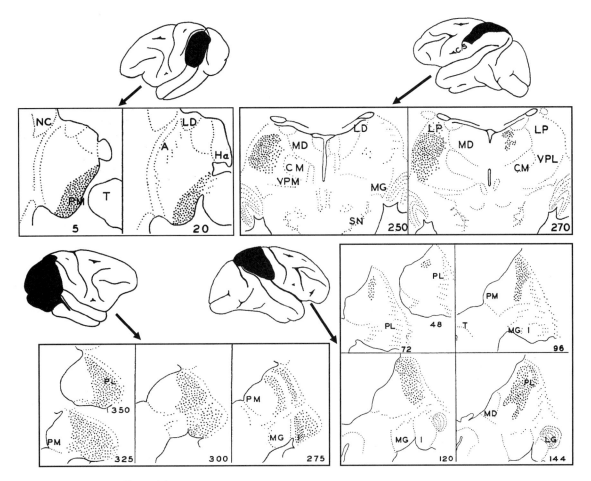

Figure 8.2
Figures from Walker (1938) to show the distribution of retrograde degeneration in the monkey thalamus after four different cortical lesions. A, pars angularis of lateral posterior nucleus; CM, center median nucleus; Ha, habenula; I, inferior pulvinar nucleus; L, nucleus limitans; LD, lateral dorsal nucleus; LG, lateral geniculate nucleus; LP, lateral posterior nucleus; MD, medial dorsal nucleus; MG, medial geniculate nucleus; NC, caudate nucleus; PL, posterior lateral nucleus; PM, medial pulvinar; SN, substantia nigra; T, tectum; VPL, ventral posterior lateral nucleus; VPM, ventral posterior medial nucleus.

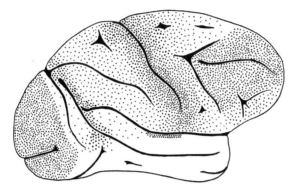

Figure 8.3
Illustration from Walker (1938) to show the density of retrograde degeneration seen in the thalamus after cortical lesions and here interpreted as the "intensity of the thalamic projection." From Walker (1938).

Walker's results for the first order thalamic nuclei. Rose and Woolsey's 1948 studies also represented a reaction to a strange but influential study published two years earlier by Lashley and Clark (1946), which had claimed that architectonic subdivisions of the cerebral cortex could not be justified on histological grounds: Lashley and Clark reported that there were essentially no architectonic borders and no separate, cyto-architectonically identifiable cortical areas in the monkey. Although this interpretation was counter to many earlier studies, it was in accord with Lashley's own previously published observations of rats learning a maze and his conclusion that cortex was equipotential for this performance (Lashley, 1931). Rose and Woolsey concluded that in terms of the demonstrable thalamocortical connections, there is a clear match between identifiable thalamic nuclei and cortical areas that could be defined by cyto-architectonic criteria. This seemed to confirm Walker's interpretation in terms of specific individual nuclei linked to identified cortical areas. It also, of course, was a strong denial of the Lashley and Clark (1946) claim.

Although their study served as a guide for many later studies of thalamocortical relationships, Rose and Woolsey's basic premise, linking single thalamic nuclei to single cortical areas, broke down not long after their 1948 studies, significantly on the basis of their own later publications described below. Today, we need to recognize complexities revealed by more recent methods, but in the 1950s these complexities were emerging slowly.

Multiple cortical areas for single sensory modalities began to appear in the 1940s and 1950s: Woolsey and Fairman (1946) for somatosensory pathways; Rose and Woolsey (1949) for auditory pathways; Thompson et al. (1950) and

Clare and Bishop (1954) for visual pathways; see Thompson et al. (1963) for an overview. The thalamic inputs to the newly defined secondary areas proved more difficult to demonstrate. Rose and Woolsey (1949) found clear evidence of thalamic cell losses in the (first order) ventral medial geniculate nucleus after lesions of the primary auditory area but reported no significant retrograde changes in any part of the medial geniculate complex of the thalamus, or in any other part of the thalamus, after damage to surrounding secondary auditory regions that also responded to cochlear stimulation. They returned to this problem in 1958 (Rose and Woolsey, 1958) and defined a second type of thalamocortical connection as a "sustaining" projection, concluding, somewhat tentatively, that a single thalamic nucleus might be innervating two separate cortical areas by branching axons. In their experiments, a lesion in one or another of the higher auditory areas produced little or no retrograde change in the thalamus, whereas lesions that involved both areas produced significant retrograde degeneration in the (higher order) dorsal medial geniculate nucleus (Rose and Woolsey, 1958). That is, one branch alone might sustain thalamic cells, whereas damage to both branches produced the classical retrograde changes. This was indirect and not entirely conclusive evidence for the presence of branching thalamocortical axons, as they recognized (Rose and Woolsey, 1958), and the method has not been widely used. However, it suggested a difference between the thalamocortical pathways for first order as opposed to higher order thalamic nuclei, and it opened a door to later studies of branching thalamocortical axons that innervate more than one cortical area (see section 8.2.2).

8.2   More Recent Views of the Thalamocortical Pathways

Walker's summary figure (figure 8.3) raised two important questions about the thalamocortical pathways. One concerns the absence of an input to the temporal lobe, and the other concerns the relative density of inputs to different cortical areas. Rose and Woolsey added a third question about the extent to which any one thalamic nucleus or cell might innervate more than one cortical area. The answer to Walker's first question, that there is good evidence for an input to the temporal lobe, is summarized in the next section, but his second question is currently unresolved and will need a method for quantitative evaluation of thalamocortical afferents, which currently does not (to our knowledge) exist (see end of next section). In response to the third question, we show in section 8.2.2 that many thalamic nuclei and many cells innervate more than one cortical area, but that here also techniques for adequate quantification of this feature are not yet available.

8.2.1 Thalamic Afferents to the Temporal Lobe

Shortly after the publication of Walker's book, Chow (1950) and Simpson (1952) used the same method as Walker and described a small zone of retrograde cell loss in the medial part of the pulvinar produced by lesions in monkeys of the anterior pole of the temporal lobe. More recently, a variety of retrogradely transported markers have been available and have been injected in several different cortical areas and in different combinations, revealing thalamic cells that send their axons to the temporal lobe, including its anterior pole. We consider these in some detail, not only to put the all-white temporal lobe in Walker's figure (figure 8.3) to rest, but also because they introduce a view of cortical areas receiving from a much wider spread of thalamic nuclei than could be revealed by the method of retrograde degeneration.

Yeterian and Pandya (1989), using retrogradely transported horseradish peroxidase, showed that all parts of the temporal lobe of monkeys receive thalamic inputs. Cells in the medial pulvinar and to a lesser extent in the lateral pulvinar send axons to the superior temporal gyrus and sulcus. In addition, cells in the suprageniculate, medial geniculate, ventral posterior lateral, central lateral, parafascicular, paracentral, and dorsal medial nuclei send axons to the temporal lobe. From the point of view of Walker's figure, it is important to note that there were thalamic projections even to the rostral parts of the temporal lobe. Webster et al. (1993) injected tritiated amino acids and horseradish peroxidase into the temporal cortex and showed reciprocal thalamocortical connections between the pulvinar and the temporal cortex, including their posterior area "TEO" and their anterior area "TE," which includes the rostral pole. They also described connections to the temporal cortex from the paracentralis, ventralis anterior, centralis, and the limitans nuclei. Evidence that the areas of the monkey's auditory cortex, including secondary areas that extend well toward the tip of the temporal lobe, receive thalamic inputs has been presented (Hackett et al., 1998), and there is also evidence that some visual areas ("MT," "MST") of the temporal lobe receive inputs from the pulvinar (Ungerleider et al., 1984; Kaas and Lyon, 2007).

It is reasonable to conclude that all parts of the temporal lobe receive thalamic inputs, and also that there are a surprising number of thalamic nuclei apart from the pulvinar that provide afferents to temporal cortex. The white areas of cortex on Walker's figure (figure 8.2) do not lack a thalamic input.

However, the second question about the relative densities of the cortical innervation remains unanswered. The inputs to the temporal lobe may well be relatively sparse. We cannot say anything rigorous about this or about the degree of overlap that may exist between any two of the thalamocortical projections that have been described. There are many relevant factors: perhaps the most important issue is that for the observations based on cortical injections (or lesions), the number of retrogradely labeled thalamic cells (or of lost cells) tells us nothing about the size of the terminal arbor of any one cell or the density of its terminals. Further, estimating the size of the injection site (or the lesion) is notoriously difficult. Anterograde methods that avoid involvement of fibers of passage, such as the autoradiographic tracing method, can provide a better view of how dense the projection from any one thalamic nucleus to a particular area of cortex is, but observations based on a single thalamic injection cannot take account of other thalamic projections to the same area. It is possible that molecular markers that label all thalamocortical axons specifically might provide a view such as Walker was aiming for in figure 8.3, but then the issues raised in chapter 5 have to be taken into account: we also need to distinguish the Class 1 drivers from the Class 2 modulators if we want to understand the functional significance of such a map of density distributions.

8.2.2 Thalamic Nuclei That Innervate More than One Cortical Area

It is now clear that, in spite of the severe and clearly marked retrograde degeneration shown in figure 8.1 after a lesion of area 17, there are some cells in the lateral geniculate nucleus of the cat and monkey projecting to areas beyond area 17. Thus, Gary and Powell (1967) showed in the cat that there are large geniculate cells that survive after a cortical lesion restricted to area 17 but degenerate when the cortical damage includes area 18. They also showed that in the cat but, contrary to more recent evidence, not in the monkey, lesions of the lateral geniculate nucleus produced anterograde fiber degeneration that extended beyond area 17 (Garey and Powell, 1971). More recently, other techniques have also been used to demonstrate that there are geniculocortical axons in cat and monkey that terminate in higher visual cortical areas beyond area 17: for the monkey (Yukie and Iwai, 1981; Sincich and Horton, 2003); for the cat (Stone and Dreher, 1973; Humphrey et al., 1985).

Conversely, pulvinar cells whose major cortical projection goes to extrastriate cortex also send some axons to area 17 itself. Thus, area 17 in the monkey receives inputs from the pulvinar (Ogren and Hendrickson, 1977; Rezak and Benevento, 1979) even though the pulvinar shows no clear retrograde changes after lesions of area 17. In the thalamocortical pathways of the monkey, these components appear to be relatively small, but for many other thalamocortical connections, such extra connections are now recognized as playing a larger role and have made the interpretation of the method of retrograde degeneration more questionable. The method has been very useful for establishing major links, providing a broad overview of major pathways, but relatively minor links revealed more recently in terms of individually labeled cells have added to the complexity of the thalamocortical connections, raising new questions and adding new complexities to the older simpler schemes.

The use of two or more distinguishable retrogradely transported markers (for example, Bentivoglio et al., 1980; Cavada et al., 1984) has demonstrated many examples of branching thalamocortical axons, some of which are considered in the next section. Whereas the observations of Rose and Woolsey (1958) seemed to show that the first order thalamocortical projections of the auditory pathway differed in their axonal branching patterns from the higher order projections, at present all we can say is that the patterns of retrograde cellular degeneration differ, but, as indicated above, this does not relate directly to knowledge of the innervation pattern. The possibility that higher order thalamic relays have more widespread connections than do first order relays merits investigation; evidence for the temporal lobe considered above and also for some of the thalamic nuclei included in the next section suggests that the first order nuclei may well not be characteristic of the rest of the thalamus in terms of the axonal branching patterns of their thalamocortical axons.

8.3   The Topography of Thalamocortical Projections

The relationships between thalamus and cortex demonstrable by retrograde degeneration or by the retrograde transport of identifiable markers consistently show a pattern of thalamic disks, laminae, sectors, or strips of thalamic cells that relate in an orderly topographic manner to the cortex. Figure 8.1 shows a sector of the lateral geniculate nucleus that receives its input from a small patch of retina and projects to a small area of cortex. Comparable but not identical arrangements can be defined in the ventral medial geniculate nucleus, where the cells show a laminar arrangement (Morest, 1965) that corresponds to the frequency tuning of the thalamic auditory cells (Aitkin and Webster, 1972). In the ventral posterior nucleus, there is an arrangement of rostrocaudally running rods or lamellae that can be revealed by cytochrome oxidase staining (Rausell and Jones, 1991) and by the arrangement of the thalamocortical projections (Jones et al., 1979; Jones and Friedman, 1982), and each of these rods appears to correspond to a small area of the body surface (Mountcastle and Henneman, 1952).

These topographic regularities are not limited to the primary sensory pathways. Asanuma et al. (1985) demonstrated a disk-like or laminar distribution of retrogradely labeled cells in the monkey's medial pulvinar after injections of retrogradely transported fluorescent marker into area 7 of the parietal cortex, with some labeled cells also in the central lateral nucleus, which appeared not to form a part of the disks. Kievit and Kuypers (1977) (figure 8.4) showed cortical areas of the frontal lobe relating to strips or bands of labeled cells extending through more than one thalamic nucleus and even crossing the internal medullary lamina. The anterior tip of the frontal lobe is shown in this figure to be connected to the most medial parts of the thalamus in strips that included the anterior ventral and medial dorsal nuclei as well as the medial pulvinar; the posterior parts of the frontal lobe are connected to the lateral and anterior parts of the thalamus, including the ventral lateral nucleus. This represents an essentially continuous strip of thalamocortical connections with a significant overall pattern of topographically organized connections stretching across the borders of several thalamic nuclei, sending inputs from several different thalamic nuclei to quite widespread cortical areas. The medial dorsal nucleus in this figure is shown projecting to the rostral parts of the lateral aspect of the frontal lobe, and the medial pulvinar contributes to some of these same cortical areas. The labeling of individual thalamic cells by the retrogradely transported label revealed a pattern of thalamocortical connections that was hidden by the early cruder method of retrograde degeneration. However, Pribram et al. (1953) had earlier used the method of retrograde cell degeneration and had shown a topographic organization of dorsoventrally arranged cell columns or strips in the medial dorsal nucleus in rough agreement with the results of Kievit and Kuypers (1977) but had limited their observations of thalamic changes to the medial dorsal nucleus, not extending them to possible connections that the medial pulvinar might be making with the more anterior parts of the frontal lobe. Later, Akert and Hartmann-von Monakow (1980) used anterograde labeling with tritiated amino acids of thalamocortical axons and confirmed the cell columns or laminae described by Kievit and Kuypers (1977) but stressed that the intralaminar nuclei were included in these projections to the frontal cortex. They did not, however, include injections of the medial pulvinar in their studies. Romanski et al. (1997) studied retrogradely transported label from the frontal lobe in monkeys and reported the distribution of thalamic label in the medial dorsal nucleus and the medial (and also the lateral) pulvinar. The continuous strips or columns illustrated in figure 8.4 were less apparent in their material,

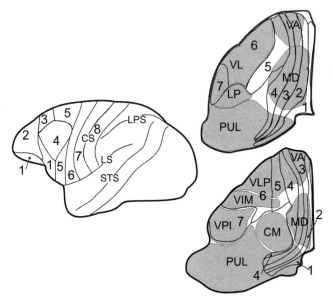

Figure 8.4
Schematic representation of the relationship between areas of frontal cortex and columns of cells that pass through several thalamic nuclei. Based on the retrograde transport of horseradish peroxidase reported by Kievit and Kuypers (1977). The numbered cortical areas in the frontal lobe on the left correspond to labeled thalamic regions identified by the same numbers in the two horizontal sections shown on the right. Some of the major thalamic nuclei are identified on the basis of the terminology used by Kievit and Kuypers. CM, center median nucleus; CS, central sulcus; LPS, lateral parietal sulcus; LA, lateral sulcus; LP, lateral posterior nucleus; LPS, lateral parietal sulcus; MD, medial dorsal nucleus; PUL, pulvinar; STS, superior temporal sulcus; VA, ventral anterior nucleus; VIM, ventral intermediate nucleus; VL, ventral lateral nucleus; VLP, ventral lateral posterior nucleus.

partly because they cut the thalamus in a different plane than that shown in figure 8.4[2] (see also Barbas and Mesulam, 1981). However, some indication of such columns can be recognized from some of their figures. Giguere and Goldman-Rakic (1988) had earlier traced anterogradely labeled thalamocorti-

2. There is an important and difficult issue relating to the plane of section that is used in the many different experiments that deal with thalamocortical connections. It would be a mistake to argue that all investigators should use the same plane, because some relationships (as indicated here) are much more evident in one plane than another. However, for many of the results reported for supposedly the same nucleus or the same part of a nucleus, comparisons are difficult to make because the section planes differ in *undefined ways*. If all section planes in any published paper could be related to defined stereotaxic planes, which in practice is fairly straightforward if fiducial section planes are established before removing the brain to serve subsequently for orienting the plane of sectioning, then the reader would have some hope of knowing whether the small group of cells described by one author is or is not the same as another group described by another author. At present, that is often impossible and makes the present account far sketchier than it ought to be.

cal axons demonstrating the orderly topography of the projection to frontal cortex from the medial dorsal nucleus and extending to the anterior limbic and the supplementary motor areas.

Experiments based on retrograde degeneration of thalamic cells or on the labeling of the reciprocally connected layer 6 corticothalamic axons show that orderly topographic projection patterns dominate in many parts of the thalamocortical projections, and one can expect this to provide clues about how the incoming messages are organized within the thalamus. For example, in the primary sensory areas, the cortical two-dimensional topography clearly relates to the distribution of sensory receptors on the relevant sensory surfaces as shown in figure 8.1, where the borders of the zone of retrograde degeneration run perpendicular to the layers of the lateral geniculate nucleus, forming lines called "lines of projection" (Bishop et al., 1962). Each line has been shown to represent a point in visual space around which the receptive fields of the geniculate cells lying along that line are scattered within a small region of visual space. Because geniculate cells lying close to that line receive their retinal inputs from several functionally distinct classes of retinal ganglion cell, and these all receive their visual inputs from that same small region in the contralateral visual field, functionally distinct geniculate cells (left eye, right eye, as well as magnocellular, parvocellular, and koniocellular in the monkey, X, Y, and W in the cat) (see chapter 3.2) are distributed to distinct geniculate layers along these lines. The borders of the degenerate geniculate segment represent the borders of the visual field loss produced by the cortical lesion, and the arrangement of cells in the layers and along the lines of projection themselves represent the distribution of the functionally distinct thalamic cells in the nucleus. That is, two dimensions of the thalamic nucleus represent the sensory surface as well as the cortical area connected to the nucleus. In the primate lateral geniculate nucleus, these are seen as separate layers, but in many other species, including the mouse and the rat, the layers are less obvious, although the lines of projection can still be defined by the retrograde degeneration or the retrogradely labeled cells produced by injections of retrogradely transported markers.

When we ask what the orderly thalamocortical projection represents in other nuclei, such as the anterior dorsal, the ventral anterior, the medial dorsal nuclei or the pulvinar, and consider the slabs or columns described earlier in this section, the only organizational rule available at present is that the two-dimensional cortex is likely to be represented repeatedly and in register in a topographic order through the three dimensions of the thalamic nuclei and that the orientation of the slabs or columns provides a hint that there are lines of projection comparable to those seen in the lateral geniculate nucleus. However, we know very little about the functional properties that characterize the three major axes of the nuclei defined by these anatomical arrangements of the thalamocortical pathways in higher order nuclei. We cannot, at present, relate them to sensory surfaces on the one hand or distinguishable functional properties on the other. It is probable, perhaps inevitable, that the orderly representation of sensory surfaces should disappear in higher cortical areas, but it is not clear what, if anything, replaces this. Goldman-Rakic and colleagues have raised the issue for the medial dorsal nucleus and the frontal cortex (O'Scalaidhe et al., 1997; Levy and Goldman-Rakic, 2000; Tanibuchi and Goldman-Rakic, 2003) and discuss how the distinct functional properties relate to the afferent pathways going to the medial dorsal nucleus. However, the possibility that the orientation of the lines of projection or the cell columns can be used to explore the distribution of functional attributes in the three major, definable axes in any one nucleus has not been fully explored. It may be reasonable to expect that an electrode passed along a column or perpendicular to a column might reveal organizational rules that show how differing functional properties are arranged in thalamocortical pathways.

Figure 6.11 illustrates the input to the lateral posterior nucleus from cortical area 18 and shows that the driver inputs from a small cortical area are distributed as small terminal nests along columns that runs obliquely through the nucleus. These columns in the lateral posterior nucleus roughly mirror those seen in the lateral geniculate nucleus about an axis representing the vertical midline of the visual field. They receive inputs from cortical areas 17, 18, and 19 (see Guillery et al., 2001), with terminals from all three areas intermingled along a column. Updyke (1977; 1981) has shown that the columns represent the retinotopic and corticotopic order in the lateral posterior nucleus, and these columns also receive afferents from other, higher visual cortical areas. The question arises what specific properties, other than visual field locus, are represented along these lines and how they are arranged along the lines of projection.

8.4   Different Types of Thalamocortical Projection

There have been several suggestions that there is more than one functionally and structurally distinct type of thalamocortical projection. The accounts all differ to a significant extent, but they all can be viewed in relation to our classification of inputs as drivers or modulators. Lorente de Nó (1938) in his account of Golgi preparations from the cerebral cortex recognized two distinct types of thalamocortical endings. One type consisted of relatively thick axons that formed dense terminals primarily in cortical layer 4 with some spill-over into layer 3. He traced these back to one of the three main classical sensory nuclei and called them "specific" axons, because they had well-localized terminals. The others were "nonspecific," thinner, apparently came from the intralaminar thalamic nuclei, and distributed sparser terminals to cortical layers 1 and 6. These spread over a greater cortical area than the specific axons. It is noteworthy that this account preceded the publications by Dempsey and Morison (Dempsey and Morison, 1941; Morison and Dempsey, 1941) in which they distinguished relatively well-localized, that is, specific "primary" cortical responses to thalamic stimulation in sensory and motor areas of the thalamus from "recruiting" responses obtained when stimuli were applied to a "rather diffuse area in the region of the internal medullary lamina," that is, nonspecific, in the intralaminar nuclei.

Frost and Caviness (Caviness and Frost, 1980; Frost and Caviness, 1980) reported the anterograde axonal degeneration produced by thalamic lesions and described thalamocortical axons going to all cortical areas in the mouse, but, like Lorente de Nó, they described two types of afferent: one that they called class I had thick axons terminating predominantly in layer 4 and the other, class II, terminated predominantly in layer 1 and had thinner axons.[3] Broadly speaking, areas of cortex in the primate that Walker (1938) had shown with a dense thalamic innervation, that is sensory and motor areas mainly, were (in the mouse) largely in receipt of class I axons, whereas other areas including those where Walker had a sparse, or absent, thalamic innervation were largely in receipt of class II thalamocortical axons. Caviness and Frost (Caviness and Frost, 1980; Frost and Caviness, 1980) made two further points about the thalamocortical projection patterns: (1) that each cortical field received afferents from many thalamic nuclei, and (2) that throughout the thalamocortical projection system, there is a topographic organization of thalamic cells and cortical terminals and that this involves both the class I and the class II axons.

Given that lesions of thalamic nuclei inevitably involve fibers of passage and that not every thalamic nucleus was included in their list of lesions, these results are hard to interpret in detail, but they raise several questions. One is about accounts of varying densities of thalamocortical innervation, because the two cell types may well produce different reports no matter what method is used, with the different fiber thicknesses producing a false impression of the relative densities. A second point concerns the extent to which any one cortical area receives from more than one thalamic nucleus, and a third

---

3. We will use Roman numerals, class I or class II, for the Frost and Caviness categories and Arabic numerals, Class 1 or Class 2, for the categories we use throughout the book.

concerns the topographic organization of the thalamocortical projection. If we treat the class II axons as comparable to Lorente de Nó's nonspecific axons, are they topographically organized everywhere as Frost and Caviness claim, or do the class II axons lack the topographic organization that Lorente de Nó described for the specific axons?

A fourth point concerns the observation that each cortical area receives inputs from several thalamic nuclei, which tends to be confirmed by experiments based on the use of retrogradely transported tracers, some of which are summarized in sections 8.2.1 and 8.2.2, and underlines the importance of defining for each cortical area not only where in the thalamus the inputs take origin, but also which inputs are Class 1 (drivers) and which are Class 2 (modulators).

Herkenham (1980), using injections of radioactive tracers into the rat thalamus, proposed a different classification of thalamocortical pathways on the basis of the cortical layer in which their axons terminated. He distinguished four distinct types of terminal pattern, with the primary sensory nuclei (visual, auditory, and somatosensory) sending afferents to layers 4 and 3, and also defined three other classes that included the intralaminar nuclei and the ventral medial nucleus in one class having terminals in cortical layers 1 or 6. Earlier, Herkenham (1979), also using thalamic injections of radioactive tracer in rats, had shown that the ventral medial thalamic nucleus has widespread projections to layer 1 of all areas of cortex, densest anteriorly with only relatively sparse projections to the deeper cortical layers that are limited to the frontal lobe rostral to the corpus callosum. Arbuthnot et al. (1990) confirmed this projection more recently, showing some topographic organization in the thalamocortical projection pattern from the ventral medial nucleus and also describing a limited termination in the motor areas of cortex extending to layer 3. Herkenham (1979), in another publication, used horseradish peroxidase as a retrogradely transported marker to show widespread ascending inputs to the ventral medial nucleus that included the globus pallidus, substantia nigra pars reticulata, the superior colliculus and periaqueductal gray, and also the brainstem reticular formation. The unusually widespread cortical projection of the ventral medial nucleus to cortical layer 1 has also been demonstrated by several other studies (Desbois and Villanueva, 2001; Monconduit and Villanueva, 2005; Kuramoto et al., 2009). Whereas Herkenham's account of the inputs to this nucleus included widespread motor connections, Villanueva and colleagues documented widespread nociceptive inputs. The extent to which the ventral medial nucleus represents a quite distinct thalamocortical relationship merits further study.

In a different approach to the classification of thalamocortical axons, Jones (1998a, 1998b, 2001, 2002) has shown that thalamic relay cells in the monkey can be distinguished on the basis of which of two calcium binding proteins, parvalbumin or calbindin, they contain. These two cell classes also differ in cell size and in the cortical distribution of their axons. The larger, "core" cells stain for parvalbumin and project primarily to layer 4 and adjacent parts of layer 3, whereas the smaller "matrix" cells stain for calbindin and project primarily to superficial cortical layers, including a dense termination in layer 1. Whereas the former have well-localized cortical terminals, the latter are described as having diffuse cortical terminals that are spread more widely across the cortex. Jones reported that the core cells are particularly concentrated in first order thalamic nuclei, whereas the matrix cells are more common in the higher order nuclei.[4] Jones, like Lorente de Nó, saw the smaller cells with the thinner axons as representing a nonspecific system concerned with linking across sensory modalities, providing a "binding" function for consciousness and cognitive events. It appears that the calcium binding proteins serve to distinguish these cell types in the monkey but do not serve in this way in other species.

More recently, Ohno et al. (2012) have distinguished two types of neuron in the posterior medial nucleus of the rat, a species that lacks parvalbumin. One neuron type is strongly immunoreactive for calbindin (in the posterior part of the posterior medial nucleus), and the other is weakly positive (anterior part of the nucleus), suggesting that the former might correspond to Jones' matrix neurons and the latter to the core neurons of monkeys. However, whereas the former projected primarily to cortical layer 1 as expected from matrix neurons, the latter projected primarily to layer 5 and were classified as neither core nor matrix by Ohno et al. (2012).

These several accounts of thalamocortical pathways not only focus on the laminar distribution of the terminals in cortex but also relate to the topographic specificity, or lack of it, in the cortex, with the clearest evidence for topographic organization in the layer 4 terminals of primary sensory areas. The evidence for a clear topography in the projections to layer 4 and its lack for at least some of the projections to layer 1 was related to the specific and nonspecific thalamocortical projections defined by physiological studies (for reviews, see Dempsey and Morison, 1941; Morison and Dempsey, 1941; Jasper and Ajmone-Marsan, 1952; Jasper, 1954; Macchi, 1993), with the

---

4. The koniocellular pathway from the lateral geniculate calbindin-containing cells appears to be a well-documented exception.

former axons providing detailed localized information about the environment through the major thalamic nuclei and the latter providing the widespread cortical arousal system through the intralaminar nuclei and the superficial layers of the cortex.

Macchi and Bentivoglio (1999) have reviewed some of these earlier accounts and have questioned the extent to which specific thalamocortical pathways can be distinguished from nonspecific pathways. They have drawn attention to the fact that the intralaminar nuclei of the thalamus do not lack a topographic projection to the cortex (see also Ullan, 1985; Royce et al., 1989; Berendse and Groenewegen, 1991). It appears that there are thalamic cell groups that have widespread projections to layer 1 of cortex, and the ventral medial thalamic nucleus appears to be the best documented of these. It also appears that the laminar distribution of thalamic terminals varies considerably not only from one thalamic nucleus to another but also for functionally distinct cell types in one nucleus such as the X, Y, and W cells in the cat or the magnocellular, parvocellular, and koniocellular groups in the monkey (for details, see Sherman and Guillery, 2006). Most of the projections that are primarily to layers 3 and 4 have some inputs to layers 1 and 6, and many of those going mainly to layer 1 also send to deeper cortical areas. The major conclusion one can draw from these and other accounts concerns our ignorance. We need to know which thalamocortical terminals are Class 1 (drivers) and which are Class 2 (modulators), and we need to learn to what extent these functions (driver vs. modulator) relate to the cortical layer of the terminations.

It is possible that the cortical layer can provide clues regarding the functional classification of the thalamocortical terminals, but this is not an interpretation to be firmly embraced until it is solidly documented for several cortical areas, because a cortical layer is merely a descriptive device, and each layer contains a great many different cell types; there is no evidence that any one layer contains the same cell types in all areas of cortex or that a number assigned to a layer by Brodmann, or by anyone else, can be a key to all of the functional properties of that layer for all of cortex. Further, because many cortical cells receive their thalamic inputs on dendrites that extend through layers other than those occupied by the cell body, the functional relevance of the layer that contains the terminals may be rather limited.

The extent to which the specific pathways show a pattern of organization that is entirely absent in a nonspecific, or "diffuse," pathway is questionable. The most strikingly widespread distribution of cortical terminals to cortical layer 1 from a single thalamic nucleus is probably that described by Herkenham for the ventral medial nucleus (Herkenham, 1979). Arbuthnot et al. (1990) describe a topographic order in that projection. As soon as there is evidence for some topographic order in a projection, the concept of a diffuse or non-specific concept is destroyed. Demonstrating that a projection is strictly "diffuse" is not easy. One projection can be more widespread than another or can show less evidence for a topographic organization, but the evidence that a projection is strictly diffuse (or nonspecific), with the implication of a random distribution, is not easy to collect and to our knowledge has never been sought.

Layer 1 is a cell-poor zone, so the likeliest target of this input is onto peripheral tufts of apical dendrites of pyramidal cells with cell bodies in layer 3 or 5. One possible insight comes from the study described in chapter 2.2 showing that activation of layer 1 inputs to a layer 5 pyramidal cell evoked no detectable response on its own but, when coupled with a separately evoked, back-propagating action potential, produced a burst of spikes (Larkum et al., 1999). Other data, which support this view, that matrix inputs to layer 1 can amplify other inputs if properly timed, include the observations that conjoint activation of core and matrix inputs to a cortical area produce a supralinear summation of responses (Llinás et al., 2002; Theyel et al., 2010a).

8.5   Cortical Outputs

To understand how thalamus and cortex relate to the body and the world, it is important to understand for each cortical area how its outputs relate to the rest of the brain. At present, the corticocortical pathways are treated as routes that can lead to the motor cortex or to memory, but in addition to these there are various other direct links from any cortical area to the subcortical motor outputs of the central nervous system. A review of these connections demonstrates that there is no known area of cortex that lacks its own (private) access to the motor apparatus by one pathway or another. Four systems of connections qualify as major motor pathways that do not go through the corticospinal tract. They are the connections established with the striatum, the tectum, the cerebellum, and the reticular cell groups and the red nucleus of the brainstem. These four regions all represent phylogenetically old neural centers present in the earliest vertebrates and contributing to the inputs to the pattern generators and lower motor centers of the brainstem and spinal cord (Murakami et al., 2005; Grillner et al., 2007b; Jones et al., 2009; Stephenson-Jones et al., 2012).

8.5.1   The Corticotectal Pathways

The corticotectal pathways come from a widespread area of cortex and distribute their terminals to varying layers of the superior colliculus, with the

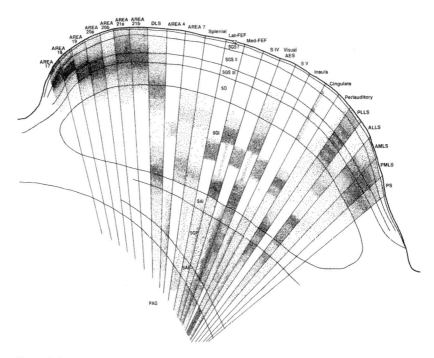

Figure 8.5
Summary of the laminar distribution of cortico-collicular inputs from 24 different cortical areas
in the cat. Based on the study of pathways labeled by cortical injections of tritiated amino acids.
From Harting et al. (1992) with permission.

terminals in the deepest layers being those most likely to be directly involved
in motor actions (figure 8.5) (Keizer et al., 1987; Harting et al., 1992; Bajo
et al., 2010; Baldwin and Kaas, 2012). The tectum represents a phylogenetically old part of the vertebrate nervous system with significant motor outputs
to the brainstem and spinal cord. In the mammalian brain, it receives afferents
from visual, auditory, and somatosensory pathways (Dräger and Hubel, 1975;
Alvarado et al., 2009). Many of the corticotectal axons that innervate the
superior colliculus have been shown to have branches that provide driver
inputs to higher order thalamic relays (see chapter 6).

8.5.2   The Corticopontine Pathways to the Cerebellum

The corticopontine pathway serves to link the phylogenetically relatively new
system of neocortical circuits with cerebellar motor controls that have a long
history in the evolution of vertebrates, longer than that of the neocortex. Most
of the cerebellar outputs in mammals feed to brainstem nuclei such as the red

nucleus, the brainstem reticular nuclei, or the vestibular nuclei, each innervating spinal motor mechanisms. The corticopontine pathways arise from extensive areas of cortex including prefrontal, posterior parietal, cingulate, and hippocampal cortical areas (see chapter 7.2.1.3). These pathways have been demonstrated in the rat (Wiesendanger and Wiesendanger, 1982a, b), in the cat (Brodal, 1981), and in the monkey (Brodal, 1978; Glickstein et al., 1985). The studies were based on methods of anterograde axon tracing or retrograde labeling of cortical cells after large injections into the pontine nuclei. When account is taken of errors that may occur in these experimental methods, it is reasonable to ignore negative results where positive results are reported. Particularly where methods based on the retrograde transport of marker molecules are used to demonstrate a projection, it is likely that corticopontine axons that also send branches to other centers, such as the thalamus, may show much less label than axons that have no non-pontine branches. Although there are some disagreements in these accounts about areas of cortex that do not project to the pons, there are few cortical areas for which positive evidence of a projection is lacking. These include extrastriate cortical areas in the temporal lobe, but not those in the parietal lobe, and also include striate cortex (area 17) itself, where relatively few cells that send their axons to the pons have been described, with those that do being limited to the areas representing peripheral visual fields. The frontal, parietal, and occipital cortical areas are well represented in the corticopontine projections and include projections from primary somatosensory and motor areas. From the point of view of assessing opportunities for higher cortical areas to contribute significant inputs to lower motor circuits, the numerically very large corticopontine pathways of the primate brain have to be seen as having the potential for a very significant contribution, even though it has to be recognized that we do not know anything about the modulatory or driving functions of these pathways on their way to cerebellum or where the cerebellar outputs reach the motor circuitry of the brainstem. Nor can a systematic search for identifying the nature of the messages be started for any part of the cortico-ponto-cerebellar-brainstem pathways until that distinction is established. At least some of the corticopontine axons have branches that go to the thalamus (see chapter 6).

8.5.3   The Corticostriatal Pathways

The corticostriatal pathways appear to arise from all parts of the neocortex (Webster, 1965; Kemp and Powell, 1970; Yeterian and Van Hoesen, 1978; Royce, 1982). The striatum sends its outputs through the globus pallidus or substantia nigra pars reticularis along GABAergic inhibitory pathways that are likely to be modulatory (see chapter 4). The pallidal input goes to the ventral anterior nucleus, thus serving to inhibit a higher order thalamic relay to motor areas of the frontal lobe, and also goes to the ventral lateral, ventral medial, parafascicular, and center median nuclei (Sidibe et al., 1997). A significant part of the nigrothalamic pathways terminates in the parafascicular and center median nuclei and also in the ventral medial nucleus of the thalamus with its widespread contribution to cortical layer 1, primarily in the rostral cortical areas described in chapter 7.4 (Di Chiara et al., 1979; Gulcebi et al., 2012). It also sends a strong output to the superior colliculus (Anderson and Yoshida, 1980; Behan et al., 1987; Smith and Bolam, 1991), opening up other possibilities for connections with the extensive tectobulbar and tectospinal pathways.

Some of the corticostriatal axons themselves appear to be branches of long descending corticofugal axons (Jinnai and Matsuda, 1979; Levesque et al., 1996; Reiner et al., 2003), and some have branches that go to the thalamus (Royce, 1983), suggesting that at least a part of striatal computations are based on information about upcoming motor instructions to the lower motor centers arising in the cortex, some of which may also be copied to the thalamus by glutamatergic afferents. The outputs of the basal ganglia are largely concerned with inhibiting (or disinhibiting) the motor outputs of the brainstem so that any cortical area that sends connections to the striatum has the capacity to play a role in the control of motor outputs, a role that is likely to be independent of that cortical area's connections to other cortical areas.

8.5.4   Direct Corticobulbar Pathways

Perhaps the most direct access that cortex has to the lower centers that connect the brain to the environment, apart from the classical corticospinal pathway that arises in the cortical areas on either side of the central fissure in primates, is through the system of corticofugal axons that terminate in the red nucleus, and in relation to the reticular nuclei of the midbrain, pons, and medulla (Zimmerman et al., 1964; Kuypers and Lawrence, 1967). In addition to this, there are corticofugal fibers that terminate in the zona incerta (Mitrofanis and Mikuletic, 1999; Bartho et al., 2007; Aronoff et al., 2010) and the subthalamic nucleus (Monakow et al., 1978; Canteras et al., 1990) and also numerous connections to sensory relay stations of the brainstem such as the inferior colliculus (Bajo et al., 2007), the cochlear nuclei (Meltzer and Ryugo, 2006), as well as the trigeminal, gracile, and cuneate nuclei (Kuypers, 1958). We include these sensory relays because through these, the corticofugal pathways can alter the relationship of the brain to the world, either reducing or increasing the information that reaches the cortex or other intermediate stations.

It is important to define the action of any one cortical area acting through its own subcortical connections, because it is extremely unlikely that we can expect to understand the functions of any area if the role(s) of these subcortical connections are ignored. The additional corticocortical connections that no doubt invariably play a role in the final behavioral or cognitive outcome are, of course, also important, but our view of cortical function for any area will inevitably be incomplete if these corticofugal pathways are ignored. It is crucial to recognize that the subcortical centers that are innervated by the corticofugal connections briefly summarized here are capable of contributing to the control of almost all aspects of behavior in a monkey except for some of the most complex manual manipulations (see Lawrence and Kuypers, 1965).

8.6   Outstanding Questions

1. Is the thalamic innervation of primary sensory areas denser than that of higher cortical areas, and if it is, what does this mean in functional terms?

2. Where a cortical area receives afferents from more than one thalamic nucleus, are the two inputs both drivers?

3. And if they are, are they processed independently (as are the inputs to the X cells and the Y cells in the cat's lateral geniculate nucleus) or do they interact?

4. How do the functional properties of cells in higher order nuclei, or in nuclei with an undefined function like the anterior medial or anterior ventral thalamic nuclei, vary along any definable line of projection?

5. There seem to be at least two distinct types of thalamocortical axon. Taking account of axon diameters, terminal distributions in different cortical layers, degree of terminal topographic order in cortex, and innervation of more than one cortical area by individual axons, as well as origin from different thalamic nuclei, can the different thalamocortical axons be classified as either Class 1 (drivers) or Class 2 (modulators) and categorized by a functionally interpretable classification of thalamocortical cell types?

6. What is the nature of the thalamic input to layer 1? Class 1, Class 2, or something entirely different?

7. As all cortical areas have the capacity to influence motor outputs, would it be useful to categorize cortical areas on the basis of what their connections with subcortical motor centers (as defined in this chapter) can do?

9 Thalamocortical Links to the Rest of the Brain and the World

9.1 A Brief Overview

We start this final chapter with a brief summary in sections 9.1.1 to 9.1.6 of the major points raised in the book that jointly lead to a number of novel and interrelated views of thalamocortical functions in cognition and behavior. Each is discussed more fully in sections 9.2 to 9.8, and some of the major implications of these several points are considered in the last three sections (9.9 to 9.11).

9.1.1 First Order and Higher Order Relays

All areas of cortex receive afferents from the thalamus bringing messages about activity in other neural centers. The best known represent inputs from peripheral sensory receptors or from subcortical neural centers, but these, defined as first order relays, represent a relatively small part of the thalamus and cortex, particularly in primates, where the major part of the thalamus forms the higher order nuclei; these relay messages to cortex about activity in other cortical areas. That is, in addition to the widely recognized direct corticocortical pathways that link cortical areas to each other, there are transthalamic corticocortical pathways, and it appears that cortical areas are often connected by both pathways in parallel.

9.1.2 Drivers and Modulators

The messages that are passed to the cortex through the thalamus involve a minority (5% to 10% or less) of the synapses in any thalamic nucleus. These inputs are the drivers, and they are glutamatergic and have the characteristic functional properties (Class 1) that define their ability to transmit the message accurately. The great majority of the inputs to the thalamus are modulators, and those that are glutamatergic have properties (Class 2) that compromise the accurate transmission of a message but provide controls for the way the message is relayed.

The importance of distinguishing the inputs that are drivers from the modulators, and for the glutamatergic pathways, distinguishing the Class 1 from the Class 2 inputs, is crucial not only for knowing where to look for the messages that the thalamus is sending to the cortex but also for understanding direct and transthalamic corticocortical communication, and, more generally, communication in any part of the nervous system. The limited information available about the functional properties of cortical and thalamic inputs as either drivers or modulators (Class 1 or Class 2) provides a serious challenge to contemporary views of corticocortical communication. Identifying the functional role for each of the many corticocortical and thalamocortical communication pathways is crucial if we are to understand the functional organization of cortical circuitry.

9.1.3   The Message and Its "Meaning"

For the first order sensory relays, the message that is transmitted to cortex can be read as signals to be understood in terms of the temporal distribution of action potentials. These relate closely to, and in that sense represent, the actions of the world or the body on sensory receptors. These messages are our link with the world. For most of the other thalamic relays, we have no clear ideas about the nature of the messages that thalamus is transmitting to cortex by any particular sequence of action potentials.

9.1.4   The Thalamic Gate

The thalamic relay can be regarded as a "gate" for the messages on their way to an area of cortex, either from the world or from other parts of the brain, including other cortical areas. The messages can either be relayed or not relayed (gate open or closed). They can be relayed either in a linear form (when the thalamic relay cells respond in tonic mode), providing the cortex with an accurate replication of the information that is received by the thalamus, or in a nonlinear form (when the thalamic relay cells respond in burst mode), with a high signal-to-noise ratio that can serve as an alert about novel or unexpected events: a wake-up call.

9.1.5   Motor Branches of the Thalamic Afferents

Essentially all of the messages that are relayed from the thalamus to the cortex reach the thalamus along axons having extrathalamic branches that pass messages to motor centers. Thus, the thalamus is receiving a single message about two distinct events: one is primarily about a sensory event or activity in another neural center, that is, an event in the past represented by a specific sequence of action potentials arriving along the thalamic branches, and the other, represented by the same pattern of action potentials passing along the extrathalamic branches, is a copy of current instruction for an upcoming action. The thalamic branches thus transmit efference copies to thalamus for relay to cortex, and these efference copies contribute to a forward model of motor activity that is available to cortex. These efference copies are richly represented in all thalamocortical pathways.

9.1.6   Cortex Acts through Lower, Subcortical Motor Centers

Much of mammalian routine behavior is controlled by pattern generators and lower motor centers in the brainstem and spinal cord. These receive inputs from higher subcortical centers that include the basal ganglia, the cerebellum, the tectum, and some higher brainstem centers primarily in the pons and midbrain. All cortical areas have outputs that relate to one or more of these motor centers. The extent to which any one cortical area depends for its role in behavior on these direct corticofugal connections and not on the more generally recognized corticocortical pathways that need to reach cortical motor areas for a role in the control of behavior is largely unexplored.

9.1.7   The Functional Capacities of Cortical Areas

Contemporary views of cortical function are commonly overly corticocentric, treating cortex as the seat of the mind and evaluating the functions of single cortical areas on the basis of: (1) the situations in which any one area is active, and (2) the other cortical areas to which it is directly or indirectly connected as part of strictly corticocortical networks. Our understanding of thalamocortical and corticofugal connections shows that a significant part of what any cortical area can do depends on the nature of its thalamic inputs and also on the corticofugal connections that it establishes through its own layer 5 outputs with phylogenetically old, lower, subcortical motor centers and pattern generators. These are the two routes upon which a cortical area depends for its most direct links to the body and the world. This account is in contrast to a common view that treats sensory areas of cortex as necessarily passing messages up a hierarchical series of cortical areas before they can either reach motor regions and issue motor commands or be stored in memory.

We will consider the seven points (summarized in sections 9.1.1 to 9.1.7) first primarily in terms of how they relate to each other, showing that the points we have listed above are closely interrelated and should all be seen as leading to a relatively unified view of cortical functions. In addition to this, we will consider where these points can stimulate new observations and finally consider some general points that arise from the material that has been included in the previous chapters.

9.2   First Order and Higher Order Relays

The evidence that significant parts of the thalamus relay information to cortex about what other cortical areas are doing raises important new questions about thalamus and cortex and reorients our understanding of corticocortical communication. The transthalamic corticocortical pathways that involve higher order thalamic relays must be distinguished from the more generally recognized direct corticocortical pathways for several reasons. One is that the relevant corticothalamic driver (layer 5) inputs to higher order thalamic nuclei are commonly, probably always, branches of descending corticofugal axons. They can provide information to higher order cortical areas not only about activity in other cortical areas, which can also be provided by the direct corticocortical pathways if they are drivers, but in addition to this, they supply information about instructions that are currently on their way to subcortical centers from cortex, contributing information about future actions. This is a new view of how any one cortical area relates to the body and the world, not only through the major incoming sensory pathways to the thalamus on the one hand and to the outgoing motor pathways on the other, but also through the receipt of information about messages that are currently on their way for motor execution (see section 9.5).

Although these connections of the higher order relays expand to a significant extent our view of what corticocortical connections can do, a probably more important property of the transthalamic corticocortical pathways that is not seen in the direct corticocortical pathways is the presence of the thalamic gate (see section 9.5). Both first and higher order cortical areas receive their inputs through this gate, and we know very little about the several different ways in which this gate may be controlled. One important feature about first and higher order nuclei and their cortical connections that needs to be studied and defined concerns the hierarchical relationships established by the transthalamic as opposed to the direct corticocortical pathways. It seems likely that they are parallel hierarchies, but we need the information necessary for this conclusion, and it should be stressed that the functional significance of this information will depend not only on knowing who is connected to whom but also on knowing for each connection whether it is a driver or a modulator (see section 9.3). This information is necessary for understanding the hierarchies of the direct corticocortical pathways, which is today based almost entirely on anatomical evidence. It is also required for understanding the transthalamic corticocortical pathways, whose exploration has only recently been started.

In section 9.7, we consider the extent to which the subcortical levels of motor control can frequently run things essentially on their own, until something unexpected happens. That is, when all is running smoothly in the periphery, information may not be relayed through the transthalamic gates (see section 9.5). Instead, the GABAergic inputs that selectively target higher order relays (see chapter 5.4.2.2) keep these gates closed, while the first order relays, with fewer GABAergic inputs, may be kept open to monitor events in the environment. However, any unexpected event, environmental or neural, will lead to a wake-up call at the thalamic gate; this will allow the recipient cortical area to register that event and send out its own instructions to lower levels, and these will, in turn, also be copied through the thalamic branches (see section 9.6) to another cortical area. That is, as one cortical area sends new instructions to lower centers, higher order cortical areas are informed about this instruction and can be recruited for further action if needed. From the point of view of distinguishing first order from higher order relays, it appears that, although an important function of all thalamocortical pathways is the job of monitoring ongoing motor instructions and confirming that they are in accord with expectations of future actions, that is, ensuring that the efference copy matches the reafference, the first order relays are monitoring subcortical motor instructions, whereas the higher order relays monitor motor instructions from cortex itself.

As indicated above, the connectional details of these links of the corticothalamocortical circuitry passing through the higher order thalamic nuclei still need to be defined to a large extent: drivers need to be distinguished from modulators, and the destinations of the extrathalamic branches of the corticofugal layer 5 axons need to be traced to their terminal sites so that the nature of their actions at those sites can be identified. It looks as though the chain of command established through the higher order thalamic nuclei is one that could be a model for any well-organized enterprise, business, military, or ecclesiastic. Each level of the organization can be alerted when it is needed and can act through its own executive powers, informing higher levels of the instructions for action that it is issuing, as do the layer 5 corticofugal outputs, which, through their thalamic branches, automatically inform a higher level that instructions for action are on their way.[1] The challenge for the neuroscientist is to find out the details of this organizational plan, to know the nature of the messages, and to understand where it might break down and, thus, where particular cognitive or behavioral phenomena can be related to the normal and abnormal functions of the system.

It is important to recognize that the unexpected event considered above may be an event in the environment or one in the organism. At every passage through the thalamus, the message that is relayed to cortex contains information that has come from the world and from the body about events that have already happened as well as information about the forward model; that is, about events that are about to happen. The subcortical driver inputs to the thalamus are not simply about any particular event in the environment, the body, or the nervous system in isolation; they are about such an event *together with* the instructions for action generated by that event in the lower centers. Any mismatch between these, that is, when the efference copy is not matching the reafference, is likely to produce a wake-up call from the thalamus to cortex (see discussion of the thalamic gate in section 9.5). Sometimes that may be a minor correction not necessarily involving higher centers, but at other times more drastic measures involving more powerful motor outputs may need to be called in. Defining the nature of the contribution that any particular transthalamic pathway makes to the forward model could well prove to be a useful way of exploring the functional capacities of that thalamic nucleus and of the cortical area(s) that it supplies.

At present, a very large part of the cortical activity that we can claim to understand is cortical activity that can be defined on the basis of the messages that are sent from the thalamus to the cortex through the major sensory relays (see section 9.4). However, these messages represent only a small proportion of the total thalamic inputs to cortex. It is probable that an exploration of the messages carried by higher order relays from thalamus to cortex will contribute very significantly to our understanding of the functions of higher cortical areas, but at present we know very little about the nature of these messages. We have made the point in earlier chapters, and repeat it here, that understanding a cortical area in terms of the messages that it receives from the thalamus

---

1. Here we are assuming that the transthalamic connection is generally a driver feedforward connection from lower to higher cortical areas. There is some evidence for this in the early visual, somatosensory, and auditory pathways, but the extent to which it can be generalized is still largely unknown.

and the instructions that it sends to lower centers is not at all the same as understanding the conditions under which a particular cortical area can be shown to be active. The aim of our approach to thalamus and cortex concerns the former rather than the latter.

9.3   Identifying the Drivers and Modulators

One issue that is vitally relevant for understanding the function of any neural circuit is a distinction between the connections that are passing on messages about activity in other centers or in the world, the drivers, and those that are modifying the transmission of such a message, the modulators (see chapter 4). The former are, so far as we know, solely the Class 1 glutamatergic pathways. The latter include Class 2 glutamatergic as well as other modulatory pathways, such as cholinergic, noradrenergic serotonergic pathways, and others, and these are either changing the way in which the message is transmitted or determining whether the message is transmitted at all on the basis of conditions that may or may not be related to the message itself. We have summarized evidence that there are such functionally distinct types of thalamocortical and corticocortical glutamatergic connections and that this distinguishes drivers, which are characterized by Class 1 functional properties, from modulators having Class 2 properties. Identifying the two classes at each of the many thalamocortical and cortical connections or possible other glutamatergic input classes represents a large area that needs detailed further study before we can claim to understand how thalamus and cortex interact.

We have seen that our knowledge of corticocortical communication is in the strange condition of including a plethora of anatomical knowledge about the existing lines of communication with extremely limited information or, more commonly, no information at all, about the nature of the communication. Even where one can be persuaded that we have clear knowledge about the activity characterizing the cells that give rise to a particular corticocortical pathway, as is sometimes the case, generally in sensory, and particularly in higher visual pathways, we lack information needed to decide whether or not that pattern of activity is being passed on as a message for further relay.

It is important to stress that the temporal pattern of action potentials carried by the drivers transmits specific information about events in the world or the nervous system, whereas the action potentials carried by the modulators only provide information that can change how the messages are sent or whether they are sent at all. The specific detailed temporal patterning of action potentials, which is characteristic of the drivers, is necessary if the drivers are to send messages that can relate closely to events. Even if the pattern of action potentials is the same, the information cannot be efficiently relayed by the modulators for reasons outlined in chapter 4.[2] That is, the Class 2 properties that have so far proved to be characteristic of modulators are such that the necessary temporal patterns of action potentials for transmitting messages cannot be transmitted faithfully; they are lost or distorted at the relevant synapses. Further, the well-localized connections that allow for a high degree of specificity in the drivers and that are an essential part of the labeled lines that transmit messages in the central nervous system are generally not available in the modulators, which have more widespread connections, except for the glutamatergic modulators, which as discussed in chapter 4.1.3.3 have topographically limited connections. We have distinguished drivers from modulators and have shown that there are functional properties that characterize a driver (Class 1 properties) or a modulator (Class 2 properties). To understand how thalamus and cortex process messages about the world, about the body, or about activity in lower centers, we need to follow the relevant pathways and to identify each input stage as either driver or modulator in function.

Almost all current textbook circuit diagrams for neural pathways are presented without these vital distinctions. In chapter 4.1.5.1, we provided examples for the medial geniculate nucleus and the anterior ventral nucleus of how the driver/modulator distinction matters for understanding the functional organization of a thalamic nucleus (see figure 4.6). Where Class 1 properties identify an input as a probable driver, it makes sense to ask about the nature of the message that is being transmitted, even though in any one instance the experimental conditions that allow a distinction between Class 1 and Class 2 may be very different from the experimental conditions that allow the identification of the message. Similarly, where Class 2 properties identify an input as a likely modulator, it becomes important to understand the nature of the modulation and the conditions that produce the modulatory action or prevent it. It also becomes important to determine whether for any one modulatory pathway there is one specific message that is modulated, or a number of different types of message, and further, as far as possible, to determine the nature of the message(s) being modulated.

Where we have data, in the few thalamic nuclei and cortical inputs so far analyzed, the modulators greatly outnumber the drivers, with the drivers providing less than 10% of the synapses. This underlines the extent to which a

2. An example of this point would be the case in which a glutamatergic pathway innervates multiple targets via branching axons and acts as a driver for one target and a modulator for another because the branching axons send identical messages along each branch. An example of this, described in chapter 4.1.6, is provided by Reyes et al. (1998).

view that treats all neural connections as equal may well be misrepresenting the lines of connection that are responsible for transmitting messages that are about details concerning other parts of the nervous system, the body, or the world.

At present, the importance of identifying the drivers and modulators in the thalamocortical pathways is largely unrecognized, and the issue is thus largely unresolved. There are comparatively few exceptions that have been summarized in chapter 4.1.5.2. Thus, it is known that in the sensory, first order thalamocortical relays of the mouse, the thalamocortical pathways to layers 4–6 are Class 1 (drivers), whereas most of those to layers 2/3 are modulators. A higher order thalamocortical pathway to primary sensory cortex can be Class 2 (probably modulator), as has been demonstrated for the connection from the posterior medial nucleus of the mouse to the primary somatosensory cortical area (see chapter 4.1.5.2), whereas the higher order thalamocortical pathways to higher order areas appear to be mostly Class 1, but more data are needed for these observations to be generalized to other sensory pathways (see chapter 5). There is likely to be a rich harvest of comparable information about thalamic and cortical functions if each of the many known pathways can now be defined as either Class 1 or Class 2.

There is some indication that the cortical layer in which a thalamocortical axon terminates may relate to one or the other of these functions, driver or modulator, but at present that is an unconfirmed hypothesis (see chapters 4 and 5). This is also an area in which more detailed evidence from several cortical areas will be extremely useful. Similarly, there are unresolved questions about the driver or modulator properties of inputs to thalamus that were discussed in chapter 4 for the dorsal medial geniculate nucleus and pulvinar, in chapter 6 for the anterior thalamic nuclei, and at several places for the superior colliculus itself.

Identification of a particular input as either Class 1 or Class 2 leads to further important potential observations. The first is that for each such identification, one can begin to question whether, indeed, the hypothesis that Class 1 inputs are drivers and Class 2 inputs, modulators can be challenged. The second is that for any candidate driver (Class 1), the nature of the message can be investigated, and for any candidate modulator (Class 2), the nature of the modulation as well as the nature of the message that is modulated can be investigated.

We have used the numerically enormous projection that goes from the layer 6 cells of visual cortex back to the lateral geniculate nucleus as an example in earlier chapters, and repeat it here, because it is such a telling example. These axons come from cells with well-defined, well-localized receptive field properties, but they do not convey information about those properties to any of the recipient geniculate cells. The axons are modulators in the lateral geniculate nucleus, changing the way in which geniculate cells pass messages from the retina to the cortex, and they also appear to be modulators where their corticocortical connections have so far been checked (Lee and Sherman, 2009a; Lee and Sherman, 2009b).

9.4   Reading the Message

Decoding the messages that are carried by the inputs to the thalamus and relayed to the cortex in the first order pathways to sensory relays has provided some clear views of how information about events in the world are conveyed to the cortex. It has also, but to a more limited extent, provided insights into the processing that occurs in the cortex. However, as we showed in chapter 7.2, of the messages that the thalamus relays to cortex, we understand the meaning for only a very small proportion of the thalamic cell groups. We have no clear ideas about the nature of the messages that the majority of thalamic relays transmit to cortex. This applies not only to most of the higher order relays but also to some of the first order relays.

Our analysis of thalamocortical pathways shows that, for many thalamic nuclei, the knowledge that is missing at present is, in the first place, an identification of the relevant drivers in the inputs on the way to the thalamus (as for the anterior medial and anterior ventral nuclei; see chapter 7.2.1.2) and, in the second place, information about the messages that are carried for relay to cortex. We summarized in chapter 7.2.1.2 how that information is available for the anterior dorsal thalamic nucleus whose cells receive information about head direction, information that has been traced back to the vestibular nuclei, whereas there is not a comparably clear identification of the drivers for the other two anterior thalamic nuclei, the anterior medial and the anterior ventral. For the higher order nuclei (see chapter 7.2.2), there is currently very limited or no information about the messages that they receive from their cortical driver inputs. To the extent that these corticothalamic drivers are branches of corticofugal axons to lower centers, and currently available evidence suggests that most or all are such branches, it makes sense to ask about the nature of the messages that the higher order thalamic nuclei receive from cortex: (1) in terms of how the message relates to the activity in the cortical area that gives origin to these axons, and (2) in terms of the activity that the descending branches are initiating at their terminations in the phylogenetically older lower centers. This may seem like a long and difficult task, but relative to the amount of skill and effort that has been invested in studying the structure and functions of the cortex and its connections in the past, it should prove tractable.

When Adrian (1928) was reporting the messages he recorded from peripheral sensory nerves, he wrote: "We will assume that the central nervous system is able to get every scrap of information out of the message, or let us say, everything that could be learnt by a physiologist who could isolate each nerve fiber and could record the impulses in it." This seems like a reasonable assumption about the nervous system that has had millions of years to evolve the best strategies for extracting that information, but it may be overestimating what it is that most physiologists are able to do after less than a century of trying. One limitation that restricted much early work was the use of anesthetics, which inevitably limits information about relationships between the nervous system and the world. Further limitations are likely to be recognized as progress is made. For example, the practical possibility that the messages on their way to the cortex might be interpreted in terms of information relevant for creating forward models of the organism in relation to the world was not considered until quite recently, nor has it seemed reasonable in the past to try to understand the nature of a message that passes along an axon to the thalamus by asking what might be the actions of the same axon's motor branches. Reading the messages that are passed through higher order thalamic nuclei is far behind what is known about sensory messages on their way to the cortex, and here, surely, the nervous system is still a long way ahead of the neurophysiologist.

A completely different approach to reading the messages in the central nervous system has recently been based on methods that exploit the possibility that the messages carried by large groups of neurons can be decoded (understood) on the basis of extensive statistical analyses that use a great many carefully controlled functional magnetic resonance imaging recordings (see Kay et al., 2008; Momennejad and Haynes, 2012). These methods, sometimes described as "brain-reading," go far beyond the decoding Adrian was discussing and raise an important question as to whether, if it can be made to work effectively, it represents a method that, in Adrian's terms, the brain can also use "to extract every scrap of information" or whether it is a method that can only be used from outside the brain itself.[3] Instead of reading the temporal codes of action potentials from individual or small groups of nerve fibers, these methods are based on the temporal changes in oxygen usage of hundreds of thousands of neural and non-neural cells in relatively large volumes of the

---

3. We leave the question unanswered.

brain recorded repeatedly under well-controlled conditions. Whether these techniques can ever provide a genuinely useful access to the messages that pass between neural centers remains an unanswered question.

9.5   The Thalamic Gate

The conditions under which messages reaching a thalamic relay cell may or may not be passed on to the cortex were considered in chapter 5.4.2.1. There are many different types of modulatory inputs to the thalamic relays, and many of them relate to all parts of the thalamus. Between them they can determine whether a message is passed to cortex or not and also determine whether that message is a "true," linear representation of the input (tonic mode for the relay cells) or is sent in a modified form with a high signal-to-noise ratio, representing what we have called a wake-up call (burst mode). We are currently not in a position to determine the precise role of each of the many modulatory inputs in this control of the thalamic relay functions, although it is clear that the dynamic changes of the potential at which the relay cell membrane is held is a crucial variable and that there are many different ways of changing this variable.

There is some evidence that the modulators do differ to some extent for different nuclei, and in chapter 5 we indicated that this is particularly noticeable for the inhibitory inputs to higher order as opposed to first order nuclei. However, the details of how they differ in the actual functioning of the nuclei has not been documented, nor can we be sure that this is actually a difference that distinguishes all of the higher order relays from all of the first order relays. Although we showed in chapters 3 and 8 that there are several different types of thalamic relay cell, definable on the basis of the branching patterns of their dendrites, the presence of axonal branches to the striatum, the calcium binding proteins that they contain, or the distribution of their axons in the several different cortical layers, at present we know of no evidence that there are differences in the action of the thalamic gate from one type to another. However, the issue merits exploration. Further, the extent to which the controls on cells having thalamocortical axons that are Class 1 at their cortical termination may differ from those that are Class 2 is not known.[4]

The functional relationships that can be produced by the thalamic gate merit further attention and raise issues for future studies. One concerns the difference between the transthalamic and the direct corticocortical pathways (see

---

4. They may well be branches from a single axon having different actions in the cortex (see chapter 4 and footnote 2), but we don't know that.

chapter 5). There is no thalamic gate for the direct corticocortical pathways, so that in these there appears not to be the same capacity to control the messages that pass from one cortical area to another. Specifically, the thalamic gate has the capacity, while not changing the messages that a cortical area is sending to lower motor centers (see section 9.6), to deal with the thalamocortical copy that is on its way to higher cortical areas in one of three different ways: (1) to prevent information about that instruction from being passed to higher cortical areas, (2) to pass that information to higher cortical areas in the form of an alert about an unexpected event, or (3) simply to send that information on to higher cortical areas where it can form a part of the forward model. This is a difference that merits further investigation because we need to learn more about how and when the thalamic gate is operating, exactly how it is controlled in any one nucleus, and the extent to which the control may differ from one nucleus to another, particularly as between first order and higher order relays.

We currently have limited information about how the function of the thalamic gate in any one thalamic nucleus relates to any relevant behavioral or cognitive situation other than indications that the levels of inhibitory inputs to the higher order posterior medial nucleus are reduced when a rat is whisking (see chapter 5.4.2.2; see also Urbain and Deschênes, 2007a and b). The details of how particular aspects of an exploratory behavior relate to the condition of the thalamic gate are largely unexplored. For example, how does control of the gate relate to a well-rehearsed and predictable navigation as opposed to one that is unfamiliar or has some surprises? The differences may well relate to the extent to which a particular behavioral pattern depends on cortical guidance and the extent to which, like walking and probably much of driving a car, it can be run by the subcortical motor centers. The role of the cortex is more likely to be one of noting unusual or unfamiliar events, recording them for future reference, and correcting them, rather than dealing with well-rehearsed behavioral patterns. We do not at present have information to demonstrate this one way or the other, but more knowledge about the condition of the thalamic gate in relation to different patterns of ongoing behavior patterns could prove instructive.

The control of the thalamic gate is raised again in section 9.6 in relation to suggestions that specific failures of the thalamic gate in some higher order nuclei to transmit information relevant for the formation of a forward model may relate to aspects of awareness of self. There is a long history of cognitive defects that are associated with higher order thalamic relays (see Means et al., 1974; Rafal and Posner, 1987; Chauveau et al., 2005), and recent evidence based on MRI and postmortem anatomy in schizophrenic patients indicates that first order nuclei appear normal but higher order nuclei (for example, the medial dorsal nucleus and pulvinar) are shrunken with neuronal loss (Danos et al., 2003; Brickman et al., 2004). It is relevant that, whereas an action that forms a part of a forward model is normally anticipated by the cortical circuits and can be registered as an action originated by the organism itself, an action that occurs when the normal formation of the forward model fails, is likely not to be registered as originating from the self. There have been proposals that failures in the production of normal forward models may play a significant role in producing some of the abnormal experiences encountered in schizo-phrenia (Frith et al., 2000; Blakemore et al., 2002; Shergill et al., 2005; Stephan et al., 2009), and these have more recently been specifically linked to the function of the thalamic gate (Vukadinovic, 2011; Vukadinovic and Rosen-zweig, 2012).

Here it is important to recognize in relation to awareness of self that the thalamic gate has the capacity to block messages on their way through the thalamus to cortex and that these blocked messages represent the copy of an instruction going to motor centers that will continue to be passed to the motor centers even when the copy fails to be transmitted to the cortex. That is, the thalamus can produce a disconnect between a motor instruction for action and a copy of that motor instruction heading for a higher cortical area where its arrival would contribute to a forward model in that area. It is this *disconnection* of information about these events—the action that occurs and its efference copy, which is blocked—that would lead to questions about assigning respon-sibility for unanticipated actions to someone other than the self; and it appears that this may be due to an abnormal action of the thalamic gate.

9.6   The Motor Branches

We have argued that the afferents to the thalamus are essentially all branches of axons that deliver messages to other, lower, motor centers. We have sum-marized some of this evidence for first order and higher order thalamic nuclei and have argued that these branched axons provide efference copies for the thalamocortical pathways. Two issues arise from these observations. One is that it looks very much as though these branched inputs to the thalamocortical pathways are common and may even prove to be truly ubiquitous, providing all cortical areas with previews of forthcoming actions. We can ask whether all cortical areas are provided with pathways that carry such previews and then raise the second issue and ask whether a branching axon is necessarily the only way in which such efference copies can be provided.

It is important to stress that we have been almost entirely concerned with the branching axons that carry inputs to the thalamus directly or trans-synaptically along one of their branches and that the occurrence of such branched axons as afferents to essentially all thalamic nuclei leads to a conclusion about the ubiquity of efference copies in the inputs to the cortex. However, there are branched axons that characterize many other pathways in the brain, pathways that do not carry inputs heading to thalamus and cortex, and we have mentioned in chapter 6 just a few of them that probably serve as efference copies, particularly where branched axons supply the cerebellum or striatum. We stress that the involvement of efference copies and the occurrence of forward models are not confined to thalamus and cortex. They are a phylogenetically ancient feature upon which much of the coordination of movements depends.[5] One can expect to find them quite unrelated to thalamocortical functions in the spinal cord, the brainstem, and in any subcortical motor center.

Where, as in the visual, somatosensory, or auditory pathways we can interpret the code about peripheral sensory events, we must retain serious doubts that this represents the whole of the message that the cortex is receiving. We have seen that there are rare glimpses of something else, as for example the forward receptive fields observed in several different visual cortical areas that shift appropriately in anticipation of an eye movement (Colby et al., 1996; Nakamura and Colby, 2002; Sommer and Wurtz, 2008; Hall and Colby, 2011) and that can be interpreted as representing information carried by the efference copies to the cortex. Gaining an insight into the roles of the efference copies that reach all parts of the thalamus through all of the branched axons that are providing driver inputs to thalamus (see chapter 6) is an important issue to be resolved. Essentially, it will involve identifying the nature of the motor actions of the extrathalamic branches and then, on the basis of knowledge about these actions, defining the extent to which the relevant recipient cortical area carries information about the identified forthcoming action. It has to be recognized that, when the pathways are functioning smoothly and the efference copy signal matches the reafference signal, nothing will happen. In a well-coordinated organism, these messages serve to deal with the unexpected. They also provide what Adrian (1928), citing Craik (1945) (see chapter 6.2.1.1), described as a model of the world. This is the forward model that contains far more information than merely the one specific motor action. If the motor action is

5. It is worth stressing that the two major proposals concerning efference copies published in 1950 (Sperry, 1950; von Holst and Mittelstaedt, 1950) were concerned with the behavior of fishes and flies.

a part of a well-established pattern generator or lower motor circuit and if that functions in relation to the environment in a consistent and repeatable manner, then the forward model can include a great deal of what the organism can reasonably expect of its regular interactions with the world (see Wolpert and Miall, 1996).

The neural details on the basis of which an efference copy is compared with the reafference signal (see figure 6.13, which is a key to understanding how an efference copy can serve in the control of movements, are often not considered but are presented schematically instead (for example, figure 6.12). Most schemes show the two messages passing along different pathways and reaching a node (presumably a nerve cell but rarely identified) where the comparison is made. The afferents to the thalamus that we have discussed in chapter 6 carry the efference copy and the reafference signal in the same pathway, with the efference copy preceding the reafference signal, as it necessarily must in any scheme. That is, there has to be a comparison of consecutive inputs. We have no information about where such comparisons are made, but it is probable that it would be in the cortex rather than in the thalamus, and the individual message in any one thalamocortical input must be read as a part of a larger set of inputs that represent the particular movement and its expected outcomes.

Theoretical considerations about the role that efference copies can contribute to establishing forward models of the future state of the motor apparatus have played an important role in understanding the necessary neural control mechanisms of motor systems (Wolpert and Miall, 1996). However, actual examples taken from neuroanatomy are uncommon. Given the anatomy of the pathways that reach the cortex (see chapter 6), it is reasonable to expect that there are many parts of the thalamocortical pathways where future states are represented in the neural circuitry comparable to the forward receptive fields in visual cortical areas. This justifies an active search for comparable evidence in other cortical regions. The fact that we very often have a clear sense of what we are going to do next is not some strange puzzle about the nature of voluntary movements; it is a direct outcome of the fact that cortical areas, almost certainly all of them, receive information about forthcoming events, and much of that information is available in higher cortical areas that have a rich overview, not just of what lower, subcortical centers are about to do, but also about what other cortical areas are doing in terms of the messages that they are sending to affect forthcoming bodily actions. This anatomically close link between higher centers concerned with perception and cognition and lower centers concerned with action throws an important light on how action and perception are interrelated in terms of actual neural connections, a relationship that has been written about by philosophers, psychologists, and roboticists (for example, Clark, 1998; O'Regan and Noë, 2001; Pfeifer and Bongard, 2006; Clark, 2011) but so far not explored in terms of the actual neural pathways that include subcortical centers. At any one moment a mammal with a large number of cortical areas will have a significant number of active forward models in its cortex and is unlikely to be initiating actions on the basis of a blank slate.

Two examples from the earlier parts of the book (for example, chapter 6.2.1.1) can be seen in the connections of the pain pathways and those of the stretch reflex. In the spinothalamic pain pathways, the activity of a small number of pain receptors will, in an unanesthetized animal, produce a flexion reflex, and stimulation of a number of muscle spindles will produce a stretch reflex. However, in terms of understanding the message that reaches the cortex, there are two separate issues. One is to record the cortical activity that relates to the inputs from the receptors, and this is the approach of the classical sensory physiologist (see Adrian, 1928; Mountcastle and Henneman, 1949; Mountcastle and Henneman, 1952). The other is to understand how the cortical activity relates to the efference copies; that is, to the messages that pass to the motor neurons. In an anesthetized preparation, the match between the cortical activity and the reaction of the muscle cannot be recorded, and in the freely moving animal, identifying the match will depend not only on knowing the relevant muscle but also on knowing the temporal interval by which the cortical activity precedes the muscle action. That is, in practice, the information about the ongoing instructions has to be understood essentially in terms of the thalamocortical pathways containing information about *forthcoming* movements. That is something that can be done readily by theoretically constructed models of neural connections (Wolpert and Miall, 1996; Perrone and Krauzlis, 2008) but cannot at present be done so easily in terms of demonstrable events in actual sensorimotor pathways. If the attempt is made on the basis of identified neural connections presented in standard textbooks and taught to medical students the world over, it may well prove quite impossible for many cortical areas, but if the attempt is based on knowledge of what action will be produced by the motor branches of the afferents to the thalamus, then the contribution that these branches make to the forward model may become accessible for study. If for any cortical area we ask what information it does carry about forthcoming actions, then clues as to the nature of the forward model can be found by looking at the thalamic inputs to that area and defining the action(s) of the motor branches given off by these thalamic inputs.

The examples of branched thalamic afferents that allow us to see the thalamic inputs as efference copies of activity in axons that innervate motor centers have been based largely on old evidence from Golgi preparations. This method is fickle in a way that no one understands, and the demonstration of a branching axon must still be a question of luck. Methods of labeling one or a few individual nerve cells and tracing their axons through serial sections allow specific afferent pathways to be selected for study (see Bourassa and Deschênes, 1995; Bourassa et al., 1995; Levesque et al., 1996; Ojima et al., 1996; Rockland, 1998) and are likely to prove more useful, but they have often failed to trace the motor branch to its final destination, where this is far from the branch point, because this involves tracing individual axons through many serial sections. Alternative methods are available: one can use two or more distinct retrogradely transported microscopically identifiable markers, inject each into different possible terminal sites, and then identify cells that contain two or more of the markers, or one can use antidromic stimulation of the axon terminals and record the responses in the suspected cells of origin. Given the apparent ubiquity of the branched inputs to the thalamus and given the available methods, it may be worth urging a far more intensive hunt for such branching patterns so that it will be possible to identify not only the thalamic cell groups that do receive such branched inputs but also identify for each such cell group the cortical or subcortical origin of these inputs to the thalamus and the specific termination of the motor branches in order to identify the possible actions of the motor branches.

The knowledge that efference copies are ubiquitous in the thalamocortical pathways, together with a general recognition that many of them are likely to represent copies of outputs from higher cortical areas concerned with complex functions, may provide a route for the easier acceptance of the fact that we very generally have a clear sense of what we are going to do next. Where we know that a branching axon is providing messages for relay to cortex, we need to understand not only the nature of the activity in the center that is giving rise to the axon that passes toward the thalamus but we also need to know the nature of the activity that is likely to be initiated by activity in the extrathalamic branch; that is, we need to know not only what the recipient cortical area can record in terms of events that have just happened but also to know what it is that the cortex can reasonably anticipate.

9.7 Cortical Areas Act through Their Connections with Lower Motor Centers

We have stressed in chapter 7 that any one cortical area is connected to the subcortical and lower motor centers and pattern generators of the brainstem and spinal cord and that these phylogenetically old vertebrate centers have a remarkable capacity to function efficiently on their own. The cortex monitors their functions and can redirect their activity patterns, but for many routine activities it does not provide the necessary individual components of the actions. If the pattern generators are working normally, it is probable that they send no message that reaches any part of the cortex. A question arises when, for one reason or another, the pattern generators are not working as expected and a message reaches cortex in the form of a wake-up call from the thalamus. Then the next question is about the message that any one cortical area sends out along its subcortical pathways. The point is important, because it leads us to ask questions about the nature of the subcortical connections established by a cortical area and to recognize that the functional capacities of a cortical area are to a significant extent determined by what its subcortical links can do; the cortical areas, as we have stressed earlier, are not limited to what they themselves can do through corticocortical connections that allow action through the classical motor outputs. The subcortical connections of the recognizably "motor" cortical areas have been studied to a significant extent, but the connections of areas far from the motor output have been studied in far less detail. We know that even the primary sensory areas have significant outputs to areas that deal with motor actions. We know that even the primary sensory areas have significant outputs to areas that deal with motor actions. How these outputs at early levels of the hierarchy relate to outputs from higher levels is not known. Motor outputs from sensory areas far from motor cortex need to do something that the motor center (that is, for vision the superior colliculus), which is receiving the same information from the retina, does not do. We need to learn what this is and how it differs from the outputs of higher cortical areas, which are in receipt of a greater variety of inputs than are the primary sensory areas.

Recognition that subcortical links exist for all cortical areas relates to the view, expressed in earlier chapters and stressed again briefly in the next section, that the functions of cortical areas cannot be understood and should not be studied in isolation from their subcortical links. The problem at present is that we have very limited information as to what those links might be providing in terms of the access that a cortical area has to action programs. For example, the many direct cortical inputs to the superior colliculus from cortical areas far from the motor cortex, including primary sensory areas (see figure 8.5), are innervating a phylogenetically old motor center that has significant direct and indirect connections with brainstem and spinal motor centers. These pathways and other corticofugal pathways to the brainstem motor centers themselves must be playing a role in the control of actions, but we know very little about these actions other than those concerned with the control of eye movements from a few visual cortical areas, including the frontal eye fields.

The actions that can be influenced through corticofugal pathways involving the basal ganglia or the cerebellum are even more mysterious, and yet most cortical areas have direct access to one or another or both of these links. They stand as challenges reminding us of our ignorance about how cortex is linked to action and perception.

9.8   The Functional Capacities of Cortical Areas

In earlier chapters of this book, we have contrasted the questions that we are asking about cortical functions with the questions that dominate much of the current literature on cortical functions. In these earlier chapters, we have been looking to understand what a cortical area may be doing on the basis of the messages that it is receiving through the thalamus and cortex and the messages that it is sending to other cortical areas and to subcortical centers. Understanding the messages and how they relate to events in the environment or in other parts of the brain forms an important part of this approach. That is, the central questions have been how any one cortical area is operating on its inputs to produce the outputs, and to what extent we can understand these inputs and outputs in terms of perception and action. The focus for us has been on the thalamic inputs, the motor outputs, and the transthalamic outputs to cortex. This is quite different from the question more generally addressed, which concerns the cognitive, perceptual, or behavioral circumstances during which a particular cortical area can be shown to be active on the basis of increased blood flow or other signs of increased activity. We have described this approach as evidence-based phrenology, because, unlike the old phrenology, it is based on the action of the brain itself and so it clearly relates to what the brain is doing and serves to localize certain definable functions. However, the functions identified for any one cortical area by these means are often not linked to any defined inputs from the world or the body nor are they linked to any defined outputs from that cortical area. They tend to separate the mental from the physical. That is, the important link between the particular activity, be it behavioral, cognitive, or perceptual, on the one hand, and the brain and the body and the world, on the other, is missing. The account of such a cortically localized function encourages the view of cortex as a neural structure that can perform its functions without any necessary links to the body or the world. Where particular functions are identified for any one cortical area, it may be possible to ask about the nature of the thalamic inputs to that area or the corticofugal or corticocortical (direct and transthalamic) outputs of that area in order to gain some insight into the role of the nervous system in the particular localized function, and this would relate the two approaches to each other. However, it is rarely done. There are important and largely unanswered questions about the extent to which we can treat the cerebral cortex as the structure that is in charge of behavior, perception, and cognition and the extent to which cortex must necessarily involve lower centers in all of its activities so that in order to understand any cortical functions, they have to be seen in terms of their relationships to the rest of the brain and through these connections to the body and world.

9.9   Comparing Two Models of Thalamocortical Functional Relationships

One way to highlight many of the ideas presented in this book is to contrast them with the accepted, textbook view for thalamocortical relationships. The two views are summarized in figure 9.1. In the textbook view (figure 9.1A), information arrives from the periphery to be relayed by a thalamic nucleus (for example, retinal input to the lateral geniculate nucleus) to sensory cortex. From there, the information is processed and sent on up the hierarchy through various stages of sensory and sensorimotor cortical areas until it reaches motor cortex, from which messages are sent to appropriate subcortical sites to affect behavior. In this scheme, there is essentially a single major entry and exit point for cortical processing, and, notably, there is no specific role for most of thalamus, which is identified in figure 9.1A with question marks and which we have identified as higher order (HO in figure 9.1B). The view based on several of the points summarized in this chapter is shown schematically in figure 9.1B; it is quite different. Right from the beginning, the messages sent to first order thalamic relays arrive over branched axons, whose extrathalamic branches target motor centers, and this logically leads to the identification of the messages passed on through the first order thalamic relays as including efference copies as well as messages about activity in peripheral sensory receptors or other neural centers. As the information is processed further through the cortical hierarchy, each area in turn produces a further motor message, and a copy of that, again an efference copy, is passed on to higher order thalamic regions for relay to cortex.

We emphasize five points from figure 9.1B, points made in earlier chapters as well. First, the information relayed through thalamus has a dual function as an efference copy and monitor of ongoing events in the body and environment. Second, all areas of cortex for which sufficient information is available have a layer 5 projection to motor centers, and thus all areas have a pronounced motor function, even cortical regions conventionally seen as primary sensory

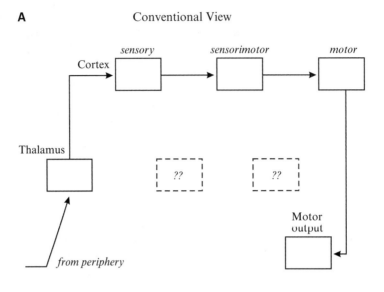

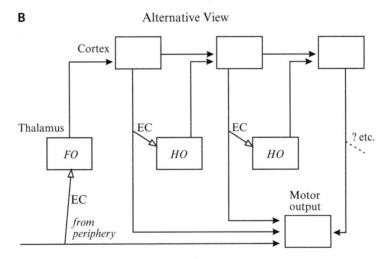

Figure 9.1
Comparison of accepted, textbook view of thalamocortical interactions (A) with the alternative view proposed here (B). See text for details. The question marks in (A) reflect the higher order thalamic relays that are not recognized in the accepted view. FO, first order; HO, higher order; EC, axons that carry information about activity at their origin in receptors or in a neural center as well as an efference copy with information about instructions for upcoming movements. Redrawn from Sherman (2005).

areas. Third, the thalamus clearly plays a role, necessarily defined by the properties of the thalamic gate in the particular nucleus,[6] in corticocortical processing. Fourth, cortical areas that communicate with each other through the direct pathways often if not always are also connected via parallel transthalamic pathways, and this raises questions about the nature of the information in each pathway (direct or transthalamic). Fifth and final, for every one of the pathways included in either scheme, the distinctions between drivers (Class 1) and modulators (Class 2) is crucial as is information about the nature of the message carried by the drivers.

9.10   Examples of How the Functions of Particular Pathways Can Be Analyzed in Terms of the Organizational Principles Summarized So Far

We can illustrate the importance of this issue by going back to some of the examples cited in earlier chapters, for example, in chapters 4.1.5.1 and 6.2.1.2. When colliculothalamic axons were first identified anatomically (Altman and Carpenter, 1961), their function was rather puzzling. At that time it was assumed that, because the superior colliculus receives visual inputs, the messages it sends via the thalamus to cortex must be a second pathway for visual messages to reach cortex, and a significant number of publications were devoted to this belief on the *implicit* assumption that these were drivers (Sprague, 1966; Schneider, 1969; Diamond, 1973; Berman and Wurtz, 2008).[7] However, at the time no one asked whether this pathway acted as a driver or a modulator in the thalamus, and to the extent that the issue has now been defined, it looks as though at least some of its components are likely to be drivers (Kelly et al., 2003). This raises a further question about the nature of the message that the pathway is carrying. Were the earlier proponents of a second visual retino-cortical pathway through the pulvinar correct to regard it as simply another sensory pathway through the thalamus to the cortex or is it more likely that these colliculothalamic axons carry copies of collicular outputs that control gaze or accommodation (Schiller and Tehovnik, 2001; Ohtsuka et al., 2002; Schiller and Tehovnik, 2005; Port and Wurtz, 2009) and that they differ from the retinal inputs to the lateral geniculate nucleus and the superior colliculus in carrying no information about retinal inputs, only information

6. These gating properties are in need of significant further exploration.

7. If the retinocollicular axons had at that time been recognized as branches of the retinogeniculate axons, the reasons why *the same* messages should be transmitted to cortex over two different routes might have attracted some speculation, but to our knowledge the issue was not raised.

about collicular outputs? The observations discussed in chapter 6.2.1., that the colliculus seems not to be sending information to cortex that is not already in cortex, suggests that the colliculothalamic axons, if they are drivers, are sending messages for relay to cortex that may be a pure efference copy (Berman and Wurtz, 2008); that is, copies of messages about instructions on their way to the extraocular muscles or the eye, with no sensory component at all. But the issue is undecided. Once the colliculothalamic cells and their axons are identified, then it should be possible to identify the nature of the message they are carrying, to identify that particular thalamic input as either a driver or a modulator, to determine whether it is an axon with extrathalamic motor branches, and finally to identify the action of the branches (if there are any) in the brainstem or spinal cord. Further, the outputs to cortex from the thalamic cells that receive the collicular inputs can be investigated to determine which ones are likely to be drivers and which ones are likely to be modulators, so that the role of the relevant recipient cortical areas can be investigated. The further question, if some of the collicular inputs to the pulvinar turn out to be modulators, would then be about the nature of the modulation and about the messages that are modulated.

Comparable approaches can be used to study some of the other areas that we have addressed in earlier chapters. For example, from chapter 4.1.5.1, where the input from the inferior colliculus to the dorsal medial geniculate nucleus has been identified as a probable modulator (Class 2), one can ask about the nature of the modulation and how it relates to the transmission of messages (from the primary to the secondary auditory area). Or where the transmission of messages from the primary to the secondary somatosensory area has been shown to pass through the posterior medial nucleus (see also chapter 4.1.5.1), the nature of the messages that are passed can be studied and compared with the messages that are passed from the first area to the second through the direct corticocortical route. The example cited in chapter 7.2.1.2 concerning the identification of a driver input to the anterior medial and anterior ventral thalamic nuclei provides another example.

9.11   Relating Thalamocortical Connectivity Patterns to Action, Perception, and Cognition

In the main parts of this book, we have tried to show some of the mechanisms of the brain that provide the interconnections needed for our interactions with the world, focusing particularly on the cerebral cortex, its inputs from the thalamus, and, crucially but more briefly, on the major cortical outputs; crucially, because these connections are certain to be playing a significant role, briefly, because many of them are so widely ignored that we know rather little about them.

We have focused specifically on these neural pathways and have seen them as providing two hierarchical systems of corticocortical connections, direct and transthalamic, that allow the cortex to form links with the phylogenetically old parts of the brain and to establish close reciprocal relationships with the body and the world. These connections can anticipate regularities in the actions of lower centers, the body, and the world and pass these anticipations to higher cognitive centers, where irregularities or unexpected events can be corrected and recorded for future reference. The ability of our cognitive centers to relate so intimately with our bodies and with the world in this manner has been recognized by many earlier writers, including psychologists, philosophers, and neuroscientists (for example, Gibson, 1979; Evans, 1982; Merleau-Ponty, 2002) and more recently others who have recognized the extent to which perception depends on action and action depends on perception, no matter whether this is viewed for a particular sensory function (for vision, see Churchland et al., 1994; for perception and action more generally, Clark, 1998; O'Regan and Noë, 2001; Clark, 2011; or for the efficient working of robots, Pfeifer and Bongard, 2006). These ideas, that cognitive and perceptual functions are "embodied," that they depend for their characteristic functions on the relationships they establish with the body and the world, are powerful proposals, but they have generally had very limited contact with knowledge of what the nervous system is like and currently lack well-defined neural correlates. These proposals work almost entirely without specifically involving the nervous system itself to any significant extent, and in this they resemble more general discussions of the mind–brain relationship rather than a piece of neuroscience.

Given what is currently published about the nervous system, where there is frequently a disconnect between the events recorded in a central nerve cell or a group of nerve cells on the one hand and the concurrent behavioral or cognitive functions of the organism on the other, this is not altogether surprising. We recognize the value of knowing how particular cognitive or behavioral patterns relate to specific neural patterns of activity in the central nervous system, particularly in the cortex, but the approach that we have presented throughout this book is aiming at something different: to show how these relationships are established in terms of neural activity, by an analysis of the neural connections that link the central nervous system, including the cortex, to the body and the world, defining the pathways and the messages that they carry or the modulatory roles that they each play.

Sensory physiology is often still dominated by a Cartesian view of an essentially passive brain upon which images of the world are projected, and cognitive neuroscientists generally treat the cortex as a surrogate mind, something that was lost to them when they stopped being psychologists; for them the functions of the cortex can readily be packaged in terms that need have no particular relation to the subcortical parts of the brain, the body, or the world, and each package can be treated and understood in its own right, with little or no concern regarding the actual links that connect the organism as a whole to the world or link the individual packages to the organism. There is broad agreement that the sensory pathways provide the way in, leading either to memory or to the motor outputs. The motor pathways provide the way out, and in between the corticocortical connectivity patterns take care of perception, cognition, and volition.

Our thoughts about the neural pathways that link the cortex to the rest of the brain have been leading us to a view of the relevant neural functions depending on closely interrelated sensory and motor functions, where the sensory pathways, as well as many of the cortical inputs to the thalamus for a transthalamic corticocortical relay, carry copies of motor instructions, and these efference copies provide information about movements that have not yet been made, actions that lie ahead. They provide information about forthcoming actions on the basis of which we can recognize that the actions originate from ourselves. This is information carried as messages in identifiable neural pathways; it is not a set of instructions emanating from one or several abstract centers for volition, nor is it necessarily information that must originate in specified areas of "motor" cortex. It is information available to the organism (to us) no matter where it comes from.

It may seem strange that in the book so far we have said almost nothing about the mind, limiting ourselves (at most) to brief references to cognitive or perceptual functions. This has been a book about nerve cells and their interconnections, and our central aim has been the provision of some ground rules for understanding how neural activities in thalamus and cortex relate to the actions of the body and to changes in the environment. We have based much of our analysis on the ground rules that can be defined in the sensory pathways, not because we regard these as more important than other systems, but because the messages that are transmitted in these pathways are easier to trace and have been studied in more detail. Also, they present many of the basic problems that need to be understood in a study of thalamus and cortex in relation to the rest of the brain.

Adrian (1928), in his introduction to the brief and lucid account of the classical experiments that revealed the neural code carried in peripheral sensory nerves, wrote: "As far as the brain is concerned the function of the sense organs, or receptors, is to construct in it a map of certain physical events occurring at the surface of the body so as to show what is taking place in the world outside us." Here, Adrian was writing about the brain as a passive recipient of messages from the external world, with the messages transmitted by electrical impulses whose patterns Adrian had demonstrated with new methods. This, as we have indicated above, is still the dominant view of sensory physiology, rooted in a dualistic approach, with the brain providing a center that can receive representations of the world much as the pineal gland did for Descartes. However, Adrian also wrote: "The final chapter deals briefly and timidly with the relation between the message in the sensory nerve and the sensation aroused in our consciousness; briefly because the relation is simple enough in a way, and timidly because the whole problem of the connection between the brain and the mind is as puzzling to the physiologist as it is to the philosopher. Perhaps some drastic revision of our systems of knowledge will explain how a pattern of nervous impulses can cause a thought, or show that the two events are really the same thing looked at from a different point of view. If such a revision is made I can only hope that I may be able to understand it."

A lifetime later, we can endorse that view, wondering whether it is not just a little optimistic, not about Adrian's understanding but about the nature of the problem. However, it is important to note that Adrian did not stop there. As we have indicated in chapter 6.2.1.1, he also quoted the work of Craik (1945, 1948), recognizing the extent to which the brain uses models of reality, the self and the world, models that allow a view of what is likely to happen next as the organism interacts with the world. We have shown some of the extent to which these models of reality can begin to be seen in neural terms and recognized as the forward models that can be created by efference copies. The efference copies themselves can also now be considered and investigated in terms of the large number of actual neural pathways that reach the cerebral cortex from the thalamus. We stress that efference copies are not limited to thalamus and cortex but that thalamus and cortex appear to be in a position to exploit them for cognitive functions. The point is that embodiment can be considered in strictly neural terms. We can ask about the specific links between action and perception in terms of the messages that are carried in identifiable neural pathways. What are the actions that the extrathalamic branches of layer 5 inputs to higher order nuclei produce at their terminals? And how do they relate to the messages carried by the transthalamic axons to cortex and the relevant perceptual processes?

If we look at the functions of the cerebral cortex from the point of view of the connections that we have described in this book, we can see that one function of any one cortical area is to monitor the activity of levels lower in the hierarchy. These lower centers can be cortical or subcortical. In a speculative sense, we can think of the cortex as large system of "listening posts" of the sort that might be used by a large national spy system. The higher echelons of the organization are not just listening directly to relevant external events or threats but can also evaluate external events on the basis of the ways in which lower levels of their own organization are reacting to the outside world. And perhaps, in a way comparable to any large intelligence organization, the highest levels can create views of what the world is really like, views of a *presumed* reality (yes, they really are making weapons of mass destruction). That is, neither the spy system nor the organism ever actually has the capacity to confirm these views but, as things stand at present, the organism has, over the course of evolution, had millions of generations of testing these "views" of reality and can rely on their *practical* validity, an advantage that most spy systems, with a much shorter history, lack. It is at this point we leave the reader to the philosophers, hoping that new insights into the workings of the nervous system will move us closer to an understanding of the puzzling ways in which our brains establish links to our bodies and to the world and allow us to interact with the world successfully without ever having certain information about what the world is really like or what we, as functioning organisms, are really like. No matter whether the brain is functioning normally or in one of its many possible abnormal ways, the neural pathways that provide our links to the world will provide the major clues we can expect for understanding the workings of the brain (and ourselves) in health and eventually also when diseased.

References

Abramson BP, Chalupa LM. 1985. The laminar distribution of cortical connections with the tecto- and cortico-recipient zones in the cat's lateral posterior nucleus. *Neuroscience* 15: 81–95.

AcunaGoycolea C, Brenowitz SD, Regehr WG. 2008. Active dendritic conductances dynamically regulate GABA release from thalamic interneurons. *Neuron* 57: 420–431.

Adams PR, Brown DA, Constanti A. 1982. Pharmacological inhibition of the M-current. *J Physiol* 332: 223–262.

Adrian ED. 1928. *The Basis of Sensation, the Action of the Sense Organs*. New York: W.W. Norton & Co.

Aggleton JP, O'Mara SM, Vann SD, Wright NF, Tsanov M, Erichsen JT. 2010. Hippocampal-anterior thalamic pathways for memory: uncovering a network of direct and indirect actions. *Eur J Neurosci* 31: 2292–2307.

Agmon A, Connors BW. 1992. Correlation between intrinsic firing patterns and thalamocortical synaptic responses of neurons in mouse barrel cortex. *J Neurosci* 12: 319–329.

Ahmed B, Anderson JC, Douglas RJ, Martin KAC, Nelson JC. 1994. Polyneuronal innervation of spiny stellate neurons in cat visual cortex. *J Comp Neurol* 341: 39–49.

Aitkin LM, Webster WR. 1972. Medial geniculate body of the cat: organization and responses to tonal stimuli of neurons in ventral division. *J Neurophysiol* 35: 365–380.

Akert K, Hartmann-von Monakow K. 1980. Relationships of precentral premotor and prefrontal cortex to the mediodorsal and intralaminar nuclei of the monkey thalamus. *Acta Neurobiol Exp (Warsz)* 40: 7–25.

Alitto HJ, Weyand TG, Usrey WM. 2005. Distinct properties of stimulus-evoked bursts in the lateral geniculate nucleus. *J Neurosci* 25: 514–523.

Allen GV, Hopkins DA. 1989. Mamillary body in the rat: topography and synaptology of projections from the subicular complex, prefrontal cortex, and midbrain tegmentum. *J Comp Neurol* 286: 311–336.

Allman JM, Kaas JH. 1971. Representation of the visual field in striate and adjoining cortex of the owl monkey (*Aotus trivirgatus*). *Brain Res* 35: 89–106.

Allman JM, Kaas JH. 1975. The dorsomedial cortical visual area: a third tier area in the occipital obe of the owl monkey (*Aotus trivirgatus*). *Brain Res* 100: 473–487.

Allman JM, Miezin F, McGuinness E. 1985. Stimulus specific responses from beyond the classical receptive field: neurophysiological mechanisms for local-global comparisons in visual neurons. *Annu Rev Neurosci* 8: 407–430.

Alonso JM, Usrey WM, Reid RC. 2001. Rules of connectivity between geniculate cells and simple cells in cat primary visual cortex. *J Neurosci* 21: 4002–4015.

Altman J, Carpenter MB. 1961. Fiber projections of the superior colliculus in the cat. *J Comp Neurol* 116: 157–177.

Alvarado JC, Stanford TR, Rowland BA, Vaughan JW, Stein BE. 2009. Multisensory integration in the superior colliculus requires synergy among corticocollicular inputs. *J Neurosci* 29: 6580–6592.

Andersen RA, Cui H. 2009. Intention, action planning, and decision making in parietal-frontal circuits. *Neuron* 63: 568–583.

Anderson CH, Van Essen DC. 1987. Shifter circuits: a computational strategy for dynamic aspects of visual processing. *Proc Natl Acad Sci USA* 84: 6297–6301.

Anderson ME, Yoshida M. 1980. Axonal branching patterns and location of nigrothalamic and nigrocollicular neurons in the cat. *J Neurophysiol* 43: 883–895.

Anderson JC, da Costa NM, Martin KA. 2009. The W cell pathway to cat primary visual cortex. *J Comp Neurol* 516: 20–35.

Andolina IM, Jones HE, Wang W, Sillito AM. 2007. Corticothalamic feedback enhances stimulus response precision in the visual system. *Proc Natl Acad Sci USA* 104: 1685–1690.

Anwyl R. 2009. Metabotropic glutamate receptor-dependent long-term potentiation. *Neuropharmacol* 56: 735–740.

Arbuthnott GW, MacLeod NK, Maxwell DJ, Wright AK. 1990. Distribution and synaptic contacts of the cortical terminals arising from neurons in the rat ventromedial thalamic nucleus. *Neuroscience* 38: 47–60.

Arcelli P, Frassoni C, Regondi MC, De Biasi S, Spreafico R. 1997. GABAergic neurons in mammalian thalamus: a marker of thalamic complexity? *Brain Res Bull* 42: 27–37.

Aronoff R, Matyas F, Mateo C, Ciron C, Schneider B, Petersen CC. 2010. Long-range connectivity of mouse primary somatosensory barrel cortex. *Eur J Neurosci* 31: 2221–2233.

Asanuma C, Andersen RA, Cowan WM. 1985. The thalamic relations of the caudal inferior parietal lobule and the lateral prefrontal cortex in monkeys: divergent cortical projections from cell clusters in the medial pulvinar nucleus. *J Comp Neurol* 241: 357–381.

Ashwell KW, McAllan BM, Mai JK, Paxinos G. 2008. Cortical cyto- and chemoarchitecture in three small Australian marsupial carnivores: Sminthopsis macroura, Antechinus stuartii and Phascogale calura. *Brain Behav Evol* 72: 215–232.

Bacci A, Huguenard JR, Prince DA. 2005. Modulation of neocortical interneurons: extrinsic influences and exercises in self-control. *Trends Neurosci* 28: 602–610.

Bacskai T, Szekely G, Matesz C. 2002. Ascending and descending projections of the lateral vestibular nucleus in the rat. *Acta Biol Hung* 53: 7–21.

Bajo VM, Nodal FR, Bizley JK, Moore DR, King AJ. 2007. The ferret auditory cortex: descending projections to the inferior colliculus. *Cereb Cortex* 17: 475–491.

Bajo VM, Nodal FR, Bizley JK, King AJ. 2010. The non-lemniscal auditory cortex in ferrets: convergence of corticotectal inputs in the superior colliculus. *Front Neuroanat* 4: 18.

Baker FH, Malpeli JG. 1977. Effects of cryogenic blockade of visual cortex on the responses of lateral geniculate neurons in the monkey. *Exp Brain Res* 29: 433–444.

Baldauf ZB, Chomsung RD, Carden WB, May PJ, Bickford ME. 2005. Ultrastructural analysis of projections to the pulvinar nucleus of the cat. I: middle suprasylvian gyrus (areas 5 and 7). *J Comp Neurol* 485: 87–107.

Baldwin MK, Kaas JH. 2012. Cortical projections to the superior colliculus in prosimian galagos (Otolemur garnetti). *J Comp Neurol* 520: 2002–2020.

Barbara JG, Auclair N, Roisin MP, Otani S, Valjent E, Caboche J, Soubrie P, Crepel F. 2003. Direct and indirect interactions between cannabinoid CB1 receptor and group II metabotropic glutamate receptor signalling in layer V pyramidal neurons from the rat prefrontal cortex. *Eur J Neurosci* 17: 981–990.

Barbas H, Mesulam MM. 1981. Organization of afferent input to subdivisions of area 8 in the rhesus monkey. *J Comp Neurol* 200: 407–431.

Barthó P, Freund TF, Acsády L. 2002. Selective GABAergic innervation of thalamic nuclei from zona incerta. *Eur J Neurosci* 16: 999–1014.

Barthó P, Slezia A, Varga V, Bokor H, Pinault D, Buzsaki G, Acsady L. 2007. Cortical control of zona incerta. *J Neurosci* 27: 1670–1681.

Bartlett EL, Smith PH. 1999. Anatomic, intrinsic, and synaptic properties of dorsal and ventral division neurons in rat medial geniculate body. *J Neurophysiol* 81: 1999–2016.

Bassett JP, Zugaro MB, Muir GM, Golob EJ, Muller RU, Taube JS. 2005. Passive movements of the head do not abolish anticipatory firing properties of head direction cells. *J Neurophysiol* 93: 1304–1316.

Behan M, Lin CS, Hall WC. 1987. The nigrotectal projection in the cat: an electron microscope autoradiographic study. *Neuroscience* 21: 529–539.

Bellone C, Luscher C, Mameli M. 2008. Mechanisms of synaptic depression triggered by metabotropic glutamate receptors. *Cell Mol Life Sci* 65: 2913–2923.

Bender DB. 1983. Visual activation of neurons in the primate pulvinar depends on cortex but not colliculus. *Brain Res* 279: 258–261.

Bentivoglio M, Kuypers HG, Catsman-Berrevoets CE, Loewe H, Dann O. 1980. Two new fluorescent retrograde neuronal tracers which are transported over long distances. *Neurosci Lett* 18: 25–30.

Berendse HW, Groenewegen HJ. 1991. Restricted cortical termination fields of the midline and intralaminar thalamic nuclei in the rat. *Neuroscience* 42: 73–102.

Berman RA, Wurtz RH. 2008. Exploring the pulvinar path to visual cortex. *Prog Brain Res* 171: 467–473.

Berman RA, Wurtz RH. 2010. Functional identification of a pulvinar path from superior colliculus to cortical area MT. *J Neurosci* 30: 6342–6354.

Berman RA, Wurtz RH. 2011. Signals conveyed in the pulvinar pathway from superior colliculus to cortical area MT. *J Neurosci* 31: 373–384.

Berman NJ, Douglas RJ, Martin KA. 1992. GABA-mediated inhibition in the neural networks of visual cortex. *Prog Brain Res* 90: 443–476.

Bezdudnaya T, Cano M, Bereshpolova Y, Stoelzel CR, Alonso JM, Swadlow HA. 2006. Thalamic burst mode and inattention in the awake LGNd. *Neuron* 49: 421–432.

Bickford ME, Carden WB, Patel NC. 1999. Two types of interneurons in the cat visual thalamus are distinguished by morphology, synaptic connections, content, and nitric oxide synthase. *J Comp Neurol* 413: 83–100.

Bishop PO, Kozak W, Levick WR, Vakkur GJ. 1962. The determination of the projection of the visual field on to the lateral geniculate nucleus in the cat. *J Physiol* 163: 503–539.

Blakemore SJ, Wolpert DM, Frith CD. 1998. Central cancellation of self-produced tickle sensation. *Nat Neurosci* 1: 635–640.

Blakemore SJ, Wolpert DM, Frith CD. 2002. Abnormalities in the awareness of action. *Trends Cogn Sci* 6: 237–242.

Bloomfield SA, Sherman SM. 1989. Dendritic current flow in relay cells and interneurons of the cat's lateral geniculate nucleus. *Proc Natl Acad Sci USA* 86: 3911–3914.

Bloomfield SA, Hamos JE, Sherman SM. 1987. Passive cable properties and morphological correlates of neurones in the lateral geniculate nucleus of the cat. *J Physiol* 383: 653–692.

Bokor H, Frere SGA, Eyre MD, Slezia A, Ulbert I, Luthi A, Acsády L. 2005. Selective GABAergic control of higher-order thalamic relays. *Neuron* 45: 929–940.

Bompas A, O'Regan JK. 2006a. Evidence for a role of action in colour perception. *Perception* 35: 65–78.

Bompas A, O'Regan JK. 2006b. More evidence for sensorimotor adaptation in color perception. *J Vis* 6: 145–153.

Bond AH. 2004. An information-processing analysis of the functional architecture of the primate neocortex. *J Theor Biol* 227: 51–79.

Bourassa J, Deschênes M. 1995. Corticothalamic projections from the primary visual cortex in rats: A single fiber study using biocytin as an anterograde tracer. *Neuroscience* 66: 253–263.

Bourassa J, Pinault D, Deschênes M. 1995. Corticothalamic projections from the cortical barrel field to the somatosensory thalamus in rats: a single-fibre study using biocytin as an anterograde tracer. *Eur J Neurosci* 7: 19–30.

Brickman AM, Buchsbaum MS, Shihabuddin L, Byne W, Newmark RE, Brand J, Ahmed S, Mitelman SA, Hazlett EA. 2004. Thalamus size and outcome in schizophrenia. *Schizophr Res* 71: 473–484.

Brodal P. 1978. The corticopontine projection in the rhesus monkey. Origin and principles of organization. *Brain* 101: 251–283.

Brodal A. 1981. *Neurological Anatomy.* Oxford: Oxford University Press.

Brodmann K. 1909. *Vergleichende Lokalisationslehre der Grosshirnrinde in ihren Prinzipien dargestellt auf Grund des Zellenbaues.* Leipzig: Johann Ambrosius Barth Verlag.

Brown SP, Brenowitz SD, Regehr WG. 2003. Brief presynaptic bursts evoke synapse-specific retrograde inhibition mediated by endogenous cannabinoids. *Nat Neurosci* 6: 1048–1057.

Bureau I, von Saint PF, Svoboda K. 2006. Interdigitated paralemniscal and lemniscal pathways in the mouse barrel cortex. *PLoS Biol* 4: e382.

Butler AB, Reiner A, Karten HJ. 2011. Evolution of the amniote pallium and the origins of mammalian neocortex. *Ann N Y Acad Sci* 1225: 14–27.

Cajal SR y 1911. *Histologie du Système Nerveux de l'Homme et des Vertébrés.* Paris: Maloine.

Callaway EM, Katz LC. 1993. Photostimulation using caged glutamate reveals functional circuitry in living brain slices. *Proc Natl Acad Sci USA* 90: 7661–7665.

Calton JL, Stackman RW, Goodridge JP, Archey WB, Dudchenko PA, Taube JS. 2003. Hippocampal place cell instability after lesions of the head direction cell network. *J Neurosci* 23: 9719–9731.

Campbell AW. 1905. *Histological Studies on the Localisation of Cerebral Function.* Cambridge, U.K.: Cambridge University Press.

Canteras NS, Shammah-Lagnado SJ, Silva BA, Ricardo JA. 1990. Afferent connections of the subthalamic nucleus: a combined retrograde and anterograde horseradish peroxidase study in the rat. *Brain Res* 513: 43–59.

Cappe C, Morel A, Rouiller EM. 2007. Thalamocortical and the dual pattern of corticothalamic projections of the posterior parietal cortex in macaque monkeys. *Neuroscience* 146: 1371–1387.

Caputi A, Rozov A, Blatow M, Monyer H. 2009. Two calretinin-positive GABAergic cell types in layer 2/3 of the mouse neocortex provide different forms of inhibition. *Cereb Cortex* 19: 1345–1359.

Carandini M, Ferster D. 1997. A tonic hyperpolarization underlying contrast adaptation in cat visual cortex. *Science* 276: 949–952.

Carden WB, Bickford ME. 2002. Synaptic inputs of class III and class V interneurons in the cat pulvinar nucleus: differential integration of RS and RL inputs. *Vis Neurosci* 19: 51–59.

Carpenter MB. 1976. *Human Neuroanatomy.* Baltimore: Williams & Wilkins.

Carpenter MB, Nakano K, Kim R. 1976. Nigrothalamic projections in the monkey demonstrated by autoradiographic technics. *J Comp Neurol* 165: 401–415.

Casagrande VA, Norton TT. 1991. Lateral geniculate nucleus: a review of its physiology and function. In: *Vision and Visual Dysfunction* (Leventhal AG, ed.), pp 41–84. London: Macmillan Press.

Catterall WA. 2010. Ion channel voltage sensors: structure, function, and pathophysiology. *Neuron* 67: 915–928.

Cavada C, Huisman AM, Kuypers HG. 1984. Retrograde double labeling of neurons: the combined use of horseradish peroxidase and diamidino yellow dihydrochloride (DY X 2HCl) compared with true blue and DY X 2HCl in rat descending brainstem pathways. *Brain Res* 308: 123–136.

Çavdar S, Hacıoğlu H, Şirvanci S, Keskinöz E, Onat F. 2011. Synaptic organization of the rat thalamus: a quantitative study. *Neurol Sci* 32: 1047–1056.

Caviness VS, Jr, Frost DO. 1980. Tangential organization of thalamic projections to the neocortex in the mouse. *J Comp Neurol* 194: 335–367.

Chalupa LM. 1991. Visual function of the pulvinar. In: *The Neural Basis of Visual Function* (Leventhal AG, ed.), pp 140–159. New York: Macmillan Press.

Chance FS, Abbott LF, Reyes A. 2002. Gain modulation from background synaptic input. *Neuron* 35: 773–782.

Chauveau F, Celerier A, Ognard R, Pierard C, Beracochea D. 2005. Effects of ibotenic acid lesions of the mediodorsal thalamus on memory: relationship with emotional processes in mice. *Behav Brain Res* 156: 215–223.

Chen C, Regehr WG. 2003. Presynaptic modulation of the retinogeniculate synapse. *J Neurosci* 23: 3130–3135.

Chow KL. 1950. A retrograde cell degeneration study of the cortical projection field of the pulvinar in the monkey. *J Comp Neurol* 93: 313–340.

Chung S, Li X, Nelson SB. 2002. Short-term depression at thalamocortical synapses contributes to rapid adaptation of cortical sensory responses in vivo. *Neuron* 34: 437–446.

Churchland PS, Ramachandran VS, Sejnowski TJ. 1994. A critique of pure vision. In: *Large-Scale Neuronal Theories in the Brain* (Koch C, Davis JL, eds.), pp 23–65. Cambridge, MA: MIT Press.

Clare MH, Bishop GH. 1954. Responses from an association area secondarily activated from optic cortex. *J Neurophysiol* 17: 271–277.

Clark A. 1998. *Being There: Putting Brain, Body, and World Together Again*. Cambridge, MA: MIT Press.

Clark A. 2011. *Supersizing the Mind*. Oxford: Oxford University Press.

Cleland BG, Lee BB. 1985. A comparison of visual responses of cat lateral geniculate nucleus neurones with those of ganglion cells afferent to them. *J Physiol* 369: 249–268.

Cleland BG, Dubin MW, Levick WR. 1971. Sustained and transient neurones in the cat's retina and lateral geniculate nucleus. *J Physiol* 217: 473–496.

Colby CL, Duhamel JR, Goldberg ME. 1996. Visual, presaccadic, and cognitive activation of single neurons in monkey lateral intraparietal area. *J Neurophysiol* 76: 2841–2852.

Conley M, Diamond IT. 1990. Organization of the visual sector of the thalamic reticular nucleus in *Galago. Eur J Neurosci* 2: 211–226.

Conn PJ, Pin JP. 1997. Pharmacology and functions of metabotropic glutamate receptors. *Annu Rev Pharmacol Toxicol* 37: 205–237.

Connor JA, Stevens CF. 1971. Prediction of repetitive firing behaviour from voltage clamp data on an isolated neurone soma. *J Physiol* 213: 31–53.

Connors BW, Gutnick MJ. 1990. Intrinsic firing patterns of diverse neocortical neurons. *Trends Neurosci* 13: 99–104.

Contreras D, Steriade M. 1995. Cellular basis of EEG slow rhythms: a study of dynamic cortico-thalamic relationships. *J Neurosci* 15: 604–622.

Covic EN, Sherman SM. 2011. Synaptic properties of connections between the primary and secondary auditory cortices in mice. *Cereb Cortex* 21: 2425–2441.

Cowey A, Stoerig P, Williams C. 1999. Variance in transneuronal retrograde ganglion cell degeneration in monkeys after removal of striate cortex: effects of size of the cortical lesion. *Vision Res* 39: 3642–3652.

Cox CL, Sherman SM. 2000. Control of dendritic outputs of inhibitory interneurons in the lateral geniculate nucleus. *Neuron* 27: 597–610.

Cox CL, Denk W, Tank DW, Svoboda K. 2000. Action potentials reliably invade axonal arbors of rat neocortical neurons. *Proc Natl Acad Sci USA* 97: 9724–9728.

Crabtree JW. 1992a. The somatotopic organization within the cat's thalamic reticular nucleus. *Eur J Neurosci* 4: 1352–1361.

Crabtree JW. 1992b. The somatotopic organization within the rabbit's thalamic reticular nucleus. *Eur J Neurosci* 4: 1343–1351.

Crabtree JW. 1996. Organization in the somatosensory sector of the cat's thalamic reticular nucleus. *J Comp Neurol* 366: 207–222.

Crabtree JW, Isaac JTR. 2002. New intrathalamic pathways allowing modality-related and cross-modality switching in the dorsal thalamus. *J Neurosci* 22: 8754–8761.

Crabtree JW, Killackey HP. 1989. The topographical organization of the axis of projection within the visual sector of the rabbit's thalamic reticular nucleus. *Eur J Neurosci* 1: 94–109.

Crabtree JW, Collingridge GL, Isaac JTR. 1998. A new intrathalamic pathway linking modality-related nuclei in the dorsal thalamus. *Nat Neurosci* 1: 389–394.

Craik KJ. 1945. *The Nature of Explanation.* Cambridge, U.K.: Cambridge University Press.

Craik KJ. 1948. Theory of the human operator in control systems; man as an element in a control system. *Br J Psychol Gen Sect* 38: 142–148.

Crandall SR, Cox CL. 2012. Local dendrodendritic inhibition regulates fast synaptic transmission in visual thalamus. *J Neurosci* 32: 2513–2522.

Crandall SR, Govindaiah G, Cox CL. 2010. Low-threshold Ca^{2+} current amplifies distal dendritic signaling in thalamic reticular neurons. *J Neurosci* 30: 15419–15429.

Crick F. 1984. Function of the thalamic reticular complex: the searchlight hypothesis. *Proc Natl Acad Sci USA* 81: 4586–4590.

Cruce JA. 1977. An autoradiographic study of the descending connections of the mammillary nuclei of the rat. *J Comp Neurol* 176: 631–644.

Cucchiaro JB, Uhlrich DJ, Sherman SM. 1991. Electron-microscopic analysis of synaptic input from the perigeniculate nucleus to the A-laminae of the lateral geniculate nucelus in cats. *J Comp Neurol* 310: 316–336.

Cushing C 1909. A note upon the Faradic stimulation of the post-central gyrus in conscious patients. *Brain* 32: 44–53.

Dacey DM, Peterson BB, Robinson FR, Gamlin PD. 2003. Fireworks in the primate retina: in vitro photodynamics reveals diverse LGN-projecting ganglion cell types. *Neuron* 37: 15–27.

Damasio A. 2006. *Descartes' Error.* London: Vintage.

Danos P, Baumann B, Kramer A, Bernstein HG, Stauch R, Krell D, Falkai P, Bogerts B. 2003. Volumes of association thalamic nuclei in schizophrenia: a postmortem study. *Schizophr Res* 60: 141–155.

Datta S, Siwek DF. 2002. Single cell activity patterns of pedunculopontine tegmentum neurons across the sleep-wake cycle in the freely moving rats. *J Neurosci Res* 70: 611–621.

Demb JB. 2002. Multiple mechanisms for contrast adaptation in the retina. *Neuron* 36: 781–783.

Dempsey EW, Morison RS. 1941. The stimulation of rhythmically recurrent cortical potentials after localized thalamic stimulation. *Am J Physiol* 135: 293–300.

DePasquale R, Sherman SM. 2011. Synaptic properties of corticocortical connections between the primary and secondary visual cortical areas in the mouse. *J Neurosci* 31: 16494–16506.

DePasquale R, Sherman SM. 2012. Modulatory effects of metabotropic glutamate receptors on local cortical circuits. *J Neurosci* 32: 7364–7372.

Desbois C, Villanueva L. 2001. The organization of lateral ventromedial thalamic connections in the rat: a link for the distribution of nociceptive signals to widespread cortical regions. *Neuroscience* 102(4): 885–898.

Descarries L, Krnjevic K, Steriade M. 2004. *Acetylcholine in the Cerebral Cortex*. Amsterdam: Elsevier.

Deschênes M, Bourassa J, Pinault D. 1994. Corticothalamic projections from layer V cells in rat are collaterals of long-range corticofugal axons. *Brain Res* 664: 215–219.

Deschênes M, Timofeeva E, Lavallée P, Dufresne C. 2005. The vibrissal system as a model of thalamic operations. *Prog Brain Res* 149: 31–40.

Desimone R, Schein SJ. 1987. Visual properties of neurons in area V4 of the macaque: sensitivity to stimulus form. *J Neurophysiol* 57: 835–868.

Desimone R, Albright TD, Gross CG, Bruce C. 1984. Stimulus-selective properties of inferior temporal neurons in the macaque. *J Neurosci* 4: 2051–2062.

Destexhe A, Contreras D, Steriade M, Sejnowski TJ, Huguenard JR. 1996. *In vivo, in vitro*, and computational analysis of dendritic calcium currents in thalamic reticular neurons. *J Neurosci* 16: 169–185.

Di Chiara G, Porceddu ML, Morelli M, Mulas ML, Gessa GL. 1979. Evidence for a GABAergic projection from the substantia nigra to the ventromedial thalamus and to the superior colliculus of the rat. *Brain Res* 176: 273–284.

Diamond IT. 1973. The evolution of the tectal-pulvinar system in mammals: structural and behavioral studies of the visual system. *Symp Zool Soc Lond* 33: 205–233.

Diamond IT, Feldman M, Galambos R, Goldberg JM, Goldstein MH, Harrison JM, Igarashi M, et al. 1973. Neuroanatomy of the audtiory system. *Arch Otolaryngol* 98: 397–413.

Diamond ME, Armstrong-James M, Budway MJ, Ebner FF. 1992. Somatic sensory responses in the rostral sector of the posterior group (POm) and in the ventral posterior medial nucleus (VPM) of the rat thalamus: dependence on the barrel field cortex. *J Comp Neurol* 319: 66–84.

Ding Y, Casagrande VA. 1997. The distribution and morphology of LGN K pathway axons within the layers and CO blobs of owl monkey V1. *Vis Neurosci* 14: 691–704.

Dittman JS, Kreitzer AC, Regehr WG. 2000. Interplay between facilitation, depression, and residual calcium at three presynaptic terminals. *J Neurosci* 20: 1374–1385.

Dixon G, Harper CG. 2001. Quantitative analysis of glutamic acid decarboxylase-immunoreactive neurons in the anterior thalamus of the human brain. *Brain Res* 923: 39–44.

Dobrunz LE, Stevens CF. 1997. Heterogeneity of release probability, facilitation, and depletion at central synapses. *Neuron* 18: 995–1008.

Dräger UC, Hubel DH. 1975. Responses to visual stimulation and relationship between visual, auditory, and somatosensory inputs in mouse superior colliculus. *J Neurophysiol* 38: 690–713.

Dubin MW, Cleland BG. 1977. Organization of visual inputs to interneurons of lateral geniculate nucleus of the cat. *J Neurophysiol* 40: 410–427.

Dubner R, Zeki SM. 1971. Response properties and receptive fields of cells in an anatomically defined region of the superior temporal sulcus in the monkey. *Brain Res* 35: 528–532.

Dugas-Ford J, Rowell JJ, Ragsdale CW. 2012. Cell-type homologies and the origins of the neocortex. *Proc Natl Acad Sci USA* 109: 16974–16979.

Dumbrava D, Faubert J, Casanova C. 2001. Global motion integration in the cat's lateral posterior-pulvinar complex. *Eur J Neurosci* 13: 2218–2226.

Dumont M, Ptito M, Cardu B, Lepore F. 1974. Collicular system as oculomotor coordination or visual perception centers. *Trans Am Neurol Assoc* 99: 23–27.

Dunlap K, Luebke JI, Turner TJ. 1995. Exocytotic Ca²⁺ channels in mammalian central neurons. *Trends Neurosci* 18: 89–98.

Elliot Smith G. 1910. On some problems relating to the evolution of the brain (Arris and Gale Lecture). *Lancet* 175: 147–153.

Erişir A, Van Horn SC, Bickford ME, Sherman SM. 1997a. Immunocytochemistry and distribution of parabrachial terminals in the lateral geniculate nucleus of the cat: a comparison with corticogeniculate terminals. *J Comp Neurol* 377: 535–549.

Erişir A, Van Horn SC, Sherman SM. 1997b. Relative numbers of cortical and brainstem inputs to the lateral geniculate nucleus. *Proc Natl Acad Sci USA* 94: 1517–1520.

Erişir A, Van Horn SC, Sherman SM. 1998. Distribution of synapses in the lateral geniculate nucleus of the cat: differences between laminae A and A1 and between relay cells and interneurons. *J Comp Neurol* 390: 247–255.

Etkin A, Egner T, Kalisch R. 2011. Emotional processing in anterior cingulate and medial prefrontal cortex. *Trends Cogn Sci* 15: 85–93.

Evans G. 1982. *The Varieties of Reference*. Oxford: Oxford University Press.

Farrow K, Masland RH. 2011. Physiological clustering of visual channels in the mouse retina. *J Neurophysiol* 105: 1516–1530.

Feig SL. 2004. Corticothalamic cells in layers 5 and 6 of primary and secondary sensory cortex express GAP-43 mRNA in the adult rat. *J Comp Neurol* 468: 96–111.

Feig SL. 2005. The differential distribution of the growth-associated protein-43 in first and higher order thalamic nuclei of the adult rat. *Neuroscience* 136: 1147–1157.

Feig S, Harting JK. 1994. Ultrastructural studies of the primate lateral geniculate nucleus: morphology and spatial relationships of axon terminals arising from the retina, visual cortex (area 17), superior colliculus, parabigeminal nucleus, and pretectum of *Galago crassicaudatus*. *J Comp Neurol* 343: 17–34.

Feldmeyer D, Lubke J, Silver RA, Sakmann B. 2002. Synaptic connections between layer 4 spiny neurone-layer 2/3 pyramidal cell pairs in juvenile rat barrel cortex: physiology and anatomy of interlaminar signalling within a cortical column. *J Physiol* 538: 803–822.

Felleman DJ, Van Essen DC. 1991. Distributed hierarchical processing in the primate cerebral cortex. *Cereb Cortex* 1: 1–47.

Ferster D. 1987. Origin of orientation-selective EPSPs in simple cells of cat visual cortex. *J Neurosci* 7: 1780–1791.

Ferster D, LeVay S. 1978. The axonal arborizations of lateral geniculate neurons in the striate cortex of the cat. *J Comp Neurol* 182: 923–944.

Ferster D, Chung S, Wheat H. 1996. Orientation selectivity of thalamic input to simple cells of cat visual cortex. *Nature* 380: 249–252.

Fornito A, Yung AR, Wood SJ, Phillips LJ, Nelson B, Cotton S, Velakoulis D, McGorry PD, Pantelis C, Yucel M. 2008. Anatomic abnormalities of the anterior cingulate cortex before psychosis onset: an MRI study of ultra-high-risk individuals. *Biol Psychiatry* 64: 758–765.

Francois C, Tande D, Yelnik J, Hirsch EC. 2002. Distribution and morphology of nigral axons projecting to the thalamus in primates. *J Comp Neurol* 447: 249–260.

Frick A, Feldmeyer D, Sakmann B. 2007. Postnatal development of synaptic transmission in local networks of L5A pyramidal neurons in rat somatosensory cortex. *J Physiol* 585: 103–116.

Friedlander MJ, Lin C-S, Stanford LR, Sherman SM. 1981. Morphology of functionally identified neurons in lateral geniculate nucleus of the cat. *J Neurophysiol* 46: 80–129.

Fries P. 2009. Neuronal gamma-band synchronization as a fundamental process in cortical computation. *Annu Rev Neurosci* 32: 209–224.

Fries W, Keizer K, Kuypers HG. 1985. Large layer VI cells in macaque striate cortex (Meynert cells) project to both superior colliculus and prestriate visual area V5. *Exp Brain Res* 58: 613–616.

Fries P, Nikolic D, Singer W. 2007. The gamma cycle. *Trends Neurosci* 30: 309–316.

Friston KJ, Ungerleider LG, Jezzard P, Turner P. 1995. Characterizing modulatory interactions between areas V1 and V2 in human cortex: a new treatment of functional MRI data. *Hum Brain Mapp* 2: 211–224.

Frith CD, Blakemore S, Wolpert DM. 2000. Explaining the symptoms of schizophrenia: abnormalities in the awareness of action. *Brain Res Brain Res Rev* 31: 357–363.

Fritsch G, Hitzig E. 1870. Über die elektrische Erregbarkeit des Grosshirns. In: *Some Papers on the Cerebral Cortex* (von Bonin G, ed.). *300-332* Springfield, IL: Charles C. Thomas.

Frost DO, Caviness VS, Jr. 1980. Radial organization of thalamic projections to the neocortex in the mouse. *J Comp Neurol* 194: 369–393.

Fukuda Y, Stone J. 1974. Retinal distribution and central projections of Y-, X-, and W-cells of the cat's retina. *J Neurophysiol* 37: 749–772.

Garcia-Cabezas MA, Rico B, Sánchez-González MA, Cavada C. 2007. Distribution of the dopamine innervation in the macaque and human thalamus. *Neuroimage* 34: 965–984.

Garey LJ, Powell TPS. 1967. The projection of the lateral geniculate nucleus upon the cortex in the cat. *Proc R Soc Lond B Biol Sci* 169: 107–126.

Garey LJ, Powell TP. 1971. An experimental study of the termination of the lateral geniculo-cortical pathway in the cat and monkey. *Proc R Soc Lond B Biol Sci* 179: 41–63.

Gasparini S, Losonczy A, Chen X, Johnston D, Magee JC. 2007. Associative pairing enhances action potential back-propagation in radial oblique branches of CA1 pyramidal neurons. *J Physiol* 580: 787–800.

Geisert EE, Langsetmo A, Spear PD. 1981. Influence of the cortico-geniculate pathway on reponse properties of cat lateral geniculate neurons. *Brain Res* 208: 409–415.

Gibson JJ. 1979. *The Ecological Approach to Visual Perception*. Hillsdale, NJ: Lawrence Erlbaum.

Giessel AJ, Sabatini BL. 2011. Boosting of synaptic potentials and spine Ca transients by the peptide toxin SNX-482 requires alpha-1E-encoded voltage-gated Ca channels. *PLoS ONE* 6: e20939.

Giguere M, Goldman-Rakic PS. 1988. Mediodorsal nucleus: areal, laminar, and tangential distribution of afferents and efferents in the frontal lobe of rhesus monkeys. *J Comp Neurol* 277: 195–213.

Gilbert CD. 1993. Circuitry, architecture, and functional dynamics of visual cortex. *Cereb Cortex* 3: 373–386.

Gilbert CD, Kelly JP. 1975. The projections of cells in different layers of the cat's visual cortex. *J Physiol* 163: 81–106.

Gilbert CD, Wiesel TN. 1979. Morphology and intracortical projections of functionally characterised neurones in the cat visual cortex. *Nature* 280: 120–125.

Glickstein M, May JG, III, Mercier BE. 1985. Corticopontine projection in the macaque: the distribution of labelled cortical cells after large injections of horseradish peroxidase in the pontine nuclei. *J Comp Neurol* 235: 343–359.

Godwin DW, Van Horn SC, Erişir A, Sesma M, Romano C, Sherman SM. 1996a. Ultrastructural localization suggests that retinal and cortical inputs access different metabotropic glutamate receptors in the lateral geniculate nucleus. *J Neurosci* 16: 8181–8192.

Godwin DW, Vaughan JW, Sherman SM. 1996b. Metabotropic glutamate receptors switch visual response mode of lateral geniculate nucleus cells from burst to tonic. *J Neurophysiol* 76: 1800–1816.

Gonzalo-Ruiz A, Romero JC, Sanz JM, Morte L. 1999. Localization of amino acids, neuropeptides and cholinergic neurotransmitter markers in identified projections from the mesencephalic tegmentum to the mammillary nuclei of the rat. *J Chem Neuroanat* 16: 117–133.

Govindaiah, Cox CL 2004. Synaptic activation of metabotropic glutamate receptors regulates dendritic outputs of thalamic interneurons. *Neuron* 41: 611–623.

Govindaiah G, Wang T, Gillette MU, Cox CL. 2012. Activity-dependent regulation of retinogeniculate signaling by metabotropic glutamate receptors. *J Neurosci* 32: 12820–12831.

Gray CM, McCormick DA. 1996. Chattering cells: Superficial pyramidal neurons contributing to the generation of synchronous oscillations in the visual cortex. *Science* 274: 109–113.

Gregoriou GG, Gotts SJ, Zhou H, Desimone R. 2009. High-frequency, long-range coupling between prefrontal and visual cortex during attention. *Science* 324: 1207–1210.

Grillner S. 2003. The motor infrastructure: from ion channels to neuronal networks. *Nat Rev Neurosci* 4: 573–586.

Grillner S, Kozlov A, Dario P, Stefanini C, Menciassi A, Lansner A, Hellgren KJ. 2007a. Modeling a vertebrate motor system: pattern generation, steering and control of body orientation. *Prog Brain Res* 165: 221–234.

Grillner S, Kozlov A, Dario P, Stefanini C, Menciassi A, Lansner A, Hellgren KJ. 2007b. Modeling a vertebrate motor system: pattern generation, steering and control of body orientation. *Prog Brain Res* 165: 221–234.

Groenewegen HJ, Berendse HW, Wolters JG, Lohman AH. 1990. The anatomical relationship of the prefrontal cortex with the striatopallidal system, the thalamus and the amygdala: evidence for a parallel organization. *Prog Brain Res* 85: 95–116.

Grüsser OJ. 1995. On the history of the ideas of efference copy and reafference. *Clio Med* 33: 35–55.

Guillery RW. 1957. Degeneration in the hypothalamic connexions of the albino rat. *Journal of Anatomy (Lond)* 91: 91–115.

Guillery RW. 1966. A study of Golgi preparations from the dorsal lateral geniculate nucleus of the adult cat. *J Comp Neurol* 128: 21–50.

Guillery RW. 1967. Patterns of fiber degeneration in the dorsal lateral geniculate nucleus of the cat following lesions in the visual cortex. *J Comp Neurol* 130: 197–222.

Guillery RW. 1969a. A quantitative study of synaptic interconnections in the dorsal lateral geniculate nucleus of the cat. *Z Zellforsch* 96: 39–48.

Guillery RW. 1969b. The organization of synaptic interconnections in the laminae of the dorsal lateral geniculate nucleus of the cat. *Z Zellforsch* 96: 1–38.

Guillery RW. 1995. Anatomical evidence concerning the role of the thalamus in corticocortical communication: a brief review. *J Anat* 187: 583–592.

Guillery RW. 2003. Branching thalamic afferents link action and perception. *J Neurophysiol* 90: 539–548.

Guillery RW. 2005. Is postnatal neocortical maturation hierarchical? *Trends Neurosci* 28: 512–517.

Guillery RW, Sherman SM. 2002. The thalamus as a monitor of motor outputs. *Philos Trans R Soc Lond B Biol Sci* 357: 1809–1821.

Guillery RW, Sherman SM. 2011. Branched thalamic afferents: what are the messages that they relay to cortex? *Brain Res Brain Res Rev* 66: 205–219.

Guillery RW, Feig SL, Van Lieshout DP. 2001. Connections of higher order visual relays in the thalamus: a study of corticothalamic pathways in cats. *J Comp Neurol* 438: 66–85.

Gulcebi MI, Ketenci S, Linke R, Hacioğlu H, Yanali H, Veliskova J, Moshé SL, Onat F, Cavdar S. 2012. Topographical connections of the substantia nigra pars reticulata to higher-order thalamic nuclei in the rat. *Brain Res Bull* 87(2-3): 312–318.

Gulledge AT, Kampa BM, Stuart GJ. 2005. Synaptic integration in dendritic trees. *J Neurobiol* 64: 75–90.

Gunning FM, Cheng J, Murphy CF, Kanellopoulos D, Acuna J, Hoptman MJ, Klimstra S, Morimoto S, Weinberg J, Alexopoulos GS. 2009. Anterior cingulate cortical volumes and treatment remission of geriatric depression. *Int J Geriatr Psychiatry* 24: 829–836.

Haas JS, Zavala B, Landisman CE. 2011. Activity-dependent long-term depression of electrical synapses. *Science* 334: 389–393.

Hackett TA, Stepniewska I, Kaas JH. 1998. Thalamocortical connections of the parabelt auditory cortex in macaque monkeys. *J Comp Neurol* 400: 271–286.

Hall NJ, Colby CL. 2011. Remapping for visual stability. *Philos Trans R Soc Lond B Biol Sci* 366: 528–539.

Hamos JE, Van Horn SC, Raczkowski D, Uhlrich DJ, Sherman SM. 1985. Synaptic connectivity of a local circuit neurone in lateral geniculate nucleus of the cat. *Nature* 317: 618–621.

Harrison JM, Warr WB. 1962. A study of the cochlear nuclei and ascending auditory pathways of the medulla. *J Comp Neurol* 119: 341–379.

Harting JK, Updyke BV, Van Lieshout DP. 1992. Corticotectal projections in the cat: anterograde transport studies of twenty-five cortical areas. *J Comp Neurol* 324: 379–414.

Hashimotodani Y, Ohno-Shosaku T, Kano M. 2007. Ca(2+)-assisted receptor-driven endocannabinoid release: mechanisms that associate presynaptic and postsynaptic activities. *Curr Opin Neurobiol* 17: 360–365.

Hayes DJ, Northoff G. 2011. Identifying a network of brain regions involved in aversion-related processing: a cross-species translational investigation. *Front Integr Neurosci* 5: 49.

Helmstaedter M, Staiger JF, Sakmann B, Feldmeyer D. 2008. Efficient recruitment of layer 2/3 interneurons by layer 4 input in single columns of rat somatosensory cortex. *J Neurosci* 28: 8273–8284.

Herkenham M. 1979. The afferent and efferent connections of the ventromedial thalamic nucleus in the rat. *J Comp Neurol* 183: 487–517.

Herkenham M. 1980. Laminar organization of thalamic projections to the rat neocortex. *Science* 207: 532–535.

Hess BJ, Blanks RH, Lannou J, Precht W. 1989. Effects of kainic acid lesions of the nucleus reticularis tegmenti pontis on fast and slow phases of vestibulo-ocular and optokinetic reflexes in the pigmented rat. *Exp Brain Res* 74: 63–79.

Higley MJ, Sabatini BL. 2008. Calcium signaling in dendrites and spines: practical and functional considerations. *Neuron* 59: 902–913.

Hilgetag CC, Kaiser M. 2004. Clustered organization of cortical connectivity. *Neuroinformatics* 2: 353–360.

Hille B. 1992. *Ionic Channels of Excitable Membranes*. Sunderland, MA: Sinauer Associates.

Hirsch JA, Martinez LM. 2006. Circuits that build visual cortical receptive fields. *Trends Neurosci* 29: 30–39.

Hodgkin AL, Huxley AF. 1952. Currents carried by sodium and potassium ions through the membrane of the giant axon of *Loligo*. *J Physiol* 116: 449–472.

Hoffman DA, Johnston D. 1999. Neuromodulation of dendritic action potentials. *J Neurophysiol* 81: 408–411.

Holmes GM. 1918. Disturbances of vision by cerebral lesions. *Br J Ophthalmol* 2: 353–384.

Holroyd CB, Yeung N. 2012. Motivation of extended behaviors by anterior cingulate cortex. *Trends Cogn Sci* 16: 122–128.

Hoover JE, Strick PL. 1999. The organization of cerebellar and basal ganglia outputs to primary motor cortex as revealed by retrograde transneuronal transport of herpes simplex virus type 1. *J Neurosci* 19: 1446–1463.

Hu B. 2003. Functional organization of lemniscal and nonlemniscal auditory thalamus. *Exp Brain Res* 153: 543–549.

Huang Y, Yasuda H, Sarihi A, Tsumoto T. 2008. Roles of endocannabinoids in heterosynaptic long-term depression of excitatory synaptic transmission in visual cortex of young mice. *J Neurosci* 28: 7074–7083.

Huang ZJ, Di CG, Ango F. 2007. Development of GABA innervation in the cerebral and cerebellar cortices. *Nat Rev Neurosci* 8: 673–686.

Hubel DH, Wiesel TN. 1962. Receptive fields, binocular interaction and functional architecture in the cat's visual cortex. *J Physiol* 160: 106–154.

Hubel DH, Wiesel TN. 1965. Receptive fields and functional architecture in two nonstriate visual areas (18 and 19) of the cat. *J Neurophysiol* 28: 229–289.

Hubel DH, Wiesel TN. 2005. *Brain and Visual Perception*. New York: Oxford University Press.

Hughlings Jackson J. 1884. Evolution and dissolution of the nervous system. *BMJ* 1: 591–593, 660–663, 703–707.

Huguenard JR, McCormick DA. 1994. *Electrophysiology of the Neuron*. New York: Oxford University Press.

Huguenard JR, Prince DA. 1992. A novel T-type current underlies prolonged Ca^{2+}-dependent burst firing in GABAergic neurons of rat thalamic reticular nucleus. *J Neurosci* 12: 3804–3817.

Hull C, Isaacson JS, Scanziani M. 2009. Postsynaptic mechanisms govern the differential excitation of cortical neurons by thalamic inputs. *J Neurosci* 29: 9127–9136.

Humphrey AL, Sur M, Uhlrich DJ, Sherman SM. 1985. Termination patterns of individual X- and Y-cell axons in the visual cortex of the cat: projections to area 18, to the 17–18 border region, and to both areas 17 and 18. *J Comp Neurol* 233: 190–212.

Inouye T. 1909. *Sehstöhrungen bei Schussverletzungen der kortikalen Sehsphäre nach Beobachtungen an Verwundeten der letzten japanischen Kriege*. Leipzig: W. Engelmann.

Isu N, Sakuma A, Kitahara M, Uchino Y, Takeyama I. 1991. Vestibulo-thalamic neurons give off descending axons to the spinal cord. *Acta Otolaryngol Suppl* 481: 216–220.

Iversen LL, Iversen SD, Dunnett SB, Björklund A. 2010. *Dopamine Handbook*. New York: Oxford University Press.

Jack JJB, Noble D, Tsien RW. 1975. *Electric Current Flow in Excitable Cells*. New York: Oxford University Press.

Jahnsen H, Llinás R. 1984. Electrophysiological properties of guinea-pig thalamic neurones: an *in vitro* study. *J Physiol* 349: 205–226.

Jasper HH. 1954. Functional properties of the thalamic reticular system. In: *Brain Mechanisms and Consciousness* (Delafresnaye JF, ed.), pp 374–401. Oxford: Blackwell.

Jasper HH, Ajmone-Marsan C. 1952. Thalamocortical integrating mechanisms. *Res Publ Assoc Res Nerv Ment Dis* 30: 493–512.

Jia Z, Lu YM, Agopyan N, Roder J. 2001. Gene targeting reveals a role for the glutamate receptors mGluR5 and GluR2 in learning and memory. *Physiol Behav* 73: 793–802.

Jinnai K, Matsuda Y. 1979. Neurons of the motor cortex projecting commonly on the caudate nucleus and the lower brain stem in the cat. *Neurosci Lett* 13: 121–126.

Johnston D, Magee JC, Colbert CM, Christie BR. 1996. Active properties of neuronal dendrites. *Annu Rev Neurosci* 19: 165–186.

Johnston D, Hoffman DA, Colbert CM, Magee JC. 1999. Regulation of back-propagating action potentials in hippocampal neurons. *Curr Opin Neurobiol* 9: 288–292.

Jones EG. 1985. *The Thalamus*. New York: Plenum Press.

Jones EG. 1998a. A new view of specific and nonspecific thalamocortical connections. *Adv Neurol* 77: 49–71.

Jones EG. 1998b. Viewpoint: The core and matrix of thalamic organization. *Neuroscience* 85: 331–345.

Jones EG. 2001. The thalamic matrix and thalamocortical synchrony. *Trends Neurosci* 24: 595–601.

Jones EG. 2002. Thalamic organization and function after Cajal. *Prog Brain Res* 136: 333–357.

Jones EG. 2007. *The Thalamus: Second Edition*. Cambridge, U.K.: Cambridge University Press.

Jones EG, Friedman DP. 1982. Projection pattern of functional components of thalamic ventrobasal complex on monkey somatosensory cortex. *J Neurophysiol* 48: 521–544.

Jones EG, Powell TP. 1969. Electron microscopy of synaptic glomeruli in the thalamic relay nuclei of the cat. *Proc R Soc Lond B Biol Sci* 172: 153–171.

Jones EG, Wise SP, Coulter JD. 1979. Differential thalamic relationships of sensory-motor and parietal cortical fields in monkeys. *J Comp Neurol* 183: 833–882.

Jones MR, Grillner S, Robertson B. 2009. Selective projection patterns from subtypes of retinal ganglion cells to tectum and pretectum: distribution and relation to behavior. *J Comp Neurol* 517: 257–275.

Kaas JH. 1995. The evolution of isocortex. *Brain Behav Evol* 46: 187–196.

Kaas JH. 2005. The evolution of isocortex. *Brain Behav Evol* 146: 187–196.

Kaas JH. 2011. Reconstructing the areal organization of neocortex of the first mammals. *Brain Behav Evol* 78: 7–21.

Kaas JH, Collins CE. 2001. The organization of sensory cortex. *Curr Opin Neurobiol* 11: 498–504.

Kaas JH, Lyon DC. 2007. Pulvinar contributions to the dorsal and ventral streams of visual processing in primates. *Brain Res Brain Res Rev* 55: 285–296.

Kaas JH, Guillery RW, Allman JM. 1972. Some principles of organization in the dorsal lateral geniculate nucleus. *Brain Behav Evol* 6: 253–299.

Kakei S, Na J, Shinoda Y. 2001. Thalamic terminal morphology and distribution of single corticothalamic axons originating from layers 5 and 6 of the cat motor cortex. *J Comp Neurol* 437: 170–185.

Kalil RE, Chase R. 1970. Corticofugal influence on activity of lateral geniculate neurons in the cat. *J Neurophysiol* 33: 459–474.

Kandel ER, Schwartz JH, Jessell TM. 2000. *Principles of Neural Science*. New York: McGraw-Hill.

Kandler K, Katz LC. 1998. Relationship between dye coupling and spontaneous activity in developing ferret visual cortex. *Dev Neurosci* 20: 59–64.

Karlen SJ, Krubitzer L. 2007. The functional and anatomical organization of marsupial neocortex: evidence for parallel evolution across mammals. *Prog Neurobiol* 82: 122–141.

Kay KN, Naselaris T, Prenger RJ, Gallant JL. 2008. Identifying natural images from human brain activity. *Nature* 452: 352–355.

Keizer K, Kuypers HG, Ronday HK. 1987. Branching cortical neurons in cat which project to the colliculi and to the pons: a retrograde fluorescent double-labeling study. *Exp Brain Res* 67: 1–15.

Keller A, White EL, Cipolloni PB. 1985. The identification of thalamocortical axon terminals in barrels of mouse SmI cortex using immunohistochemistry of anterogradely transported lectin (Phaseolus vulgaris-leucoagglutinin). *Brain Res* 343: 159–165.

Kelly LR, Li J, Carden WB, Bickford ME. 2003. Ultrastructure and synaptic targets of tectothalamic terminals in the cat lateral posterior nucleus. *J Comp Neurol* 464: 472–486.

Kemp JM, Powell TP. 1970. The cortico-striate projection in the monkey. *Brain* 93: 525–546.

Khasnis A, Gokula RM. 2003. Romberg's test. *J Postgrad Med* 49: 169–172.

Kievit J, Kuypers HG. 1977. Organization of the thalamo-cortical connexions to the frontal lobe in the rhesus monkey. *Exp Brain Res* 29: 299–322.

Killackey HP, Sherman SM. 2003. Corticothalamic projections from the rat primary somatosensory cortex. *J Neurosci* 23: 7381–7384.

Kiritoshi T, Sun H, Ren W, Stauffer SR, Lindsley CW, Conn PJ, Neugebauer V. 2013. Modulation of pyramidal cell output in the medial prefrontal cortex by mGluR5 interacting with CB1. *Neuropharmacol* 66: 170–178.

Kölliker A. 1896. *Handbuch der Gewebelehre des Menschen. Nervensystemen des Menschen und der Thiere*. Leipzig: Engelmann.

Kolling N, Behrens TE, Mars RB, Rushworth MF. 2012. Neural mechanisms of foraging. *Science* 336: 95–98.

Koralek KA, Jensen KF, Killackey HP. 1988. Evidence for two complementary patterns of thalamic input to the rat somatosensory cortex. *Brain Res* 463: 346–351.

Kullmann DM, Lamsa K. 2008. Roles of distinct glutamate receptors in induction of anti-Hebbian long-term potentiation. *J Physiol* 586: 1481–1486.

Kultas-Ilinsky K, Ilinsky I. 1991. Fine structure of the ventral lateral nucleus (VL) of the *Macaca mulatta* thalamus: cell types and synaptology. *J Comp Neurol* 314: 319–349.

Kultas-Ilinsky K, Sivan-Loukianova E, Ilinsky IA. 2003. Reevaluation of the primary motor cortex connections with the thalamus in primates. *J Comp Neurol* 457: 133–158.

Kuramoto E, Furuta T, Nakamura KC, Unzai T, Hioki H, Kaneko T. 2009. Two types of thalamo-cortical projections from the motor thalamic nuclei of the rat: a single neuron-tracing study using viral vectors. *Cereb Cortex* 19: 2065–2077.

Kuramoto E, Fujiyama F, Nakamura KC, Tanaka Y, Hioki H, Kaneko T. 2011. Complementary distribution of glutamatergic cerebellar and GABAergic basal ganglia afferents to the rat motor thalamic nuclei. *Eur J Neurosci* 33: 95–109.

Kuroda M, Murakami K, Kishi K, Price JL. 1992b. Distribution of the piriform cortical terminals to cells in the central segment of the mediodorsal thalamic nucleus of the rat. *Brain Res* 595: 159–163.

Kuypers HG. 1958. An anatomical analysis of cortico-bulbar connexions to the pons and lower brain stem in the cat. *J Anat* 92: 198–218.

Kuypers HG, Lawrence DG. 1967. Cortical projections to the red nucleus and the brain stem in the Rhesus monkey. *Brain Res* 4: 151–188.

Lachica EA, Casagrande VA. 1993. The morphology of collicular and retinal axons ending on small relay (W-like) cells of the primate lateral geniculate nucleus. *Vis Neurosci* 10: 403–418.

Lam YW, Sherman SM. 2007. Different topography of the reticulothalmic inputs to first- and higher-order somatosensory thalamic relays revealed using photostimulation. *J Neurophysiol* 98: 2903–2909.

Lam YW, Sherman SM. 2010. Functional organization of the somatosensory cortical layer 6 feedback to the thalamus. *Cereb Cortex* 20: 13–24.

Lam YW, Nelson CS, Sherman SM. 2006. Mapping of the functional interconnections between reticular neurons using photostimulation. *J Neurophysiol* 96: 2593–2600.

Lamme VA. 2003. Recurrent corticocortical interactions in neural disease. *Arch Neurol* 60: 178–184.

Lamme VA, Spekreijse H. 2000. Modulations of primary visual cortex activity representing attentive and conscious scene perception. *Front Biosci* 5: D232–D243.

Landisman CE, Connors BW. 2005. Long-term modulation of electrical synapses in the mammalian thalamus. *Science* 310: 1809–1813.

Landisman CE, Connors BW. 2007. VPM and PoM nuclei of the rat somatosensory thalamus: Intrinsic neuronal properties and corticothalamic feedback. *Cereb Cortex* 17: 2853–2865.

Landisman CE, Long MA, Beierlein M, Deans MR, Paul DL, Connors BW. 2002. Electrical synapses in the thalamic reticular nucleus. *J Neurosci* 22: 1002–1009.

Landò L, Zucker RS. 1994. Ca^{2+} cooperativity in neurosecretion measured using photolabile Ca^{2+} chelators. *J Neurophysiol* 72: 825–830.

Lanska DJ, Goetz CG. 2000. Romberg's sign: development, adoption, and adaptation in the 19th century. *Neurology* 55: 1201–1206.

Larkum ME, Zhu JJ, Sakmann B. 1999. A new cellular mechanism for coupling inputs arriving at different cortical layers. *Nature* 398: 338–341.

Larkum ME, Senn W, Luscher HR. 2004. Top-down dendritic input increases the gain of layer 5 pyramidal neurons. *Cereb Cortex* 14: 1059–1070.

Larkum ME, Waters J, Sakmann B, Helmchen F. 2007. Dendritic spikes in apical dendrites of neocortical layer 2/3 pyramidal neurons. *J Neurosci* 27: 8999–9008.

Lashley KS. 1931. Mass action in cerebral function. *Science* 73: 245–254.

Lashley KS, Clark G. 1946. The cytoarchitecture of the cerebral cortex of Ateles; a critical examination of architectonic studies. *J Comp Neurol* 85: 223–305.

Lassek AM. 1954. Motor deficits produced by posterior rhizotomy versus section of the dorsal funiculus. *Neurology* 4: 120–123.

Lavallée P, Urbain N, Dufresne C, Bokor H, Acsády L, Deschênes M. 2005. Feedforward inhibitory control of sensory information in higher-order thalamic nuclei. *J Neurosci* 25: 7489–7498.

Lawrence DG, Kuypers HG. 1965. Pyramidal and non-pyramidal pathways in monkeys: anatomical and functional correlation. *Science* 148: 973–975.

Lee CC, Sherman SM. 2008. Synaptic properties of thalamic and intracortical inputs to layer 4 of the first- and higher-order cortical areas in the auditory and somatosensory systems. *J Neurophysiol* 100: 317–326.

Lee CC, Sherman SM. 2009a. Glutamatergic inhibition in sensory neocortex. *Cereb Cortex* 19: 2281–2289.

Lee CC, Sherman SM. 2009b. Modulator property of the intrinsic cortical projection from layer 6 to layer 4. *Front Syst Neurosci* 3: 1–5.

Lee CC, Sherman SM. 2010. Topography and physiology of ascending streams in the auditory tectothalamic pathway. *Proc Natl Acad Sci USA* 107: 372–377.

Lee CC, Sherman SM. 2012. Intrinsic modulators of auditory thalamocortical transmission. *Hear Res* 287(1-2): 43–50.

Lee Y, Lopez DE, Meloni EG, Davis M. 1996. A primary acoustic startle pathway: obligatory role of cochlear root neurons and the nucleus reticularis pontis caudalis. *J Neurosci* 16: 3775–3789.

Lesica NA, Stanley GB. 2004. Encoding of natural scene movies by tonic and burst spikes in the lateral geniculate nucleus. *J Neurosci* 24: 10731–10740.

LeVay S, Ferster D. 1977. Relay cell classes in the lateral geniculate nucleus of the cat and the effects of visual deprivation. *J Comp Neurol* 172: 563–584.

Leventhal AG, Rodieck RW, Dreher B. 1981. Retinal ganglion cell classes in the old world monkey: morphology and central projections. *Science* 213: 1139–1142.

Leventhal AG, Rodieck RW, Dreher B. 1985. Central projections of cat retinal ganglion cells. *J Comp Neurol* 237: 216–226.

Levesque M, Charara A, Gagnon S, Parent A, Deschênes M. 1996. Corticostriatal projections from layer V cells in rat are collaterals of long-range corticofugal axons. *Brain Res* 709: 311–315.

Levy R, Goldman-Rakic PS. 2000. Segregation of working memory functions within the dorsolateral prefrontal cortex. *Exp Brain Res* 133: 23–32.

Li J, Guido W, Bickford ME. 2003. Two distinct types of corticothalamic EPSPs and their contribution to short-term synaptic plasticity. *J Neurophysiol* 90: 3429–3440.

Lisman JE. 1997. Bursts as a unit of neural information: making unreliable synapses reliable. *Trends Neurosci* 20: 38–43.

Liu XB, Jones EG. 2003. Fine structural localization of connexin-36 immunoreactivity in mouse cerebral cortex and thalamus. *J Comp Neurol* 466: 457–467.

Liu X-B, Honda CN, Jones EG. 1995. Distribution of four types of synapse on physiologically identified relay neurons in the ventral posterior thalamic nucleus of the cat. *J Comp Neurol* 352: 69–91.

Liu XB, Munoz A, Jones EG. 1998. Changes in subcellular localization of metabotropic glutamate receptor subtypes during postnatal development of mouse thalamus. *J Comp Neurol* 395: 450–465.

Liu X, Robertson E, Miall RC. 2003. Neuronal activity related to the visual representation of arm movements in the lateral cerebellar cortex. *J Neurophysiol* 89: 1223–1237.

Llano DA, Sherman SM. 2008. Evidence for nonreciprocal organization of the mouse auditory thalamocortical-corticothalamic projection systems. *J Comp Neurol* 507: 1209–1227.

Llano DA, Sherman SM. 2009. Differences in intrinsic properties and local network connectivity of identified layer 5 and layer 6 adult mouse auditory corticothalamic neurons support a dual corticothalamic projection hypothesis. *Cereb Cortex* 19: 2810–2826.

Llinás R. 1988. The intrinsic electrophysiological properties of mammalian neurons: insights into central nervous system. *Science* 242: 1654–1664.

Llinás RR, Steriade M. 2006. Bursting of thalamic neurons and states of vigilance. *J Neurophysiol* 95: 3297–3308.

Llinás RR, Leznik E, Urbano FJ. 2002. Temporal binding via cortical coincidence detection of specific and nonspecific thalamocortical inputs: a voltage-dependent dye-imaging study in mouse brain slices. *Proc Natl Acad Sci USA* 99: 449–454.

Lorente de Nó R. 1938. Cerebral cortex: architecture, intracortical connections, motor projections. In: *Physiology of the Nervous System* (Fulton J, ed.), pp 291–340. Oxford: Oxford University Press.

Lu SM, Lin RC. 1993. Thalamic afferents of the rat barrel cortex: a light- and electron-microscopic study using Phaseolus vulgaris leucoagglutinin as an anterograde tracer. *Somatosens Mot Res* 10: 1–16.

Lu GW, Willis WD, Jr. 1999. Branching and/or collateral projections of spinal dorsal horn neurons. *Brain Res Brain Res Rev* 29: 50–82.

Lujan R, Nusser Z, Roberts JD, Shigemoto R, Somogyi P. 1996. Perisynaptic location of metabotropic glutamate receptors mGluR1 and mGluR5 on dendrites and dendritic spines in the rat hippocampus. *Eur J Neurosci* 8: 1488–1500.

Ma W, Peschanski M, Ralston HJ, III. 1987. The differential synaptic organization of the spinal and lemniscal projections to the ventrobasal complex of the rat thalamus. Evidence for convergence of the two systems upon single thalamic neurons. *Neuroscience* 22: 925–934.

Ma Y, Hioki H, Konno M, Pan S, Nakamura H, Nakamura KC, Furuta T, Li JL, Kaneko T. 2011. Expression of gap junction protein connexin36 in multiple subtypes of GABAergic neurons in adult rat somatosensory cortex. *Cereb Cortex* 21: 2639–2649.

Macchi G. 1993. The intralaminar system revisited. In: *Somatosensory Integration in the Thalamus* (Minciacchi D, Molinari M, Macchi G, Jones EG, eds.), pp 175–184. Oxford: Pergamon Press.

Macchi G, Bentivoglio M. 1999. Is the "nonspecific" thalamus still "nonspecific"? *Arch Ital Biol* 137: 201–226.

Magee JC, Cook EP. 2000. Somatic EPSP amplitude is independent of synapse location in hippocampal pyramidal neurons. *Nat Neurosci* 3: 895–903.

Magee JC, Johnston D. 2005. Plasticity of dendritic function. *Curr Opin Neurobiol* 15: 334–342.

Mao LM, Zhang GC, Liu XY, Fibuch EE, Wang JQ. 2008. Group I metabotropic glutamate receptor-mediated gene expression in striatal neurons. *Neurochem Res* 33: 1920–1924.

Mao T, Kusefoglu D, Hooks BM, Huber D, Petreanu L, Svoboda K. 2011. Long-range neuronal circuits underlying the interaction between sensory and motor cortex. *Neuron* 72: 111–123.

Marinelli S, Pacioni S, Bisogno T, Di M, Prince V, Huguenard DA, Jr, Bacci A. 2008. The endocannabinoid 2-arachidonoylglycerol is responsible for the slow self-inhibition in neocortical interneurons. *J Neurosci* 28: 13532–13541.

Markram H, Wang Y, Tsodyks M. 1998. Differential signaling via the same axon of neocortical pyramidal neurons. *Proc Natl Acad Sci USA* 95: 5323–5328.

Marlinski V, McCrea RA. 2009. Self-motion signals in vestibular nuclei neurons projecting to the thalamus in the alert squirrel monkey. *J Neurophysiol* 101: 1730–1741.

Marrocco RT, McClurkin JW, Alkire MT. 1996. The influence of the visual cortex on the spatiotemporal response properties of lateral geniculate nucleus cells. *Brain Res* 737: 110–118.

Martinez A, Di RF, Anllo-Vento L, Sereno MI, Buxton RB, Hillyard SA. 2001. Putting spatial attention on the map: timing and localization of stimulus selection processes in striate and extrastriate visual areas. *Vision Res* 41: 1437–1457.

Masri R, Trageser JC, Bezdudnaya T, Li Y, Keller A. 2006. Cholinergic regulation of the posterior medial thalamic nucleus. *J Neurophysiol* 96: 2265–2273.

Masri R, Bezdudnaya T, Trageser JC, Keller A. 2008. Encoding of stimulus frequency and sensor motion in the posterior medial thalamic nucleus. *J Neurophysiol* 100: 681–689.

Mastronarde DN. 1987a. Two classes of single-input X-cells in cat lateral geniculate nucleus. I. Receptive field properties and classification of cells. *J Neurophysiol* 57: 357–380.

Mastronarde DN. 1987b. Two classes of single-input X-cells in cat lateral geniculate nucleus. II. Retinal inputs and the generation of receptive-field properties. *J Neurophysiol* 57: 381–413.

Mateo Z, Porter JT. 2007. Group II metabotropic glutamate receptors inhibit glutamate release at thalamocortical synapses in the developing somatosensory cortex. *Neuroscience* 146: 1062–1072.

Mathers LH. 1972. The synaptic organization of the cortical projection to the pulvinar of the squirrel monkey. *J Comp Neurol* 146: 43–60.

Matthews G. 1996. Neurotransmitter release. *Annu Rev Neurosci* 19: 219–233.

Matyas F, Sreenivasan V, Marbach F, Wacongne C, Barsy B, Mateo C, Aronoff R, Petersen CC. 2010. Motor control by sensory cortex. *Science* 330: 1240–1243.

Maunsell JH, Newsome WT. 1987. Visual processing in monkey extrastriate cortex. *Annu Rev Neurosci* 10: 363–401.

Maunsell JHR, Van Essen DC. 1983. Functional properties of neurons in middle temporal visual area of the macaque monkey. I. Selectivity for stimulus direction, speed, and orientation. *J Neurophysiol* 49: 1127–1147.

McClurkin JW, Marrocco RT. 1984. Visual cortical input alters spatial tuning in monkey lateral geniculate nucleus cells. *J Physiol* 348: 135–152.

McClurkin JW, Optican LM, Richmond BJ. 1994. Cortical feedback increases visual information transmitted by monkey parvocellular lateral geniculate nucleus neurons. *Vis Neurosci* 11: 601–617.

McCormick DA. 1992. Neurotransmitter actions in the thalamus and cerebral cortex and their role in neuromodulation of thalamocortical activity. *Prog Neurobiol* 39: 337–388.

McCormick DA, Pape H-C. 1990. Properties of a hyperpolarization-activated cation current and its role in rhythmic oscillation in thalamic relay neurones. *J Physiol* 431: 291–318.

McCormick DA, Von Krosigk M. 1992. Corticothalamic activation modulates thalamic firing through glutamate "metabotropic" receptors. *Proc Natl Acad Sci USA* 89: 2774–2778.

McCrea RA, Bishop GA, Kitai ST. 1978. Morphological and electrophysiological characteristics of projection neurons in the nucleus interpositus of the cat cerebellum. *J Comp Neurol* 181: 397–419.

McFarland NR, Haber SN. 2002. Thalamic relay nuclei of the basal ganglia form both reciprocal and nonreciprocal cortical connections, linking multiple frontal cortical areas. *J Neurosci* 22: 8117–8132.

Means LW, Harrell TH, Mayo ES, Alexander GB. 1974. Effects of dorsomedial thalamic lesions on spontaneous alternation, maze, activity and runway performance in the rat. *Physiol Behav* 12: 973–979.

Melcher D, Colby CL. 2008. Trans-saccadic perception. *Trends Cogn Sci* 12: 466–473.

Meltzer NE, Ryugo DK. 2006. Projections from auditory cortex to cochlear nucleus: a comparative analysis of rat and mouse. *Anat Rec A Discov Mol Cell Evol Biol* 288: 397–408.

Mendez-Lopez M, Mendez M, Lopez L, Arias JL. 2009. Spatial working memory learning in young male and female rats: involvement of different limbic system regions revealed by cytochrome oxidase activity. *Neurosci Res* 65: 28–34.

Meng H, May PJ, Dickman JD, Angelaki DE. 2007. Vestibular signals in primate thalamus: properties and origins. *J Neurosci* 27: 13590–13602.

Merleau-Ponty M. 2002. *The Phenomenology of Perception*. New York: Routledge.

Meyer HS, Wimmer VC, Hemberger M, Bruno RM, de Kock CP, Frick A, Sakmann B, Helmstaedter M. 2010. Cell type-specific thalamic innervation in a column of rat vibrissal cortex. *Cereb Cortex* 20: 2287–2303.

Miall RC, King D. 2008. State estimation in the cerebellum. *Cerebellum* 7: 572–576.

Miller RJ. 1998. Presynaptic receptors. *Annu Rev Pharmacol Toxicol* 38: 201–227.

Milner AD, Goodale MA. 2008. Two visual systems re-viewed. *Neuropsychologia* 46: 774–785.

Minkowski M. 1914. Untersuchungen über die Beziehungen der Grosshirnrinde und der Netzhaut zu den primären optischen Zentren, besonders zum Corpus geniculatum externum. *Arbeiten Hirnanat Inst Zurich* 7: 259–362.

Mishkin M, Ungerleider LG, Macko KA. 1983. Object vision and spatial vision: two cortical pathways. *Trends Neurosci* 4: 414–417.

Mitrofanis J, Mikuletic L. 1999. Organisation of the cortical projection to the zona incerta of the thalamus. *J Comp Neurol* 412: 173–185.

Momennejad I, Haynes JD. 2012. Human anterior prefrontal cortex encodes the 'what' and 'when' of future intentions. *Neuroimage* 61: 139–148.

Monakow KH, Akert K, Kunzle H. 1978. Projections of the precentral motor cortex and other cortical areas of the frontal lobe to the subthalamic nucleus in the monkey. *Exp Brain Res* 33: 395–403.

Monconduit L, Villanueva L. 2005. The lateral ventromedial thalamic nucleus spreads nociceptive signals from the whole body surface to layer I of the frontal cortex. *Eur J Neurosci* 21: 3395–3402.

Moore T, Armstrong KM. 2003. Selective gating of visual signals by microstimulation of frontal cortex. *Nature* 421: 370–373.

Moran J, Desimone R. 1985. Selective attention gates visual processing in the extrastriate cortex. *Science* 229: 782–784.

Morest DK. 1965. The laminar structure of the medial geniculate body of the cat. *J Anat* 99: 143–160.

Morison RS, Dempsey EW. 1941. A study of thalamo-cortical relations. *Am J Physiol* 135: 281–292.

Mott DD, Lewis DV. 1994. The pharmacology and function of central GABAB receptors. *Int Rev Neurobiol* 36: 97–223.

Mountcastle VB, Henneman E. 1949. Pattern of tactile representation in thalamus of cat. *J Neurophysiol* 12: 85–100.

Mountcastle VB, Henneman E. 1952. The representation of tactile sensibility in the thalamus of the monkey. *J Comp Neurol* 97: 409–439.

Movshon JA, Adelson EH, Gizzi MS, Newsome WT. 1985. The analysis of visual moving patterns. In: *Pattern Recognition Mechanisms* (Chagas C, Gattass R, Gross CG, eds.), pp 117–151. New York: Springer.

Murakami Y, Uchida K, Rijli FM, Kuratani S. 2005. Evolution of the brain developmental plan: insights from agnathans. *Dev Biol* 280: 249–259.

Nakamura K, Colby CL. 2002. Updating of the visual representation in monkey striate and extrastriate cortex during saccades. *Proc Natl Acad Sci USA* 99: 4026–4031.

Nhan HL, Callaway EM. 2012. Morphology of superior colliculus- and middle temporal area-projecting neurons in primate primary visual cortex. *J Comp Neurol* 520: 52–80.

Nicoll RA, Malenka RC, Kauer JA. 1990. Functional comparison of neurotransmitter receptor subtypes in mammalian central nervous system. *Physiol Rev* 70: 513–565.

Niwa M, Johnson JS, O'Connor KN, Sutter ML. 2012. Activity related to perceptual judgment and action in primary auditory cortex. *J Neurosci* 32: 3193–3210.

Norrsell U, Finger S, Lajonchere C. 1999. Cutaneous sensory spots and the "law of specific nerve energies": history and development of ideas. *Brain Res Bull* 48: 457–465.

Nowak LG, Azouz R, Sanchez-Vives MV, Gray CM, McCormick DA. 2003. Electrophysiological classes of cat primary visual cortical neurons in vivo as revealed by quantitative analyses. *J Neurophysiol* 89: 1541–1566.

Nusser Z, Mulvihill E, Streit P, Somogyi P. 1994. Subsynaptic segregation of metabotropic and ionotropic glutamate receptors as revealed by immunogold localization. *Neuroscience* 61: 421–427.

O'Regan JK, Noë A. 2001. A sensorimotor account of vision and visual consciousness. *Behav Brain Sci* 24: 939–973.

O'Scalaidhe SP, Wilson FA, Goldman-Rakic PS. 1997. Areal segregation of face-processing neurons in prefrontal cortex. *Science* 278: 1135–1138.

Ogren MP, Hendrickson AE. 1977. The distribution of pulvinar terminals in visual areas 17 and 18 of the monkey. *Brain Res* 137: 343–350.

Ogren MP, Hendrickson AE. 1979. The morphology and distribution of striate cortex terminals in the inferior and lateral subdivisions of the Macaca monkey pulvinar. *J Comp Neurol* 188: 179–199.

Ohno S, Kuramoto E, Furuta T, Hioki H, Tanaka YR, Fujiyama F, Sonomura T, Uemura M, Sugiyama K, Kaneko T. 2012. A morphological analysis of thalamocortical axon fibers of rat posterior thalamic nuclei: a single neuron tracing study with viral vectors. *Cereb Cortex* 22: 2840–2857.

Ohtsuka K, Maeda S, Oguri N. 2002. Accommodation and convergence palsy caused by lesions in the bilateral rostral superior colliculus. *Am J Ophthalmol* 133: 425–427.

Ojima H. 1994. Terminal morphology and distribution of corticothalamic fibers originating from layers 5 and 6 of cat primary auditory cortex. *Cereb Cortex* 4: 646–663.

Ojima H, Murakami K. 2011. Triadic synaptic interactions of large corticothalamic terminals in non-lemniscal thalamic nuclei of the cat auditory system. *Hear Res* 274: 40–47.

Ojima H, Murakami K, Kishi K. 1996. Dual termination modes of corticothalamic fibers originating from pyramids of layers 5 and 6 in cat visual cortical area 17. *Neurosci Lett* 208: 57–60.

Oswald AM, Reyes AD. 2008. Maturation of intrinsic and synaptic properties of layer 2/3 pyramidal neurons in mouse auditory cortex. *J Neurophysiol* 99: 2998–3008.

Otani S, Daniel H, Takita M, Crepel F. 2002. Long-term depression induced by postsynaptic group II metabotropic glutamate receptors linked to phospholipase C and intracellular calcium rises in rat prefrontal cortex. *J Neurosci* 22: 3434–3444.

Padgett CL, Slesinger PA. 2010. GABAB receptor coupling to G-proteins and ion channels. *Adv Pharmacol* 58: 123–147.

Pape HC, McCormick DA. 1995. Electrophysiological and pharmacological properties of interneurons in the cat dorsal lateral geniculate nucleus. *Neuroscience* 68: 1105–1125.

Pasik T, Pasik P, Bender MB. 1966. The superior colliculi and eye movements. An experimental study in the monkey. *Arch Neurol* 15: 420–436.

Passingham RE, Stephan KE, Kotter R. 2002. The anatomical basis of functional localization in the cortex. *Nat Rev Neurosci* 3: 606–616.

Penfield W, Boldrey E. 1937. Somatic motor and sensory representation in the cerebral cortex of man as studied by electrical stimulation. *Brain* 60: 389–443.

Perrone JA, Krauzlis RJ. 2008. Vector subtraction using visual and extraretinal motion signals: a new look at efference copy and corollary discharge theories. *J Vis* 8: 24.1–14.

Pesaran B, Nelson MJ, Andersen RA. 2008. Free choice activates a decision circuit between frontal and parietal cortex. *Nature* 453: 406–409.

Peschanski M, Lee CL, Ralston HJ, III. 1984. The structural organization of the ventrobasal complex of the rat as revealed by the analysis of physiologically characterized neurons injected intracellularly with horseradish peroxidase. *Brain Res* 297: 63–74.

Peschanski M, Roudier F, Ralston HJ, III, Besson JM. 1985. Ultrastructural analysis of the terminals of various somatosensory pathways in the ventrobasal complex of the rat thalamus: an electron-microscopic study using wheatgerm agglutinin conjugated to horseradish peroxidase as an axonal tracer. *Somatosens Res* 3: 75–87.

Petrof I, Sherman SM. 2009. Synaptic properties of the mammillary and cortical afferents to the anterodorsal thalamic nucleus in the mouse. *J Neurosci* 29: 7815–7819.

Petrof I, Viaene AN, Sherman SM. 2012. Two populations of corticothalamic and interareal corticocortical cells in the subgranular layers of the mouse primary sensory cortices. *J Comp Neurol* 520: 1678–1686.

Pfeifer R, Bongard J. 2006. *How the Body Shapes the Way We Think: A New View of Intelligence.* Cambridge, MA: MIT Press.

Pierrot-Deseilligny C, Rosa A, Masmoudi K, Rivaud S, Gaymard B. 1991. Saccade deficits after a unilateral lesion affecting the superior colliculus. *J Neurol Neurosurg Psychiatry* 54: 1106–1109.

Pin JP, Duvoisin R. 1995. The metabotropic glutamate receptors: structure and functions. *Neuropharmacol* 34: 1–26.

Pinault D, Bourassa J, Deschênes M. 1995a. Thalamic reticular input to the rat visual thalamus: A single fiber study using biocytin as an anterograde tracer. *Brain Res* 670: 147–152.

Pinault D, Bourassa J, Deschênes M. 1995b. The axonal arborization of single thalamic reticular neurons in the somatosensory thalamus of the rat. *Eur J Neurosci* 7: 31–40.

Polyak S. 1957. *The Vertebrate Visual System.* Chicago: University of Chicago Press.

Port NL, Wurtz RH. 2009. Target selection and saccade generation in monkey superior colliculus. *Exp Brain Res* 192: 465–477.

Power BD, Kolmac CI, Mitrofanis J. 1999. Evidence for a large projection from the zona incerta to the dorsal thalamus. *J Comp Neurol* 404: 554–565.

Powers RK, Sawczuk A, Musick JR, Binder MD. 1999. Multiple mechanisms of spike-frequency adaptation in motoneurones. *J Physiol Paris* 93: 101–114.

Pribram KH, Chow KL, Semmes J. 1953. Limit and organization of the cortical projection from the medial thalamic nucleus in monkey. *J Comp Neurol* 98: 433–448.

Price JL. 1986. Subcortical projections from the amygdaloid complex. *Adv Exp Med Biol* 203: 19–33.

Purushothaman G, Marion R, Li K, Casagrande VA. 2012. Gating and control of primary visual cortex by pulvinar. *Nat Neurosci* 15: 905–912.

Purves D, Augustine GJ, Fitzpatrick D, Hall WC, LaMantia A-S, McNamara JO, White LE. 2008. *Neuroscience.* Sunderland, MA: Sinauer.

Raastad M, Shepherd GM. 2003. Single-axon action potentials in the rat hippocampal cortex. *J Physiol* 548: 745–752.

Rafal RD, Posner MI. 1987. Deficits in human visual spatial attention following thalamic lesions. *Proc Natl Acad Sci USA* 84: 7349–7353.

Rall W. 1977. Core conductor theory and cable properties of neurons. In: *Handbook of Physiology: The Nervous System I* (Kandel ER, Geiger S, eds.), pp 39–97. New York: Oxford University Press.

Ralston HJ, III. 1969. The synaptic organization of lemniscal projections to the ventrobasal thalamus of the cat. *Brain Res* 14: 99–116.

Ralston HJ. 1971. Evidence for presynaptic dendrites and a proposal for their mechanism of action. *Nature* 230: 585–587.

Ramcharan EJ, Gnadt JW, Sherman SM. 2005. Higher-order thalamic relays burst more than first-order relays. *Proc Natl Acad Sci USA* 102: 12236–12241.

Rausell E, Jones EG. 1991. Histochemical and immunocytochemical compartments of the thalamic VPM nucleus in monkeys and their relationship to the representational map. *J Neurosci* 11: 210–225.

Reichova I, Sherman SM. 2004. Somatosensory corticothalamic projections: Distinguishing drivers from modulators. *J Neurophysiol* 92: 2185–2197.

Reiner A, Jiao Y, Del MN, Laverghetta AV, Lei WL. 2003. Differential morphology of pyramidal tract-type and intratelencephalically projecting-type corticostriatal neurons and their intrastriatal terminals in rats. *J Comp Neurol* 457: 420–440.

Reuter H. 1996. Diversity and function of presynaptic calcium channels in the brain. *Curr Opin Neurobiol* 6: 331–337.

Reyes A, Lujan R, Rozov A, Burnashev N, Somogyi P, Sakmann B. 1998. Target-cell-specific facilitation and depression in neocortical circuits. *Nat Neurosci* 1: 279–285.

Reynolds JH, Chelazzi L. 2004. Attentional modulation of visual processing. *Annu Rev Neurosci* 27: 611–647.

Reynolds JH, Chelazzi L, Desimone R. 1999. Competitive mechanisms subserve attention in macaque areas V2 and V4. *J Neurosci* 19: 1736–1753.

Rezak M, Benevento LA. 1979. A comparison of the organization of the projections of the dorsal lateral geniculate nucleus, the inferior pulvinar and adjacent lateral pulvinar to primary visual cortex (area 17) in the macaque monkey. *Brain Res* 167: 19–40.

Rockland KS. 1994. Further evidence for two types of corticopulvinar neurons. *Neuroreport* 5: 1865–1868.

Rockland KS. 1996. Two types of corticopulvinar terminations: round (type 2) and elongate (type 1). *J Comp Neurol* 368: 57–87.

Rockland KS. 1998. Convergence and branching patterns of round, type 2 corticopulvinar axons. *J Comp Neurol* 390: 515–536.

Rockland KS, Knutson T. 2001. Axon collaterals of Meynert cells diverge over large portions of area V1 in the macaque monkey. *J Comp Neurol* 441: 134–147.

Romanski LM, Giguere M, Bates JF, Goldman-Rakic PS. 1997. Topographic organization of medial pulvinar connections with the prefrontal cortex in the rhesus monkey. *J Comp Neurol* 379: 313–332.

Rose JE, Woolsey CN. 1948a. Structure and relations of limbic cortex and anterior thalamic nuclei in rabbit and cat. *J Comp Neurol* 89: 279–347.

Rose JE, Woolsey CN. 1948b. The orbitofrontal cortex and its connections with the mediodorsal nucleus in rabbit, sheep and cat. *Res Publ Assoc Res Nerv Ment Dis* 27: 210–232.

Rose JE, Woolsey CN. 1949. The relations of thalamic connections, cellular structure and evocable electrical activity in the auditory region of the cat. *J Comp Neurol* 91: 441–466.

Rose JE, Woolsey CN. 1958. Cortical connections and functional organization of the thalamic auditory system of the cat. In: *Biological and Biochemical Bases of Behavior* (Harlow HF, Woolsey CN, eds.), pp 127–150. Madison: University of Wisconsin Press.

Rossignol S, Frigon A. 2011. Recovery of locomotion after spinal cord injury: some facts and mechanisms. *Annu Rev Neurosci* 34: 413–440.

Rouiller EM, Tanné J, Moret V, Kermadi I, Boussaoud D, Welker E. 1998. Dual morphology and topography of the corticothalamic terminals originating from the primary, supplementary motor, and dorsal premotor cortical areas in macaque monkeys. *J Comp Neurol* 396: 169–185.

Rowell JJ, Mallik AK, Dugas-Ford J, Ragsdale CW. 2010. Molecular analysis of neocortical layer structure in the ferret. *J Comp Neurol* 518: 3272–3289.

Royce GJ. 1982. Laminar origin of cortical neurons which project upon the caudate nucleus: a horseradish peroxidase investigation in the cat. *J Comp Neurol* 205: 8–29.

Royce GJ. 1983. Cortical neurons with collateral projections to both the caudate nucleus and the centromedian-parafascicular thalamic complex: a fluorescent retrograde double labeling study in the cat. *Exp Brain Res* 50: 157–165.

Royce GJ, Bromley S, Gracco C, Beckstead RM. 1989. Thalamocortical connections of the rostral intralaminar nuclei: an autoradiographic analysis in the cat. *J Comp Neurol* 288: 555–582.

Saito H, Yukie M, Tanaka K, Hikosaka K, Fukada Y, Iwai E. 1986. Integration of direction signals of image motion in the superior temporal sulcus of the macaque monkey. *J Neurosci* 6: 145–157.

Sakai ST, Inase M, Tanji J. 1996. Comparison of cerebellothalamic and pallidothalamic projections in the monkey (Macaca fuscata): a double anterograde labeling study. *J Comp Neurol* 368: 215–228.

Salin PA, Bullier J. 1995. Corticocortical connections in the visual system: structure and function. *Physiol Rev* 75: 107–154.

Salt TE. 1987. Excitatory amino acid receptors and synaptic transmission in the rat ventrobasal thalamus. *J Physiol* 391: 490–510.

Salt TE, Eaton SA. 1989. Function of non-NMDA receptors and NMDA receptors in synaptic responses to natural somatosensory stimulation in the ventrobasal thalamus. *Exp Brain Res* 77: 646–652.

Sánchez-González MA, Garcia-Cabezas MA, Rico B, Cavada C. 2005. The primate thalamus is a key target for brain dopamine. *J Neurosci* 25: 6076–6083.

Sanchez-Vives MV, Nowak LG, McCormick DA. 2000. Membrane mechanisms underlying contrast adaptation in cat area 17 *in vivo*. *J Neurosci* 20: 4267–4285.

Sandell JH, Schiller PH. 1982. Effect of cooling area 18 on striate cortex cells in the squirrel monkey. *J Neurophysiol* 48: 38–48.

Saunders RC, Vann SD, Aggleton JP. 2012. Projections from Gudden's tegmental nuclei to the mammillary body region in the cynomolgus monkey (Macaca fascicularis). *J Comp Neurol* 520: 1128–1145.

Scheibel ME, Scheibel AB. 1966. The organization of the nucleus reticularis thalami: a Golgi study. *Brain Res* 1: 43–62.

Schell GR, Strick PL. 1984. The origin of thalamic inputs to the arcuate premotor and supplementary motor areas. *J Neurosci* 4: 539–560.

Schiller PH, Tehovnik EJ. 2001. Look and see: how the brain moves your eyes about. *Prog Brain Res* 134: 127–142.

Schiller PH, Tehovnik EJ. 2005. Neural mechanisms underlying target selection with saccadic eye movements. *Prog Brain Res* 149: 157–171.

Schmielau F, Singer W. 1977. The role of visual cortex for binocular interactions in the cat lateral geniculate nucleus. *Brain Res* 120: 354–361.

Schneider GE. 1969. Two visual systems. *Science* 163: 895–902.

Schofield BR. 2009. Projections to the inferior colliculus from layer VI cells of auditory cortex. *Neuroscience* 159: 246–258.

Schwartz ML, Dekker JJ, Goldman-Rakic PS. 1991. Dual mode of corticothalamic synaptic termination in the mediodorsal nucleus of the rhesus monkey. *J Comp Neurol* 309: 289–304.

Schwarz C, Horowski A, Mock M, Thier P. 2005. Organization of tectopontine terminals within the pontine nuclei of the rat and their spatial relationship to terminals from the visual and somatosensory cortex. *J Comp Neurol* 484: 283–298.

Sclar G, Lennie P, DePriest DD. 1989. Contrast adaptation in striate cortex of macaque. *Vision Res* 29: 747–755.

Scollo-Lavizzari G, Akert K. 1963. Cortical area 8 and its thalamic projection in Macaca mulatta. *J Comp Neurol* 121: 259–269.

Shergill SS, Samson G, Bays PM, Frith CD, Wolpert DM. 2005. Evidence for sensory prediction deficits in schizophrenia. *Am J Psychiatry* 162: 2384–2386.

Sherman SM. 1982. Parallel pathways in the cat's geniculocortical system: W-, X-, and Y-cells. In: *Changing Concepts of the Nervous System* (Morrison AR, Strick PL, eds.), pp 337–359. New York: Academic Press.

Sherman SM. 2001. Tonic and burst firing: dual modes of thalamocortical relay. *Trends Neurosci* 24: 122–126.

Sherman SM. 2004. Interneurons and triadic circuitry of the thalamus. *Trends Neurosci* 27: 670–675.

Sherman SM. 2005. Thalamic relays and cortical functioning. *Prog Brain Res* 149: 107–126.

Sherman SM, Guillery RW. 1998. On the actions that one nerve cell can have on another: Distinguishing "drivers" from "modulators". *Proc Natl Acad Sci USA* 95: 7121–7126.

Sherman SM, Guillery RW. 2001. *Exploring the Thalamus*. San Diego: Academic Press.

Sherman SM, Guillery RW. 2006. *Exploring the Thalamus and Its Role in Cortical Function*. Cambridge, MA: MIT Press.

Sherman SM, Guillery RW. 2011. Distinct functions for direct and transthalamic corticocortical connections. *J Neurophysiol* 106: 1068–1077.

Sherman SM, Koch C. 1986. The control of retinogeniculate transmission in the mammalian lateral geniculate nucleus. *Exp Brain Res* 63: 1–20.

Sherman SM, Spear PD. 1982. Organization of visual pathways in normal and visually deprived cats. *Physiol Rev* 62: 738–855.

Sherrington CS. 1906. Observations on the scratch-reflex in the spinal dog. *J Physiol* 34: 1–50.

Shinoda Y, Futami T, Mitoma H, Yokota J. 1988. Morphology of single neurones in the cerebellorubrospinal system. *Behav Brain Res* 28: 59–64.

Shlosberg D, Amitai Y, Azouz R. 2006. Time-dependent, layer-specific modulation of sensory responses mediated by neocortical layer 1. *J Neurophysiol* 96: 3170–3182.

Shostak Y, Ding Y, Casagrande VA. 2003. Neurochemical comparison of synaptic arrangements of parvocellular, magnocellular, and koniocellular geniculate pathways in owl monkey (*Aotus trivirgatus*) visual cortex. *J Comp Neurol* 456: 12–28.

Shushruth S, Ichida JM, Levitt JB, Angelucci A. 2009. Comparison of spatial summation properties of neurons in macaque V1 and V2. *J Neurophysiol* 102: 2069–2083.

Sidibe M, Bevan MD, Bolam JP, Smith Y. 1997. Efferent connections of the internal globus pallidus in the squirrel monkey: I. Topography and synaptic organization of the pallidothalamic projection. *J Comp Neurol* 382: 323–347.

Simms ML, Kemper TL, Timbie CM, Bauman ML, Blatt GJ. 2009. The anterior cingulate cortex in autism: heterogeneity of qualitative and quantitative cytoarchitectonic features suggests possible subgroups. *Acta Neuropathol* 118: 673–684.

Simola J, Stenbacka L, Vanni S. 2009. Topography of attention in the primary visual cortex. *Eur J Neurosci* 29: 188–196.

Simpson DA. 1952. The projection of the pulvinar to the temporal lobe. *J Anat* 86: 20–28.

Sincich LC, Horton JC. 2003. Independent projection streams from macaque striate cortex to the second visual area and middle temporal area. *J Neurosci* 23: 5684–5692.

Sincich LC, Park KV, Wohlgemuth MJ, Horton JC. 2004. Bypassing V1: a direct geniculate input to area MT. *Nat Neurosci* 7: 1123–1128.

Smith GD, Cox CL, Sherman SM, Rinzel J 2001. A firing-rate model of spike-frequency adaptation in sinusoidally-driven thalamocortical relay neurons. *Thalamus Related Sys* 1: 135–156.

Smith GD, Sherman SM. 2002. Detectability of excitatory versus inhibitory drive in an integrate-and-fire-or-burst thalamocortical relay neuron model. *J Neurosci* 22: 10242–10250.

Smith MA, Ellis-Davies GC, Magee JC. 2003. Mechanism of the distance-dependent scaling of Schaffer collateral synapses in rat CA1 pyramidal neurons. *J Physiol* 548: 245–258.

Smith Y, Bolam JP. 1991. Convergence of synaptic inputs from the striatum and the globus pallidus onto identified nigrocollicular cells in the rat: a double anterograde labelling study. *Neuroscience* 44: 45–73.

Solomon SG, Peirce JW, Dhruv NT, Lennie P. 2004. Profound contrast adaptation early in the visual pathway. *Neuron* 42: 155–162.

Sommer MA, Wurtz RH. 2004a. What the brain stem tells the frontal cortex. I. Oculomotor signals sent from superior colliculus to frontal eye field via mediodorsal thalamus. *J Neurophysiol* 91: 1381–1402.

Sommer MA, Wurtz RH. 2004b. What the brain stem tells the frontal cortex. II. Role of the SC-MD-FEF pathway in corollary discharge. *J Neurophysiol* 91: 1403–1423.

Sommer MA, Wurtz RH. 2008. Brain circuits for the internal monitoring of movements. *Annu Rev Neurosci* 31: 317–338.

Somogyi G, Hajdu F, Tömböl T. 1978. Ultrastructure of the anterior ventral and anterior medial nuclei of the cat thalamus. *Exp Brain Res* 31: 417–431.

Sorra KE, Harris KM. 2000. Overview on the structure, composition, function, development, and plasticity of hippocampal dendritic spines. *Hippocampus* 10: 501–511.

Sperry RW. 1950. Neural basis of the spontaneous optokinetic response produced by visual inversion. *J Comp Neurol* 43: 482–489.

Spillane JD. 1981. *Chapters in the History of Neurology*. Oxford: Oxford University Press.

Sprague JM. 1966. Interaction of cortex and superior colliculus in mediation of visually guided behavior in the cat. *Science* 153: 1544–1547.

Sprague JM. 1972. The superior colliculus and pretectum in visual behavior. *Invest Ophthalmol* 11: 473–482.

Spreafico R, Frassoni C, Arcelli P, De Biasi S. 1994. GABAergic interneurons in the somatosensory thalamus of the guinea-pig: a light and ultrastructural immunocytochemical investigation. *Neuroscience* 59: 961–973.

Spruston N. 2008. Pyramidal neurons: dendritic structure and synaptic integration. *Nat Rev Neurosci* 9: 206–221.

Stafstrom CE, Schwindt PC, Crill WE. 1984. Cable properties of layer V neurons from cat sensorimotor cortex in vitro. *J Neurophysiol* 52: 278–289.

Stanford LR, Friedlander MJ, Sherman SM. 1983. Morphological and physiological properties of geniculate W-cells of the cat: a comparison with X- and Y-cells. *J Neurophysiol* 50: 582–608.

Stanley GB, Li FF, Dan Y. 1999. Reconstruction of natural scenes from ensemble responses in the lateral geniculate nucleus. *J Neurosci* 19: 8036–8042.

Stanton GB. 1980. Topographical organization of ascending cerebellar projections from the dentate and interposed nuclei in *Macaca mulatta*: and anterograde degeneration study. *J Comp Neurol* 190: 699–731.

Stanton GB. 2001. Organization of cerebellar and area "y" projections to the nucleus reticularis tegmenti pontis in macaque. *J Comp Neurol* 432: 169–183.

Steenland HW, Li XY, Zhuo M. 2012. Predicting aversive events and terminating fear in the mouse anterior cingulate cortex during trace fear conditioning. *J Neurosci* 32: 1082–1095.

Stein JF, Glickstein M. 1992. Role of the cerebellum in visual guidance of movement. *Physiol Rev* 72: 967–1017.

Stephan KE, Friston KJ, Frith CD. 2009. Dysconnection in schizophrenia: from abnormal synaptic plasticity to failures of self-monitoring. *Schizophr Bull* 35: 509–527.

Stephenson-Jones M, Samuelsson E, Ericsson J, Robertson B, Grillner S. 2011. Evolutionary conservation of the basal ganglia as a common vertebrate mechanism for action selection. *Curr Biol* 21: 1081–1091.

Stephenson-Jones M, Floros O, Robertson B, Grillner S. 2012. Evolutionary conservation of the habenular nuclei and their circuitry controlling the dopamine and 5-hydroxytryptophan (5-HT) systems. *Proc Natl Acad Sci USA* 109: E164–E173.

Stone J. 1983. *Parallel Processing in the Visual System*. New York: Plenum Press.

Stone J, Dreher B. 1973. Projection of X- and Y-cells of the cat's lateral geniculate nucleus to areas 17 and 18 of visual cortex. *J Neurophysiol* 36: 551–567.

Sun HY, Lyons SA, Dobrunz LE. 2005. Mechanisms of target-cell specific short-term plasticity at Schaffer collateral synapses onto interneurones versus pyramidal cells in juvenile rats. *J Physiol* 568: 815–840.

Sun YG, Wu CS, Lu HC, Beierlein M. 2011. Target-dependent control of synaptic inhibition by endocannabinoids in the thalamus. *J Neurosci* 31: 9222–9230.

Sung CH, Chuang JZ. 2010. The cell biology of vision. *J Cell Biol* 190: 953–963.

Swadlow HA, Gusev AG. 2001. The impact of 'bursting' thalamic impulses at a neocortical synapse. *Nat Neurosci* 4: 402–408.

Swadlow HA, Gusev AG, Bezdudnaya T. 2002. Activation of a cortical column by a thalamocortical impulse. *J Neurosci* 22: 7766–7773.

Szentágothai J. 1963. The structure of the synapse in the lateral geniculate nucleus. *Acta Anat (Basel)* 55: 166–185.

Takahashi KA, Castillo PE. 2006. The CB1 cannabinoid receptor mediates glutamatergic synaptic suppression in the hippocampus. *Neuroscience* 139: 795–802.

Tamietto M, Cauda F, Corazzini LL, Savazzi S, Marzi CA, Goebel R, Weiskrantz L, de Gelder B. 2010. Collicular vision guides nonconscious behavior. *J Cogn Neurosci* 22: 888–902.

Tan ZJ, Hu H, Huang ZJ, Agmon A. 2008. Robust but delayed thalamocortical activation of dendritic-targeting inhibitory interneurons. *Proc Natl Acad Sci USA* 105: 2187–2192.

Tang J, Suga N. 2008. Modulation of auditory processing by cortico-cortical feed-forward and feedback projections. *Proc Natl Acad Sci USA* 105: 7600–7605.

Tanibuchi I, Goldman-Rakic PS. 2003. Dissociation of spatial-, object-, and sound-coding neurons in the mediodorsal nucleus of the primate thalamus. *J Neurophysiol* 89: 1067–1077.

Tanibuchi I, Kitano H, Jinnai K. 2009. Substantia nigra output to prefrontal cortex via thalamus in monkeys. I. Electrophysiological identification of thalamic relay neurons. *J Neurophysiol* 102: 2933–2945.

Taube JS. 2007. The head direction signal: origins and sensory-motor integration. *Annu Rev Neurosci* 30: 181–207.

Tehovnik EJ, Slocum WM, Schiller PH. 2003. Saccadic eye movements evoked by microstimulation of striate cortex. *Eur J Neurosci* 17: 870–878.

Teune TM, van der Burg J, van der Moer J, Voogd J, Ruigrok TJ. 2000. Topography of cerebellar nuclear projections to the brain stem in the rat. *Prog Brain Res* 124: 141–172.

Thach WT. 1975. Timing of activity in cerebellar dentate nucleus and cerebral motor cortex during prompt volitional movement. *Brain Res* 88: 233–241.

Theyel BB, Lee CC, Sherman SM. 2010a. Specific and nonspecific thalamocortical connectivity in the auditory and somatosensory thalamocortical slices. *Neuroreport* 21: 861–864.

Theyel BB, Llano DA, Sherman SM. 2010b. The corticothalamocortical circuit drives higher-order cortex in the mouse. *Nat Neurosci* 13: 84–88.

Thompson JM, Woolsey CN, Talbot SA. 1950. Visual areas I and II of cerebral cortex of rabbit. *J Neurophysiol* 13: 277–288.

Thompson RF, Johnson RH, Hoopes JJ. 1963. Organization of auditory, somatic sensory, and visual projection to association fields of cerebral cortex in the cat. *J Neurophysiol* 26: 343–364.

Thomson AM. 2000. Facilitation, augmentation and potentiation at central synapses. *Trends Neurosci* 23: 305–312.

Thomson AM, Deuchars J. 1994. Temporal and spatial properties of local circuits in neocortex. *Trends Neurosci* 17: 119–126.

Thomson AM, Deuchars J. 1997. Synaptic interactions in neocortical local circuits: dual intracellular recordings in vitro. *Cereb Cortex* 7: 510–522.

Thomson AM, West DC. 2003. Presynaptic frequency filtering in the gamma frequency band; dual intracellular recordings in slices of adult rat and cat neocortex. *Cereb Cortex* 13: 136–143.

Tokita K, Inoue T, Boughter JD, Jr. 2010. Subnuclear organization of parabrachial efferents to the thalamus, amygdala and lateral hypothalamus in C57BL/6J mice: a quantitative retrograde double labeling study. *Neuroscience* 171: 351–365.

Tömböl T. 1967. Short neurons and their synaptic relations in the specific thalamic nuclei. *Brain Res* 3: 307–326.

Tömböl T. 1969. Two types of short axon (Golgi 2nd) interneurons in the specific thalamic nuclei. *Acta Morphol Acad Sci Hung* 17: 285–297.

Torigoe Y, Blanks RH, Precht W. 1986. Anatomical studies on the nucleus reticularis tegmenti pontis in the pigmented rat. II. Subcortical afferents demonstrated by the retrograde transport of horseradish peroxidase. *J Comp Neurol* 243: 88–105.

Torrealba F, Partlow GD, Guillery RW. 1981. Organization of the projection from the superior colliculus to the dorsal lateral geniculate nucleus of the cat. *Neuroscience* 6: 1341–1360.

Trageser JC, Keller A. 2004. Reducing the uncertainty: gating of peripheral inputs by zona incerta. *J Neurosci* 24: 8911–8915.

Trageser JC, Burke KA, Masri R, Li Y, Sellers L, Keller A. 2006. State-dependent gating of sensory inputs by zona incerta. *J Neurophysiol* 96: 1456–1463.

Trettel J, Fortin DA, Levine ES. 2004. Endocannabinoid signalling selectively targets perisomatic inhibitory inputs to pyramidal neurones in juvenile mouse neocortex. *J Physiol* 556: 95–107.

Tsanov M, Manahan-Vaughan D. 2009. Synaptic plasticity in the adult visual cortex is regulated by the metabotropic glutamate receptor, mGLUR5. *Exp Brain Res* 199: 391–399.

Tseng K-Y, Atzori M. 2007. *Monoaminergic Modulation of Cortical Excitability*. New York: Springer.

Turner JP, Salt TE. 2000. Synaptic activation of the group I metabotropic glutamate receptor mGlu1 on the thalamocortical neurons of the rat dorsal lateral geniculate nucleus *in vitro*. *Neuroscience* 100: 493–505.

Tzounopoulos T, Rubio ME, Keen JE, Trussell LO. 2007. Coactivation of pre- and postsynaptic signaling mechanisms determines cell-specific spike-timing-dependent plasticity. *Neuron* 54: 291–301.

Uhlrich DJ, Cucchiaro JB, Sherman SM. 1988. The projection of individual axons from the parabrachial region of the brainstem to the dorsal lateral geniculate nucleus in the cat. *J Neurosci* 8: 4565–4575.

Uhlrich DJ, Cucchiaro JB, Humphrey AL, Sherman SM. 1991. Morphology and axonal projection patterns of individual neurons in the cat perigeniculate nucleus. *J Neurophysiol* 65: 1528–1541.

Ullan J. 1985. Cortical topography of thalamic intralaminar nuclei. *Brain Res* 328: 333–340.

Ungerleider LG, Desimone R, Galkin TW, Mishkin M. 1984. Subcortical projections of area MT in the macaque. *J Comp Neurol* 223: 368–386.

Updyke BV. 1977. Topographic organization of the projections from cortical areas 17, 18, and 19 onto the thalamus, pretectum and superior colliculus in the cat. *J Comp Neurol* 173: 81–122.

Updyke BV. 1981. Projections from visual areas of the middle suprasylvian sulcus onto the lateral posterior complex and adjacent thalamic nuclei in cat. *J Comp Neurol* 201: 477–506.

Urbain N, Deschênes M. 2007a. A new thalamic pathway of vibrissal information modulated by the motor cortex. *J Neurosci* 27: 12407–12412.

Urbain N, Deschênes M. 2007b. Motor cortex gates vibrissal responses in a thalamocortical projection pathway. *Neuron* 56: 714–725.

Usrey WM, Reppas JB, Reid RC. 1999. Specificity and strength of retinogeniculate connections. *J Neurophysiol* 82: 3527–3540.

van Atteveldt N, Roebroeck A, Goebel R. 2009. Interaction of speech and script in human auditory cortex: insights from neuro-imaging and effective connectivity. *Hear Res* 258: 152–164.

Van Buren JM. 1963. Trans-synaptic retrograde degeneration in the visual system of primates. *J Neurol Neurosurg Psychiatry* 26: 402–409.

Van Essen DC, Anderson CH. 1990. Information processing strategies and pathways in the primate retina and visual cortex. In: *Introduction to Neural and Electronic Networks* (Zornetzer SF, Davis JF, Lau C, eds.), pp 43–72. Orlando: Academic.

Van Essen DC, DeYoe EA, Olavarria JF, Fox JM, Sagi D, Julesz B. 1989. Neural responses to static and moving texture patterns in visual cortex of the macaque monkey. In: *Neural Mechanisms of Visual Perception* (Lam DM-K, ed.), pp 137–154. Woodlands, TX: Portfolio.

Van Essen DC, Anderson CH, Felleman DJ. 1992. Information processing in the primate visual system: an integrated systems perspective. *Science* 255: 419–423.

Van Horn SC, Sherman SM. 2004. Differences in projection patterns between large and small corticothalamic terminals. *J Comp Neurol* 475: 406–415.

Van Horn SC, Sherman SM. 2007. Fewer driver synapses in higher order than in first order thalamic relays. *Neuroscience* 475: 406–415.

Van Horn SC, Erişir A, Sherman SM. 2000. The relative distribution of synapses in the A-laminae of the lateral geniculate nucleus of the cat. *J Comp Neurol* 416: 509–520.

Varela C, Sherman SM. 2007. Differences in response to muscarinic agonists between first and higher order thalamic relays. *J Neurophysiol* 98: 3538–3547.

Varela C, Sherman SM. 2008. Differences in response to serotonergic activation between first and higher order thalamic nuclei. *Cereb Cortex* 19: 1776–1786.

Varela JA, Song S, Turrigiano GG, Nelson SB. 1999. Differential depression at excitatory and inhibitory synapses in visual cortex. *J Neurosci* 19: 4293–4304.

Veinante P, Lavallée P, Deschênes M. 2000. Corticothalamic projections from layer 5 of the vibrissal barrel cortex in the rat. *J Comp Neurol* 424: 197–204.

Viaene AN, Petrof I, Sherman SM. 2011a. Properties of the thalamic projection from the posterior medial nucleus to primary and secondary somatosensory cortices in the mouse. *Proc Natl Acad Sci USA* 108: 18156–18161.

Viaene AN, Petrof I, Sherman SM. 2011b. Synaptic properties of thalamic input to layers 2/3 in primary somatosensory and auditory cortices. *J Neurophysiol* 105: 279–292.

Viaene AN, Petrof I, Sherman SM. 2011c. Synaptic properties of thalamic input to the subgranular layers of primary somatosensory and auditory cortices in the mouse. *J Neurosci* 31: 12738–12747.

Viaene AN, Petrof I, Sherman SM 2013. Activation requirements for metabotropic glutamate receptors. *Neurosci Lett* 541: 67–72.

Vidnyanszky Z, Gorcs TJ, Negyessy L, Borostyankio Z, Knopfel T, Hamori J. 1996. Immunocytochemical visualization of the mGluR1a metabotropic glutamate receptor at synapses of corticothalamic terminals originating from area 17 of the rat. *Eur J Neurosci* 8: 1061–1071.

Von Economo C, Koskinas GN. 1929. *The Cytoarchitectonics of the Human Cerebral Cortex.* London: Oxford University Press.

von Holst E. 1954. Relations between the central nervous system and the peripheral organs. *Br J Anim Behav* 2: 89–94.

von Holst E, Mittelstaedt H. 1950. The reafference principle. Interaction between the central nervous system and the periphery. In: *Selected Papers of Erich von Holst: The Behavioural Physiology of Animals and Man* (Transl. by Martin R, ed.), pp 139–173. Coral Gables, FL: University of Miami Press.

Voshart K, Van der Kooy D. 1981. The organization of the efferent projections of the parabrachial nucleus of the forebrain in the rat: a retrograde fluorescent double-labeling study. *Brain Res* 212: 271–286.

Vukadinovic Z. 2011. Sleep abnormalities in schizophrenia may suggest impaired trans-thalamic cortico-cortical communication: towards a dynamic model of the illness. *Eur J Neurosci* 34: 1031–1039.

Vukadinovic Z, Rosenzweig I. 2012. Abnormalities in thalamic neurophysiology in schizophrenia: could psychosis be a result of potassium channel dysfunction? *Neurosci Biobehav Rev* 36: 960–968.

Walker AE. 1938. *The Primate Thalamus.* Chicago: University of Chicago Press.

Walzl EM, Woolsey CN. 1946. Effects of cochlear lesions on click responses in the auditory cortex of the cat. *Bull Johns Hopkins Hosp* 79: 309–319.

Wang C, Waleszczyk WJ, Burke W, Dreher B. 2000. Modulatory influence of feedback projections from area 21a on neuronal activities in striate cortex of the cat. *Cereb Cortex* 10: 1217–1232.

Wang S, Bickford ME, Van Horn SC, Erişir A, Godwin DW, Sherman SM. 2001. Synaptic targets of thalamic reticular nucleus terminals in the visual thalamus of the cat. *J Comp Neurol* 440: 321–341.

Wang S, Eisenback MA, Bickford ME. 2002. Relative distribution of synapses in the pulvinar nucleus of the cat: implications regarding the "driver/modulator" theory of thalamic function. *J Comp Neurol* 454: 482–494.

Wang C, Waleszczyk WJ, Burke W, Dreher B. 2007a. Feedback signals from cat's area 21a enhance orientation selectivity of area 17 neurons. *Exp Brain Res* 182: 479–490.

Wang X, Wei Y, Vaingankar V, Wang Q, Koepsell K, Sommer FT, Hirsch JA. 2007b. Feedforward excitation and inhibition evoke dual modes of firing in the cat's visual thalamus during naturalistic viewing. *Neuron* 55: 465–478.

Webster KE. 1965. The cortico-striatal projection in the cat. *J Anat* 99: 329–337.

Webster MJ, Bachevalier J, Ungerleider LG. 1993. Subcortical connections of inferior temporal areas TE and TEO in macaque monkeys. *J Comp Neurol* 335: 73–91.

Wei H, Bonjean M, Petry HM, Sejnowski TJ, Bickford ME. 2011. Thalamic burst firing propensity: a comparison of the dorsal lateral geniculate and pulvinar nuclei in the tree shrew. *J Neurosci* 31: 17287–17299.

Weisenhorn DMV, Illing RB, Spatz WB. 1995. Morphology and connections of neurons in area 17 projecting to the extrastriate areas MT and 19DM and to the superior colliculus in the monkey *Callithrix jacchus*. *J Comp Neurol* 362: 233–255.

White EL. 1978. Identified neurons in mouse SmI cortex which are postsynaptic to thalamocortical axon terminals: a combined Golgi-electron microscopic and degeneration study. *J Comp Neurol* 181: 627–661.

Whittle S, Chanen AM, Fornito A, McGorry PD, Pantelis C, Yucel M. 2009. Anterior cingulate volume in adolescents with first-presentation borderline personality disorder. *Psychiatry Res* 172: 155–160.

Wiesendanger R, Wiesendanger M. 1982a. The corticopontine system in the rat. I. Mapping of corticopontine neurons. *J Comp Neurol* 208: 215–226.

Wiesendanger R, Wiesendanger M. 1982b. The corticopontine system in the rat. II. The projection pattern. *J Comp Neurol* 208: 227–238.

Williams SR, Stuart GJ. 2000. Action potential backpropagation and somato-dendritic distribution of ion channels in thalamocortical neurons. *J Neurosci* 20: 1307–1317.

Wilson JR, Friedlander MJ, Sherman SM. 1984. Fine structural morphology of identified X- and Y-cells in the cat's lateral geniculate nucleus. *Proc R Soc Lond B Biol Sci* 221: 411–436.

Wirtshafter D, Stratford TR. 1993. Evidence for GABAergic projections from the tegmental nuclei of Gudden to the mammillary body in the rat. *Brain Res* 630: 188–194.

Wise SP, Boussaoud D, Johnson PB, Caminiti R. 1997. Premotor and parietal cortex: corticocortical connectivity and combinatorial computations. *Annu Rev Neurosci* 20: 25–42.

Wolfart J, Debay D, Le Masson G, Destexhe A, Bal T. 2005. Synaptic background activity controls spike transfer from thalamus to cortex. *Nat Neurosci* 8: 1760–1767.

Wolpert DM, Ghahramani Z. 2000. Computational principles of movement neuroscience. *Nat Neurosci* 3(Suppl): 1212–1217.

Wolpert DM, Miall RC. 1996. Forward models for physiological motor control. *Neural Netw* 9: 1265–1279.

Womelsdorf T, Fries P, Mitra PP, Desimone R. 2006. Gamma-band synchronization in visual cortex predicts speed of change detection. *Nature* 439: 733–736.

Woolsey CN, Fairman D. 1946. Contralateral, ipsilateral, and bilateral representation of cutaneous receptors in somatic areas I and II of the cerebral cortex of pig, sheep, and other mammals. *Surgery* 19: 684–702.

Woolsey TA, Van der Loos H. 1970. The structural organization of layer IV in the somatosensory region (SI) of mouse cerebral cortex. The description of a cortical field composed of discrete cytoarchitectonic units. *Brain Res* 17: 205–242.

Wright NF, Erichsen JT, Vann SD, O'Mara SM, Aggleton JP. 2010. Parallel but separate inputs from limbic cortices to the mammillary bodies and anterior thalamic nuclei in the rat. *J Comp Neurol* 518: 2334–2354.

Wu LG, Borst JG, Sakmann B. 1998. R-type Ca^{2+} currents evoke transmitter release at a rat central synapse. *Proc Natl Acad Sci USA* 95: 4720–4725.

Wurtz RH. 2008. Neuronal mechanisms of visual stability. *Vision Res* 48: 2070–2089.

Yen C-T, Conley M, Jones EG. 1985. Morphological and functional types of neurons in cat ventral posterior thalamic nucleus. *J Neurosci* 5: 1316–1338.

Yeterian EH, Pandya DN. 1989. Thalamic connections of the cortex of the superior temporal sulcus in the rhesus monkey. *J Comp Neurol* 282: 80–97.

Yeterian EH, Van Hoesen GW. 1978. Cortico-striate projections in the rhesus monkey: the organization of certain cortico-caudate connections. *Brain Res* 139: 43–63.

Yu C, Derdikman D, Haidarliu S, Ahissar E. 2006. Parallel thalamic pathways for whisking and touch signals in the rat. *PLoS Biol* 4: e124.

Yukie M, Iwai E. 1981. Direct projection from the dorsal lateral geniculate nucleus to the prestriate cortex in macaque monkeys. *J Comp Neurol* 201: 81–97.

Zeki SM. 1969. Representation of central visual fields in prestriate cortex of monkey. *Brain Res* 14: 271–291.

Zeki S. 1983. Colour coding in the cerebral cortex: the reaction of cells in monkey visual cortex to wavelengths and colours. *Neuroscience* 9: 741–765.

Zhan XJ, Cox CL, Rinzel J, Sherman SM. 1999. Current clamp and modeling studies of low threshold calcium spikes in cells of the cat's lateral geniculate nucleus. *J Neurophysiol* 81: 2360–2373.

Zhou Q, Godwin DW, O'Malley DM, Adams PR. 1997. Visualization of calcium influx through channels that shape the burst and tonic firing modes of thalamic relay cells. *J Neurophysiol* 77: 2816–2825.

Zimmerman EA, Chambers WW, Liu CN. 1964. An experimental study of the anatomical organization of the cortico-bulbar system in the albino rat. *J Comp Neurol* 123: 301–323.

Index

Note: f after a number indicates a figure.

Accommodation, 183, 239
Action potential, 23, 24f, 26f, 28f, 86, 91, 98
 backpropagation, 25, 35, 36, 46
 at branch point, 149
Action and perception linked, 145, 232
Auditory pathways/functions, 163, 93
Autonomic nervous system, 5
Autoradiographic tracing methods, 201
Axon, 17, 31, 36, 49, 50f, 82, 92–94, 104
 branches, 3f, 9, 15, 144, 149, 164, 172, 183,
  202, 219, 233–4
 diameter, 27, 94
 initial segment, 18, 19, 23
 myelinated, 25, 28f
 terminal arbors, 81, 85, 95
 terminals, 66, 66f, 67f, 68, 113
 unmyelinated, 28f

Basal ganglia, 5, 106f, 111, 114–115, 144, 236
Binding functions, 209
Blindsight, 156
"Brain-reading," 227
Brainstem, 84, 98, 113, 145f
Breathing, 144
Burst mode, 27, 97, 109, 124, 137, 228

Cable modeling, 17, 20, 21f
 of cortical cells, 22
 of dendrites, 35
Calbindin, 73, 209
Calcium binding proteins, 73, 209
Central pattern generators, 5, 143
Cerebellar pathways, 157, 190
Cerebellum, 5, 189, 211, 236
Channels, 18
 high-threshold Ca^{2+} channels, 35, 43
Chewing, 144
Cholinergic inputs, 62, 73f, 77, 100, 113,
 115–116, 123, 134

Cholinergic modulators, 223
Cilia, 141, 142f
Class 1/Class 2 glutamatergic inputs, 14, 70,
 85, 87f, 88f, 89f, 90–116, 101f, 103f, 106f,
 109f, 120, 185, 191, 201, 223
 Class 1A, 1B, 1C, 100, 101f, 103f
Classification of afferents, 83
 of glutamatergic afferents, 85
 of possible mixtures of input classes, 96
Cochlear nerve, 163
Cochlear root neurons, 163
Coincidence detection, 46
Colliculo-thalamo-cortical pathway, 156
Conductance. *See* Membrane conductances
Consciousness, 243
Contrast adaptation, 78
Convergence of synaptic inputs, 65, 85, 90,
 92, 99, 112
Corollary discharge, 8, 170
Cortex, 5, 49, 50f, 112
 functional localization, 54
 neocortex, 49
Cortical areas, 8, 49, 52, 53f, 197f
 architectonically defined, 57, 196
 anterior limbic, 183, 184f, 187
 auditory, 52, 53f, 109f, 129f
 connections to motor centers, 126, 132,
  234
 in early mammals, 141
 extrastriate, 157
 functional capacities, 219, 236
 functional localization, 55f, 56f
 limbic, 196
 motor, 52, 53f, 189
 MT, MST, 200
 multiple areas of single modalities, 198
 orbitofrontal, 196
 posterior limbic, 183, 184f
 retrosplenial, 160, 183, 184f

Cortical areas (cont.)
  somatosensory, 52, 53f, 57, 61, 109f, 110,
    132
  visual, 52, 53f, 55, 56f, 57, 59, 110, 129f
Cortical cells, 22, 50f, 51f, 57, 86, 94
  layer 5 cells/axons, 64, 67–70, 94, 95, 105
  layer 6 cells/axons, 49, 68–70, 79, 83
  thalamic recipient cells, 94
Cortical connections, 2f, 59, 112. *See also*
  Corticocortical connections
  conventional view, 2f
  our view, 2f
Cortical hierarchies, 59, 132, 221
Cortical layers, 49, 68, 108
  thalamic inputs to, 109f, 110, 206–207, 210
Cortical maps, 53f, 55f, 56f, 57, 58
Cortical outputs, 180–181, 211
Cortical recruiting response, 207
Corticobulbar pathways, 214–5
Corticocentric view, 59
Corticocortical connections, 112, 223, 229
  between areas, 110
  branches to cortical layer 4 from layer 6, 94
  direct, 127
  direct and transthalamic conjoint, 137
  direct vs. transthalamic, 126, 131
  drivers in, 85, 106f, 225
  functional linking of cortical areas, 138
  laminar origins and targets, 129f
  local, 112
  modulators in, 225
  for monkey, 129f, 130
  for mouse, 129f, 130
  transthalamic, 132
Corticopontine pathways, 2012–3
Corticopontocerebellar pathways, 190
Corticostriatal pathways, 213–4
Corticotectal pathways, 2011, 2012f
Corticothalamic axons, 66, 67f, 68, 120, 167f
  from areas 17, 18, 19, 168
  to lateral geniculate nucleus, 97, 109, 225
  from layer 5, 68, 69, 165, 166f
  from layer 6, 69, 81, 97, 109
  from temporal and parietal cortex, 168
  from visual, somatosensory, and motor
    areas, 168
Cytoarchitectonics, 57, 198f

Decoding messages in the Brain, 227
Defecating, 144
Dendrite, 17, 35, 46, 49, 51f, 57, 62, 63f, 67,
  70–82, 80f
  dendritic arbors, 21f, 70, 71, 76, 78
  diameter, 20, 21f
Developmental considerations, 49, 52, 59–61,
  124
Diencephalon, 60

Dopaminergic inputs, 79, 113
Dorsal roots, 143, 146f
  dual role of, 150
Drivers, 10, 59, 62, 66f, 83, 120, 144, 179,
  184f, 185, 191, 217, 223
  topography, 100

Efference copy, 8, 133f, 151, 169, 169f, 171f,
  231–2
Embodied perception, 241
EPSP, 31–35, 40, 42f, 81, 86, 87f, 91
Evidence based phrenology, 180, 236
Evolutionary considerations, 52–54, 124

Feedback, 59, 128
Feedforward, 59, 128
Feeding, 144
Firing rates, 25, 30f, 32, 33, 37, 40, 45
Firing modes. *See* Thalamic gate; Burst
  mode; Tonic mode
First order relays, 62, 68–71, 119, 122, 144,
  217, 220
Fornix, 184f, 185
Forward models, 169, 182, 189
Forward receptive fields, 170, 231
Frontal lobe/cortex, 53f, 156, 174, 190, 192,
  203

GABAergic inputs, 61–62, 63f, 65, 72f, 73f,
  78–80, 113, 114, 134, 122, 221
  from basal ganglia, 114
  in thalamic gating, 135
Gap junctions, 37, 81
Gate. *See* Thalamic gate
Gaze control, 155, 160, 239
Globus palidus, 64, 65, 79, 137
Glutamate receptors. *See* Receptors
Glutamatergic inputs, 62, 64, 67f, 73f, 81.
  See also Class 1, Class 2
  from cortical layer 6, 79
  as inhibitory inputs, 104
Golgi method, 49, 164, 189
Gustatory pathways, 163

Head direction cells, 160
Hedgehog, 52
Higher order relays, 65, 69–71, 82, 122, 133f,
  137, 217, 220
  afferents to, 67f, 68, 164, 191
  anatomical evidence for, 119
  physiological evidence for, 121
  role in corticocortical connections, 126
Hippocampus, 160, 162f, 183
Hypothalamus, 163, 182
Hyperpolarization-activated cation
  conductance. *See* Sag current
Hypothalamus, 145f

Inferior colliculus, 106f, 163, 240
Inferior olive, 182
Intelligence organization, 244
Interneurons, 20, 62, 74, 75f
Ion channels. *See* Channels
IPSP, 40, 98

Korsakoff's syndrome, 188

Labeled line (Müller's law), 98, 152
Lamprey, 144
Lemniscal pathways, 190
Levels of organization, 221
Lines of projection, 205
Lower motor mechanisms, 145f
Low-threshold Ca^{2+} spike, 26f, 28f, 31, 43

Mamillary bodies, 162f
Mamillary peduncle, 184f, 185
Mamillotegmental pathway/tract, 160, 162f, 183, 184f
Mamillothalamic tract, 160, 162f, 184f
Maps. *See* Cortical maps
  topographical maps, 81
Marchi method, 195
Medial mamillary nucleus, 184f
Membrane conductances, 22–37, 47
Membrane properties, 19f, 22, 23, 47
Memory functions, 188
Message, 179–183, 218, 226
Method of retrograde degeneration, 195, 196f, 198f, 202
Meynert cell, 131
Midbrain, 5
Midbrain reticular formations, 182
Mind, 241–242
Modulators, 10, 59, 79, 83, 120, 137, 179, 184f, 217, 223. *See also under transmitters (glutamatergic, cholinergic, etc.)*
  topography, 100
Modulatory inputs to first and higher order relays, 123
Motor branches, 190, 230
Movement control, 125
Myeloarchtectonics, 57

Nigrothalamic pathways, 214
Nissl method, 49, 50f, 58
Nociceptive afferents to thalamus, 208
Noradrenergic inputs, 113, 223
Nucleus. *See also* Thalamic nuclei
  amygdaloid, 163
  anterior pretectal, 166f
  caudate, 8
  cochlear, 163
  deep tegmental, 184f, 185
  dentate, 189
  dorsal tegmental, 160, 161f, 184f
  facial nucleus, 148f
  first order, 3
  glossopharyngeal, 148f
  higher order, 3
  interpositus, 189
  lateral mamillary, 160, 161f, 184f, 185
  medial mamillary, 184f, 185
  medial pontine reticular, 160, 161f, 183
  posterior pretectal, 165f
  of the solitary tract, 163
  vestibular, 161f

Ocular movements, 192. *See also* Saccades
Ontogeny, 125
Optic tract, 154f
Opossum, 52
Organizational principles, 239

Pain pathways, 233
Paired-pulse depression/facilitation, 41, 42f, 88f, 89f, 91
Paleocortex, 49
Pallidothalamic axons, 214
Parvalbumen, 73, 209
Periaqueductal gray, 182
Place cells, 160
Posterior columns, 149, 151
Postsynaptic receptors, 72f, 77
  cannabinoid, 41
  $GABA_A$, 40
  $GABA_B$, 37, 40
  glutamate, 37, 40, 41, 77
  ionotropic, 37, 39f, 40, 85, 88f
  M1–M5, 40
  metabotropic, 37, 39f, 40, 85, 88f, 103
  muscarinic, 40
  NMDA, 40, 45
  postsynaptic, 37, 86, 88f
  presynaptic, 41
  sensory, 141
  serotonergic, 41
Postsynaptic responses, 88f
Pretectal region/nucleus, 79, 135, 137, 155
Principal mamillary tract, 162f
Probability of transmitter release (p), 43, 84, 91
"Pure sensation," 153

Reafference, 169f, 171f, 231
Receptive fields, 32, 68, 96, 97, 99
Receptors. *See* Postsynaptic receptors
Recruiting responses, 207
Reticular nucleus/formation, 5, 60, 80, 80f, 81
  medial pontine, 160, 161f, 183
Reticular cells, 20

Retinal afferents, 66f, 72f, 75, 77, 154
Retinal rod development, 142f
Retrogradely transported markers, use of, 200
Robots, 241
Romberg's sign, 150

Sag current, 35
saltatory conduction, 28f
Schizophrenia, 153, 229–30
Scratch reflex, 143, 144
Self (the Self), 153, 230
Sensory code, 71
Serotonergic inputs, 79, 113, 123, 223
 receptors, 41
Signal-to-noise ratio, 34
Short-term plasticity, 41
Soma, 18, 23, 25, 31, 36
Somatosensory pathways, 52, 54, 59, 68, 93,
 145
Somatosensory perception, 152
Spinal cord, 145f, 146f
Spinocerebellar pathways, 190
Striatum, 211
Stretch reflex, 233
Subcortical motor centers, 219
Subiculum, 185
Substantia nigra, 137
Superior cerebellar peduncle, 158f
Superior colliculus, 5, 144, 155, 157, 166f,
 182, 192, 239
Sustaining projections, 199
Swallowing, 144
Swimming, 144
Synaptic inputs. See under transmitter; Class
 1; Class 2
Synaptic integration, 46
Synaptic properties, 17, 37
Synaptic terminals, 63f, 66f, 94

Tectobulbar pathways, 214
Tectospinal pathways, 214
Tectum, 144, 211
Temporal lobe, 195, 199
 afferents to, 200
Tenrec, 52
Thalamic cells. See also Relay cells; Drivers;
 Modulators; Class 1; Class 2
 core and matrix, 209
Thalamic gate, 10, 132, 137, 218, 228
 burst mode, 30f, 27
 tonic mode, 30f, 29
Thalamic glomerulus, 63f, 66f, 67f, 72f
Thalamic inputs, 62, 181
 connectivity patterns, 72f, 76
 F terminals, 62, 63f, 66f, 67f, 72f, 75f, 77
 RL terminals, 62, 63f, 66f, 67f, 72f, 75f, 77
 RS terminals, 62, 63f, 66f, 67f, 72f

Thalamic interneurons, 74, 75f
Thalamic nucleus, 8, 61, 196, 197f
 anterior dorsal nucleus, 8f, 160, 161f, 183,
  184f, 225
 anterior medial nucleus, 8f, 183, 1184f,
  225
 anterior pulvinar, 13f
 anterior thalamic (inputs to), 183
 anterior ventral nucleus, 8f, 183, 184f, 203,
  225
 center medial nucleus, 8f, 191, 214
 dorsal medial geniculate nucleus, 8f, 66f,
  67f, 106f, 107, 187, 240
 first order nuclei, 63, 119
 intralaminar nuclei, 8f, 203, 208, 210
 habenular nuclei, 8f
 higher order nuclei, 65, 119
 lateral dorsal nucleus, 8f, 166f, 191
 lateral geniculate nucleus, 8f, 63f, 154f, 155,
  167f, 196f, 239
 lateral posterior nucleus, 8f, 157, 166f, 167f,
  206
 medial dorsal nucleus, 8f, 107, 157, 191,
  196, 203, 206
 parafascicular nucleus, 21
 posterior medial nucleus, 107
 posterior nucleus, 8f, 166f, 167f
 pulvinar, 8f, 157, 201, 203
 reticular nucleus, 8f
 ventral anterior nucleus, 8f, 106f, 114, 187,
  190
 ventral lateral nucleus, 8f, 106f, 114, 188,
  214
 ventral medial geniculate nucleus, 8f, 106f,
  199
 ventral medial nucleus, 208, 214
 ventral posterior inferior nucleus, 8f
 ventral posterior medial nucleus, 8f, 87f,
  163
 ventral posterior nucleus, 181
Thalamic nuclei that receive branched
  afferents, 174
Thalamic discs and lamellae, 202, 204f
Thalamocortical axons
 drivers and modulators, 225
 branches to cortical layers, 4, 94
 specific and nonspecific, 207
 terminals in cortical layers, 109f, 201, 208
 topographic organization, 208
Thalamocortical relationships/interactions,
  195
 conventional view, 237, 238f
 distinct types of projection, 206
 our view, 237, 238f
 quantification, 199, 201
 related to action and perception, 240
 topography, 1196f, 202, 208

Thalamus, 60. *See also preceding classified entries*
T-type Ca^{2+} spikes. *See* Low-threshold Ca^{2+} spike
Tonic mode, 124, 137, 228
Triads, 62, 63f, 71, 72f, 75f, 76, 77–79, 191
Trigeminal nerve, 147, 148f
  motor nucleus of, 148f

Unexpected events, 221

Vestibular nuclei, 144
Vestibular pathways, 157, 159f
Vigilance, 124
Visual cortex. *See also* Cortical areas
  lesions of, 157
Visual pathways, 70, 93
  magno, parvo, konio pathways, 70, 205, 210
  X, Y, W pathways, 70, 205, 210
Visually guided movements, 189

Wake-up call, 134, 222, 235
Walking, 143
Whisking, 229

Zona incerta, 61, 79, 135, 136